AN INTRODUCTION TO POPULATION

Second Edition

Helen Ginn Daugherty
Kenneth C. W. Kammeyer

THE GUILFORD PRESS
New York   London

© 1995 The Guilford Press
A Division of Guilford Publication, Inc.
72 Spring Street, New York, NY 10012
Marketed and distributed outside North America by
Longman Group Limited.

Printed in the United States of America

Last digit is print number: 9 8 7 6 5 4 3 2 1

Library of Congress Cataloging-in-Publication Data
Daugherty, Helen Ginn.
    An introduction to population / Helen Ginn Daugherty,
Kenneth C. W. Kammeyer.—2nd ed.
        p.   cm.
    Kammeyer's name appears first on the earlier edition.
    Includes bibliographical references and index.
    ISBN 0-89862-616-1
    1. Population. 2. Population—Study and teaching.
I. Kammeyer, Kenneth C. W. II. Title.
HB871.K25 1995
304.6—dc20   95-6657
                                                    CIP

PREFACE

The guiding thought behind this book continues to be the same as it was for the previous two editions: It is, as the title says, an introduction to the study of population. Many college and university students opt to take a course that has *population* someplace in its title, but their reasons for choosing it are many and varied. Most students are not aiming to be professional demographers, though a few are, and, happily, some find the topic of enduring interest. The majority of those who take a first population course are best described as open-minded, or perhaps curious, about the study of population. We hope this book will prove to be, for the vast majority of readers, a satisfying combination of interesting, informative, and sometimes provocative ideas and facts about the study of population.

One of the important features of this book is the way it emphasizes that the study of population is not simply a presentation of descriptive demographic statistics. The study of population, at its best, poses questions and then tries to answer them. And the questions are extremely varied, as a few selected examples from the text will illustrate:

- Is population growth a negative or positive factor in economic development and the improvement of human lives?
- Are some occupations related in special ways to migration?
- Do left-handed people have shorter life-spans than right-handed people do, and if so, why?
- Do people in lower socioeconomic classes always have the highest level of childbearing?
- Should population growth be slowed down and, if so, is this possible?

These are just some samples of the many questions that we take up in this book and for which we will try to provide answers based on the best available scientific data. Not every question raised, of course, is definitively answered. In any science there is always more work to be done to resolve ambiguities and even contradictions.

We continue, as we did in the previous edition of this book, to present the study of population in a framework that distinguishes the different ways of accomplishing this (Chapter 3). This analytic scheme is summarized in a table near the end of the chapter. In this edition, for the first time, we have introduced a similar analytic scheme for describing the burgeoning field of *family demography*. Most population textbooks confound and conflate demography and family demography, but we make a clear distinction between the two.

Demography in its most developed form is a statistically sophisticated science. At an introductory level, we like to introduce readers to the rudimentary statistical measurements and methods of demography. We do that by beginning with an introduction to *rates* in the very first chapter. Later in the book we introduce a few more statistical approaches, including a sample of life tables, which give us measures of life expectancy. However, the statistical demographic techniques presented in this text do not make the "compleat" demographer; this is only an introduction.

While the statistical techniques of demography are important, it must also be noted that the quality of the data is equally important. For that reason we devote Chapter 4 to the *data of demography*.

The study of population leads quite naturally, and persistently, to the study of population policy. The scientific information about population both shapes and informs policy questions. In Chapter 10 we take up population policy, and show that it is almost always influenced by political considerations.

We hope that you will share our view that demography is an exciting field of study. The chapters that follow attempt to heighten your awareness and curiosity about the dynamics of population.

HELEN GINN DAUGHERTY
KENNETH C.W. KAMMEYER

ACKNOWLEDGMENTS

Our colleagues in demography, sociology, and related disciplines have been of great help in making suggestions, providing us with articles, and sharing their research. In particular, we would like to acknowledge the following: Frank Bean, University of Texas; Suzanne Bianchi, University of Maryland; Andrew Cherlin, Johns Hopkins University; Tom Espenshade, Princeton University; Bill Falk, University of Maryland; Lea Keil Garson; Brenda Gilliam, National Center for Health Statistics; Joan Hermson, University of Maryland; Jan Hoem, University of Stockholm; Joan Kahn, University of Maryland; Sunita Kishor, Macro International; Anju Malhotra, University of Maryland; Joe McFalls, University of Pennsylvania; Wayne McVey, University of Alberta; Bill Mosher, National Center for Health Statistics; Dudley Posten, Cornell University; John Relethford, SUNY-Oneonta; Richard Rogers, University of Colorado; J. Mayone Stycos, Cornell University; Jay Teachman, Washington State University; and Reeve Vanneman, University of Maryland. We also appreciate the anonymous reviewers who made many useful suggestions. Most of the time, we took their advice.

The Sociology Department Staff at the University of Maryland, College Park, deserve recognition, and we specifically wish to thank Dorothy Bowers, Cyndi Mewborn, Cass O'Toole, Gerry Todd, and Agnes Zane. At St. Mary's College of Maryland, we would like to thank Lucy Myers, Jeanne Cain, Vicki Spalding, Michelle Vanisko, and Shannon O'Hara. Special appreciation goes to Joann Klear, who typed the first draft, and to Trenna Solomon, whose research assistance was valuable in preparing the manuscript in the final stages. The Provost at St. Mary's College, Melvin Endy, provided a course release and awarded a faculty development grant in support of this project.

Our spouses, Sonia Kammeyer and Tom Daugherty, spent many weekend hours waiting for us to finish our work before we could turn to more convivial activities. We are grateful for their patience and their encouragement throughout. Lastly, to our families, near and far, we say thank you.

HELEN GINN DAUGHERTY
KENNETH C.W. KAMMEYER

v

CONTENTS

vii

1

THE STUDY OF POPULATION:
AN INTRODUCTION

D emography is the name given to the scientific study of human populations. The most fundamental concerns of demography are childbearing and birthrates, dying and death rates, and migration and migration rates, the three fundamental processes that produce and change populations. Births, deaths, and migration account for the growth, size, density, and distribution of populations. Demographers also study the characteristics of human populations, especially the way people are distributed in terms of different ages and the proportions who are male and female.

An important subfield of demography is called family demography. *Family demography* is concerned with the composition and sizes of households and families, and with the processes related to family life, such as marriage and divorce (and, increasingly, cohabitation and remarriage).

This brief description provides only a glimpse of the many different kinds of demographic analysis and research. There are many different ways of approaching the study of population, including, in addition to the purely demographic approach, the application of economic, biological, medical, geographic, political, sociological, and psychological factors. We will describe and illustrate these various approaches to the study of population in Chapter 3, showing that demography is unquestionably an interdisciplinary science. Through this book we plan to show you the wide variety of approaches used to study population. We will begin by examining the fundamental demographic variables and how they are related to each other.

DEMOGRAPHIC VARIABLES

Demographers, like other scientists, deal most of the time with variables. A *variable* is any event, occurrence, or thing that can change or take on different values. The fundamental demographic processes mentioned above—births, deaths, and migration—are key variables of demography because their levels do often change from one time to another, and are very different from one population to another.

1

Demographic variables are most often expressed as rates. A *rate* is a measure that reflects the frequency of some event (e.g., deaths) relative to some population that may experience that event. By using rates, demographers can make comparisons between two or more populations, even when the populations are very different in size, or they can compare rates from year to year to discover trends occurring in a specific population. Some examples of commonly used demographic rates will provide more details about how rates are calculated and why they are useful.

Fertility

The United Kingdom has a population of 58.4 million people, while Thailand has a population of 59.4 million people (1994 data).[1] Even though these two countries have nearly the same size populations, Thailand experiences many more births each year than the United Kingdom. In the United Kingdom there are about 759,000 births, while in Thailand there are about 1,188,000. These figures may be used to calculate the *crude birthrate*, which is the measure of fertility most often used, especially for international comparisons.

The crude birthrate is calculated by comparing the number of births in a year to the total number of people in the population (at midyear). This produces a fraction, as is shown in the cases of the United Kingdom and Thailand:

United Kingdom
$$\frac{\text{No. of births, 1994}}{\text{Total pop., United Kingdom, 1994}} \times 1{,}000 = \frac{759{,}000}{58{,}400{,}000} \times 1{,}000 = 13$$

Thailand
$$\frac{\text{No. of births, 1994}}{\text{Total pop., Thailand, 1994}} \times 1{,}000 = \frac{1{,}188{,}000}{59{,}400{,}000} \times 1{,}000 = 20$$

It is customary when calculating rates to multiply the resulting decimal fraction by a constant number. For demographic rates the constant number is usually 1,000. In the case of the United Kingdom, the decimal fraction is 0.013, which when multiplied by 1,000 yields a crude birthrate of 13. For Thailand the decimal fraction is 0.020, which when multiplied by 1,000 produces a crude birthrate of 20. These crude birthrates are interpreted in the following way: In the United Kingdom, there are 13 births for every 1,000 people in the population, while in Thailand there are 20 births for every 1,000 people. The crude birthrate of Thailand is nearly half again as large as that of the United Kingdom.

---

[1]All numbers are estimates for 1994, and are based on a recent census, or on an official national estimate, or on United Nations or World Bank Projections (Population Reference Bureau, 1994).

This comparison between Thailand and the United Kingdom is an illustration of how rates can be used to make meaningful comparisons. While we intentionally chose two nations with very similar population sizes, we could have chosen nations of very different sizes, and still the resulting comparison would have been valid. The population of the United States, for example, is about 260 million (1994 data). The U.S. crude birthrate is 16 per 1,000 people in the population (1994 estimate). When we compare the crude birthrate of the United States with the other two countries, we can say that the birthrate of the United States is between those of the United Kingdom and Thailand, but somewhat closer to the former.

As the term "crude rate" suggests, much more refined rates for measuring fertility do exist. They are "refined" in the sense that the number of births are compared only to the number of those people in a population who are likely to be "at risk." Obviously, when the number of births is compared to the total population, as in the example above, many people included as part of the total population are not able to bear children (e.g., males, young children, postmenopausal women, and others). But for the time being, it is enough for us to consider only the crude rates for births, deaths, and migration. More refined measures will be presented in Chapters 6, 7, and 8.

Mortality

The second major demographic process is related to death and dying. In demographic terms, this is *mortality*. The principal and most commonly used mortality measure is the *crude death rate*, which is calculated in much the same way as the crude birthrate: The number of deaths in a year is related to the size of the total population. In 1993, there were 2,268,000 deaths in the United States, while the total population was 257,900,000 (National Center for Health Statistics, 1994). Once again, we calculate a decimal fraction and multiply by 1,000 to arrive at the crude death rate for that year:

$$\frac{\text{No. of deaths in 1993}}{\text{Total pop., United States, 1993}} \times 1,000 = \frac{2,268,200}{257,900,000} \times 1,000 = 9$$

Thus, in the United States there were 9 deaths[2] for every 1,000 people. After calculating this rate, we could compare it to the death rates of earlier years to see what the mortality trend has been in the United States. We could also compare the death rate of the United States with the death rates of other countries.

While the data required to calculate the crude death rate are usually

[2]The actual decimal fraction was 0.00879, which multiplied by 1,000 produces 8.79. While we have rounded to the nearest whole number (9), birthrates and death rates are often reported to one decimal place.

available for most populations, the resulting measure has some of the same disadvantages as the crude birthrate. Older people and infants in the population have a greater risk of dying, and thus populations with very different age compositions may produce misleading comparisons. More refined measures of mortality, where the age of the population is taken into account, will be presented at the end of Chapter 7.

Another widely used measure that reflects mortality is life expectancy at birth. *Life expectancy at birth* can be thought of as the average number of years newborns may be expected to live under the death rates that prevail at different ages (age 1, age 2, age 3, etc.). Life expectancy is a measure of mortality because there is a general inverse, or negative, relationship between the death rate and life expectancy—the higher the death rate, the lower the life expectancy of a population.

While the method for calculating life expectancy at birth is quite detailed, most people seem to have an intuitive sense of its meaning. For example, most people find it significant that since 1920, in the United States, life expectancy at birth has increased by over 20 years. In 1920, life expectancy at birth was 54 years, but by 1990, life expectancy at birth had increased to approximately 75 years. These life expectancies were calculated for the total U.S. population, males and females combined. It is well known, however, that life expectancy for males in the United States is considerably lower than that for females (because males have higher death rates at all ages). Appendix B illustrates that difference, as well as the difference in life expectancy between whites and blacks in the United States. Appendix B also provides the details about how life expectancy is calculated.

Migration

The third major demographic process is migration, the act of changing one's residence from one place to another. There are two basic ways of calculating the migration rates of populations. Migration rates can focus on the amount of migration *into* an area or on the amount of migration *out of* an area. These rates are calculated in much the same way as crude birthrates and death rates (see Chapter 6). And, like the other rates, migration rates allow us to compare migration differences between two or more populations, or the migration rates of a single population over time.

The three demographic variables we have been discussing—fertility, mortality, and migration—are the immediate demographic causes of all population changes. At a fundamental level, only these three demographic variables can change the size, rate of growth, density, and characteristics of populations.

To round out this introductory discussion of the basic population variables, we will discuss briefly the idea of population characteristics, and the population outcomes of size, density, and growth.

Population Characteristics

When demographers speak of the characteristics of a population, they may be referring to many different things, depending on the particular population under consideration. In the United States, for example, the characteristics of the population may include race (blacks, whites, Asians, and others), or religion (Protestants, Catholics, Jews, and others), or residence (rural and urban). These characteristics may or may not be found in the populations of other countries of the world, and even if they are found elsewhere the cultural definitions may vary greatly. But there are two population characteristics that are found in every population, and their meanings are relatively consistent from one society to another: age and sex.

Age and sex are biological attributes common to every person in every population. Therefore, every population can be described in terms of its age and sex composition. Some populations are relatively young and some populations are relatively old. Some populations will have a higher proportion of males and some will have a higher proportion of females.

The age and sex characteristics of populations are shaped by the combined influences of births, deaths, and migration. A high proportion of young people usually means that the population has a high birthrate, but it can also mean that life expectancy is low. A youthful population can also be produced by a high level of in-migration since migrants tend to be young adults, and young adults tend to have young children.

The age of a population may be measured by using some measure that reflects the average age. The *median age* of a population is the most commonly used measure of average age, though the *mean age* (an arithmetic average) could also be used. The median age of a population is the age at which half the population is older and half is younger. The median age of the U.S. population in 1993 was 32. In 1970 the median age was 28, while in 1980 it was 30. Clearly, the median age of the U.S. population has been increasing over the last quarter century.

A second way of measuring the age of a population is to measure the proportion or percentage in some significant age category. The percentage of a population under 15 years of age will reflect its relative youthfulness. The percentage of the population aged 65 and over will reflect its relative elderliness. The percentage of a population that is female and in the childbearing years of life (generally ages 15–49) reflects a significant age and sex category. In Chapter 5 we will present more elaborate measures of the age composition of populations.

The sex composition of populations can be shaped by all three demographic processes, but is most likely to be influenced by mortality and migration. While the numbers of males and females at the time of birth are nearly equal, slightly more males than females are born. In the United States today there are about 106 males born for every 100 females, a ratio similar to that of many other countries. In some populations, that ratio is closer to equality, since male fetuses are not as able to survive under harsh living conditions. In yet other populations, as in contemporary China, the ratio of males to females at birth is often extremely imbalanced in favor of males. The high proportion of males at birth is believed to be related to the use of abortion (or even infanticide) among Chinese parents who prefer a male child and who have agreed to have only one child. (China's one-child policy and its implications will be discussed in Chapters 7 and 10.)

Migration often influences the sex composition of populations since males and females are likely to have different migration patterns. The state of Alaska, for example, which has a relatively high percentage of population resulting from migration from other states, has historically had more males than females in its population. In 1990, the Alaskan population had 111.4 males for every 100 females (U.S. Census, 1992).

The sex composition of populations is typically expressed in the form of a ratio called the sex ratio. The *sex ratio* is calculated by using the following formula:

$$\frac{\text{No. of males}}{\text{No. of females}} \times 100 = \text{Sex ratio}$$

Population Size, Density, and Growth/Decline

A fundamental demographic variable is *population size*, the number of people in a population. The population in question may be that of a society, a nation, or, even the world. A related, but distinctly different, demographic variable is *density*, which is the number of people living in some areal unit. Commonly used measures of population density include the number of people per square mile or the number of people per block (in urban places). Density and size are different measures of population because even countries with very large populations can have some areas with very low density. China, with a population size of over 1 billion people, is such a country; its cities and river valleys are densely settled, while its mountainous and desert areas are sparsely settled.

Populations, over time, can grow, they can decline, and at least hypothetically they can remain unchanged. Population growth is usually the focus of attention among demographers, since almost every population in the contemporary world is growing, although some only very slightly. *Population growth* is the number of people added to a population in a given year, and is expressed as a percentage of the base population.

The world's population, for example, is currently growing at a rate[3] of

---

[3]The world *rate* is often used in connection with population growth; demographers conventionally speak of a population's "growth rate." But they are not using the word *rate* in the same way as it was introduced above. Growth rate, as the text says, is best understood as a *percentage rate of growth*.

1.6% per year. The growth of the world's population may be calculated if we know the number of births in a year, the number of deaths in that same year, and the size of the world's population at the beginning of the year. Migration can be ignored when we are dealing with the world's population, since there cannot be a net loss or gain due to migration. The calculation formula for the world's population is:

$$\frac{\text{No. of births} - \text{No. of deaths}}{\text{Total pop.}} \times 100 =$$

The resulting number, since we have multiplied by 100, can be expressed as a percentage. In the case of the world's population, the *addition* of about 90 million more people each year, with a total world population of 5.6 billion, yields the rate of growth previously mentioned: 1.6% (Population Reference Bureau, 1994).

While a population growth rate of less than 2% in a year may seem like a small amount, the cumulative effects of this growth can be surprisingly high. Since individual countries, and even whole continents, have population growth rates as high as 2, 3, and in some cases nearly 4%, the numbers can be even more impressive.[4]

One easy way to get a sense of how fast a population grows at these percentages is to ask how long it would take for a population to double. A population growing at 1% per year will double in 70 years. Growing at 1.6%—the world's present rate of growth—a population will double in about 44 years. At 3% growth each year a population will double in 23 years.

A convenient method for calculating *doubling time* is to divide the number 70 by the percentage growth rate. As an example, a country such as the Ivory Coast, with its growth rate of 3.5%, will double its population in 20 years (70/3.5 = 20). The U.S. population is growing at 0.7% per year and will double in approximately 98 years (Population Reference Bureau, 1994).

There is great variation in population growth from the more economically developed parts of the world to the less-developed parts. Among the more-developed countries, the doubling time is about 264 years, while among the less-developed countries, the doubling time is about 36 years (Population Reference Bureau, 1994).

WORLD POPULATION GROWTH

The population of the earth is growing at a rate that leads to its doubling in about 40 years. In numerical terms this means the addition of about

---

[4]If we are dealing with the population growth of a country or continent, the net migration (in-migration minus out-migration) will have to be added to, or subtracted from, the birth and death numbers. The number of people added to the population would then be: number of births − number of deaths ± net migration.

5.5 billion people during the lifetimes of most readers of this book. This percentage growth rate, as well as the numbers being added yearly, are far, far greater than the earth's population growth has been through most of human existence.

Before we begin our consideration of contemporary demography, we will briefly review the history of world population growth. We will see that before the modern era, the estimates of world population are neither precise, nor based on actual population counts. Nonetheless, the estimates, even though not firmly based on demographic data, will show us how extraordinary human population growth has been in the last 300 years and especially in the 20th century.

The Population of the Past

In the long record of human life on earth—a record that by some archaeological accounts may go back as much as 2 million years—the "population problem" has generally not been one of overpopulation. While there were very likely many instances where the resources of an area were overwhelmed by the numbers of humans living there, it is probably the case that through most of human existence the most basic struggle has been to *maintain* the population size of human groupings.

The last assertion is difficult to document, since prehistory data are scant and inconclusive, but we do know that after 2 million years of existence as a species, the total number of human beings was not very large. It is estimated that even after the world was well into the medieval period, about 1000 A.D., there were fewer than 300 million people on earth (Durand, 1977). In all the preceding thousands of years, the population of the human species had reached a size that was about as large as the combined populations of the United States and Canada today (approximately 290 million people).

A world population of 300 million in the year 1000 may be viewed in another way. With the present growth rate of the world's population, it now takes less than 4 years to add 300 million more people. This comparison begins to give us some perspective on how rapidly the world's population is growing today.

If one goes back in time more than 2,000 years, estimates of the world's population are very uncertain since they are based largely on circumstantial evidence. Durand (1977) has studied the few estimates that have been made for the beginning of the Age of Agriculture (10,000–8000 B.C.), and he believes that the population at that time may have numbered between five and 10 million people. The uncertainty of our knowledge is reflected in the fact that the figures 5 and 10 million are two quite different estimates, but we have no basis for preferring one over the other (Durand, 1977, p. 260). These estimates are not based on any records, of course, not even on archaeological evidence. They are instead inferences made from information we have about primitive nonagricultural populations and their population densities (Durand, 1977).

For the period from the beginning of agriculture and the year 0 A.D., we again know very little about the world's population size, except that it probably grew at a relatively brisk pace as agriculture spread around the world. By the years 0–14 A.D., Durand judges that the world population was probably between 270 and 330 million.

The next 1,000 years did not add to those numbers substantially, since, as we have already mentioned, the best estimates for the year 1000 are between 275 and 345 million. After the year 1000, and especially after 1650, the size of the world's population started increasing fairly rapidly. And during the 20th century the size of the world's population has grown at a rate that is truly remarkable.

In all likelihood, the world first achieved a population of 1 billion in the early 1800s. By 1930, the second billion had been added. While this addition took somewhat more than a century, the third billion was added in only 30 years, by 1960. The fourth billion was added in about 13 years, coming in 1973. Since then, the rate of growth has slowed down a little, but with a larger base population the world now takes only slightly more than 10 years to add 1 billion people. Current population projections indicate there will be over 6 billion people by the year 2000, and 7 billion by the year 2010.

Recent Growth of the World's Population

This brief review of the history of the world's population growth reveals that in the last 300 years, and especially in the 20th century, population growth has taken on a new and very different character. When the growth of the world's population is shown graphically, as in Figure 1.1, it is difficult to depict accurately the enormity of this change. It requires a population scale that accumulates to only 5 or 10 million people in more than 1 million years (up to the beginning of agriculture, 10,000–8000 A.D.) and yet now grows by nearly 1 billion people every decade. Such rapid growth clearly calls for explanations of the cause. There are some well-developed answers to the question of why population growth in the last 300 years has been so rapid, but there are also some mysteries. Some of the most rapid population growth of the world occurred in Europe, especially in the 18th and 19th centuries, and is generally linked to declines in mortality rates then that came with economic development. This topic will be discussed in Chapter 10. For some other parts of the world, according to Durand (1977), there were also increases in population size, even though economic development had not begun in those areas. China, for example, for which there are fairly high-quality population data during this period, shows population increases in the 18th century comparable to those in Europe. The same is true for Russia. In the 19th century, there were also apparent (the statistical data are less complete)

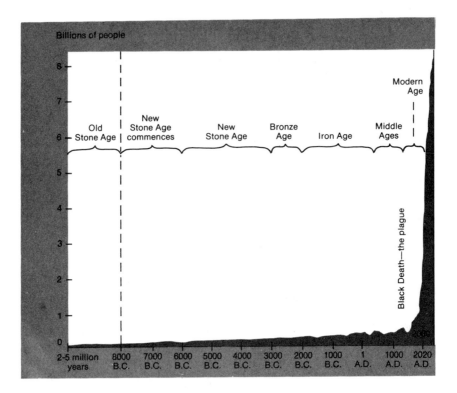

FIGURE 1.1. World population growth through history. From Population Reference Bureau (1984).

population increases in North Africa, southwest Asia, India–Pakistan–Bangladesh, southeast Asia, and Middle and South America (Durand, 1977). These non-European population increases in the 18th and 19th centuries present a major demographic mystery for historical demographers to solve in the future.

Currently, the overall *rate* of population growth is declining, and as in the past the pace of growth is not uniform throughout the world. Great portions of the world's population, especially in Africa and Latin America, are growing at a very rapid rate. Ignoring net migration, the Latin American population could double in the next 35 years and the African population in 24 years. Some countries, for example, Iran and Iraq, are currently growing at rates that will double their populations in 19 years (Population Reference Bureau, 1994).

The European population, at its current rate of growth, will require over 1,000 years to double. Indeed, a number of European countries are currently losing population: Bulgaria, Estonia, Germany, Hungary, Latvia, Romania, Russia, and Ukraine (Population Reference Bureau, 1994).

What should be the response to the rapid rate of growth in many countries? If the objective is to lower the rate of growth in many countries of the world, how can this be accomplished? Indeed, can it be accomplished? Do countries with declining populations, or low growth, even have a population problem? These questions will be considered in the remainder of this book.

SUMMARY

Demography is the scientific study of population. Three key demographic variables are fertility, or childbearing; mortality, or death; and migration, changing place of residence. These three basic demographic processes are often measured as rates: the frequency of an event relative to some population that may experience that event.

In addition to their interest in these three processes, demographers also are concerned with the characteristics of populations, especially age and sex characteristics. The age composition of populations is influenced by fertility, mortality, and migration. The age of a population is typically measured by an average, such as the median, or by the percentage of the population in some significant age category (e.g., youth, the elderly, childbearing women).

The sex composition of populations is influenced primarily by differences in mortality and migration, but there is a sex imbalance at birth, with more males generally being born. The sex composition of populations is measured by the sex ratio, which is expressed as the number of males per 100 females.

The remaining basic population variables are population size, density, and growth/decline. Population size is simply a count of the total population, while density is the number of people living in some areal unit (e.g., number per square mile). Population growth (or decline) is the number of people added to (or subtracted from) a population in a given year, and is expressed as a percentage of the base population.

The population of the world grew very slowly over the first two million years of human existence. Even the population spurt produced by the advent of agriculture around 10,000 B.C. led to a population of no more than some 300 million people in the year 1000 A.D.

The world's population did not reach 1 billion until the early 1800s. By 1930 a second billion had been added. Thirty years later, in 1960, the population was at 3 billion. Since then more than 2.5 billion more people have been added to the world's population. Currently, 1 billion people are being added to the world's population in slightly over 10 years, but some countries of the world are growing rapidly, while others are growing hardly at all.

2
THE BEGINNINGS
OF POPULATION STUDY

I t is rare that the beginning date of a science can be precisely identified, but the beginnings of the scientific study of population, the science we now call demography, can be logically assigned to January 25, 1662. On that date John Graunt, a London shopkeeper who also had a taste for scholarship, published a small volume titled *Natural and Political Observations . . . Made upon the Bills of Mortality* (Graunt, 1662/1939), in which he reported his systematic observations about the population in and around the city of London. Graunt's observations were in the form of generalizations about the patterns and regularities in births, deaths, and migration in the English population. His observations were based largely on city and church records that had been kept on deaths (the bills of mortality) and births (church christenings).

Graunt's key innovation was to apply the scientific method to the study of population. In Graunt's time science was largely limited to the observation and description of naturally occurring events. That is what Graunt did with regard to births, deaths, and other aspects of population, and it is this work that gives Graunt a claim to having started the scientific field we now call demography.

While Graunt can be given credit for founding the science of demography, many other people before his time had thought about and even written about population. We may even speculate that concerns about population, and in particular about population size, predate the historical period of human existence. As soon as early humans began to identify with their groups, tribes, or bands, they probably started to have some concerns about the numbers in their population. Certainly, the historical record, about which we can be more certain, includes examples of interest in human populations that go back as far as we have written records.

There are two distinct strands in the history of population study. One strand reflects a concern about the way in which population relates to or affects the well-being or survival of the group, tribe, or society. Concerns about declining populations and overpopulation illustrate this strand. The other strand reveals an interest in making scientific statements or generali-

12

zations about how and why populations grow and decline. This is the strand of population study John Graunt started when he wrote his treatise on population in the 17th century. Both strands in this history are important, and as we will see once we reach the scientific era, they are frequently intertwined. Even today, the scientific study of population is often justified in terms of the significance it has for the welfare of human groups and societies.

We will return later in the chapter to the work of John Graunt and his contemporaries, and to the beginnings of the scientific study of population. But first we will go back in history to the earliest writings that reveal an interest in population.

EARLY POPULATION CONCERNS

Early writings on population are typically contained in the works of social and political philosophers who were primarily concerned with the effects of population size on political or economic processes. Some of the earliest discussions addressing the relationship between societal well-being and population appear in the writings of the ancient Chinese philosophers.

The Early Chinese

Confucius and other Chinese scholars gave consideration to the ideal proportion between land area and the number of people. Confucius held that it was the responsibility of the government to move people from overpopulated to underpopulated areas. He also noted several factors he believed would act as checks on population growth, naming among them such things as insufficient food supply, war, premature marriage (resulting in high infant mortality), and costly marriage ceremonies. It is clear from these latter concerns that Confucius and other early Chinese sages had an awareness of the way in which economic, biological, and social factors shaped and influenced population. Their statements about population were, of course, only speculative and nonscientific; we possess no record that indicates that any empirical or statistical studies were conducted by the sages to check or verify the hypothesized relationships (United Nations, 1953).

The Greeks

If we turn to the classical period in Greece, we find that two of the major philosophers of the Western world, Plato and Aristotle, discussed population. They were both concerned with the ways in which population size could affect the political workings of the city-state, the principal political unit of Greece in classical times. Both Plato and Aristotle devoted attention to what would be the best, or optimum, size of the city-state. The optimum-size city-state, in their view, would have a population in which the well-being and the security of the citizens could best be maintained. They were also interested in determining a population size that would allow the most efficient administration by the government. Plato was more specific with regard to population size, setting the optimum number of citizens at 5,040. However, since women, children, and slaves were not counted as citizens, the actual number of people living in Plato's ideal city-state would have been much larger than 5,040. The total population of such a political unit would have ranged up to 10 times that number (United Nations, 1953).

While Plato allowed that the number of citizens might possibly vary according to local circumstances and conditions (e.g., the availability of land and relations with neighboring groups), he nevertheless held firmly to the exact number of 5,040 citizens and believed that the state should make an effort to maintain that number. The primary mechanism he suggested for retaining that precise number of citizens was the inheritance system. The land of a newly formed city-state was to be divided equally among the 5,040 citizens. Each family was then to pass its allotment on to a single, preferred male heir. In Plato's words:

> Let the possessor of a lot leave the one of his children who is best loved, and only one, to be the heir of his dwelling . . . but of his other children, if he have more than one, he shall give the females in marriage . . . and the males he shall distribute as sons of those citizens who have no children. (quoted in Jowett, 1937, pp. 531–532)

If some citizen families appeared to be having too many or too few children, Plato considered it reasonable that the government and the elders of the society exert the pressures necessary to keep their numbers of offspring at the desired level. Such pressures were to include rewards, social stigma, advice, and rebuke. If all else failed, the size of the population was to be held at 5,040 citizens by sending the surplus numbers to a colony.

Aristotle, while less explicit than Plato about the exact number of citizens who should live in a city-state, clearly supported the notion that the size of the population should be stabilized and controlled by the government. He argued that a population could become too large to be properly governed. Aristotle noted that the number of citizens of a state might be used to judge the quality of that state, and he cautioned that "a great state is not the same thing as a state with a large population . . . it is difficult and perhaps impossible for a state with too large a population to have a good legal government" (Aristotle, 1932).

Aristotle was also concerned with controlling the size of the population and warned about the dangers of a population that was too great for its resources or too large to allow for an orderly life. He advocated government control of the number of births per family, approved of abortion under certain circumstances, and recommended the exposure of any infant who was born deformed (i.e., infanticide).

Both Plato and Aristotle were interested in population size because they believed that the size of a population would affect the well-being of the society and the effectiveness of the governing processes. Population size, from this point of view, influences and affects the social and political order of society. Both Plato and Aristotle also were interested in the societal methods by which population size could be controlled (e.g., inheritance systems, rewards and stigmas, abortion and infanticide). It was their surmise that these different economic and social measures would influence people's choices concerning the number of children they would have and thus total population size. These hypothesized relationships between economic/social variables and demographic variables were probably not tested with any empirical research, since the Greek philosophers were more inclined toward logical deduction than empirical demonstration. Scientific inquiry generally, and scientific demography specifically, were still about 2,000 years in the future.

The Romans

Following the classical Greek period, the focus of history in Western civilization shifts to the Roman world. Roman scholars, philosophers, and politicians were generally in favor of an increasing population. Early in Roman history a larger population was needed to provide protection from threatening neighbors. Later, as the Roman Empire expanded to areas ranging from Britain in the north to Africa in the south, more and more people were needed to occupy the conquered lands. The need for military manpower was a strong factor influencing Roman views on population. Since Rome's rulers favored a growing population, they tended to disapprove of celibacy, while they gave support to marriage and childbearing. Legislation was passed during the period of the Roman Empire that aimed at stimulating both marriage rates and birthrates (Hutchinson, 1967; United Nations, 1953).

Mercantilist Economic Philosophers

European writings on population began to appear more frequently by the 17th century. Many political and social philosophers offered their ideas on the influence that population might have on society. Increasingly, the emphasis came to be placed on the relationship between population and the economy. Characteristic of this emphasis were the views of the mercantilists of the 17th and 18th centuries.

The mercantilist philosophers argued that a large and growing population would lead to economic and military advantages for the state blessed with such a population. They held that the accumulation of wealth and power by the state—particularly by amassing large supplies of precious metals—could be enhanced by a large and growing population. The aim of the mercantilists was not to increase per capita income but to increase the aggregate national income. They argued that a growing population, which would produce a growing labor force, would depress the wages of workers, make competition among them keener, and thus increase the gap between the cost of wages and the national income. Following this line of reasoning, the mercantilists advocated a high rate of population growth to enhance the economic goals of the state.

THE EMERGENCE OF SCIENTIFIC POPULATION STUDY

Early population writings, as we have seen, focused mainly on the way in which population can affect the well-being of the state or economy. Research on population, conducted in a scientific manner, was virtually nonexistent until the 17th century when John Graunt and others started their work. It is now time to take a closer look at his work and the work of a group of scholars who called themselves *political arithmeticians*.

The scholars known as political arithmeticians were so called because they used arithmetic methods to study the numbers of people in various political entities. The political arithmeticians were greatly influenced by the rise of science during the Renaissance. Scientific research in astronomy, physics, and physiology, which was based mostly on observations of the natural world, was firmly established by the 17th century. Scientists of the stature of Bacon, Galileo, Harvey, and Kepler had proven the usefulness of scientific observation for understanding the physical world. The political arithmeticians were contemporaries of Sir Isaac Newton, who was establishing scientific inquiry even more firmly as the best means of verifying the uniformities and the laws of nature. Greatly influenced by these intellectual currents, the political arithmeticians recognized the potential for a science based on observations of human populations.

However, applying the scientific method to establish facts about human populations was only one part of the motivation of the political arithmeticians. The second motive was related to the goals of mercantilism, for many political arithmeticians acknowledged they were setting out to measure the resources of the state, or the Crown, in terms of land, capital, wealth, and population.

Along with John Graunt, a second eminent English political arithmetician was Sir William Petty. It was Petty who actually coined the term "political arithmetic" (Hull, 1899/1963). Graunt and Petty knew each other well; when Graunt's home was destroyed by the Great Fire of London in 1666, Petty provided the financial assistance to rebuild it (Sutherland, 1963).

The works of Graunt and Petty reveal the two sides of political arithmetic—scientific research and political service—but with different emphasis. Graunt's work gave greater prominence to the scientific study of population, while Petty's work was more heavily weighted toward measuring the resources of the state or Crown.

John Graunt (1620–1674)

The "bills of mortality" referred to in the title of Graunt's book were the weekly reports of all deaths that had occurred in the parishes of London. These reports, which were actually reports of burials, had been started in London in the early years of the 16th century for the purpose of keeping an account of plague deaths. The severity of plagues could thus be judged from these accounts since plague-related deaths were separated from other causes of death. Even after the plagues had ended, the bills of mortality continued to be kept, and by 1563 they were regularly published (Sutherland, 1963).

On the base of information found in the bills of mortality, augmented by christening data collected from England's parishes, Graunt made some very insightful observations about the population of 17th-century England. To appreciate Graunt's accomplishment, it is necessary to understand that in his time almost nothing was known about the population or population processes; indeed, even the population sizes of England and its largest city, London, were unknown.

Although many people in the 17th century believed they knew something about the size of London, Graunt was very skeptical about the accuracy of their beliefs. Many Londoners estimated the population of their city as being in the millions. Graunt mentioned that he was especially provoked when one day he heard one of the eminent men of the city say that London's population had grown by 2 million people since the year 1625 (the year of an epidemic plague). That amount of growth would have meant a London population of 6 or 7 million people (Graunt, 1662). Graunt believed the number to be much less, and set about to support his view.

First, by using the average number of christenings over a period of years, Graunt estimated the number of women who could bear children. From this number, he estimated the total number of adult women, and from this, he estimated the total number of families living in the city. He guessed that the average family (or household) had eight members: a husband, wife, three children, and three additional servants or lodgers. By multiplying the number of families by eight, Graunt arrived at a figure of 384,000 people in London proper. To this number, Graunt added another 20% because of some outlying parishes not covered by the christening data. His total number of inhabitants thus came to 460,000.

But Graunt was not satisfied with just one estimate. He also used data on the yearly number of deaths to estimate the total number of families. He believed there were probably three deaths for every 11 families. Since he knew there were about 13,000 deaths yearly, he was able to calculate how many people, at eight per family, must have been living in the city of London. As a third procedure, Graunt used a map of London, which had been drawn to scale in yards, and estimated that 54 families were likely to live in a 100-square-yard area. Both of these latter methods of estimation indicated that the number of families was about 48,000. Using the same family size as before, and adding the outlying parishes, Graunt concluded on the basis of three separate methods of estimation that the population of London in his time was 460,000 people. His use of three different methods, now called *triangulation*, is commonly used in many kinds of scientific research to make estimates or to reach conclusions.

Graunt's estimate of the size of the London population may not have been precisely correct, but it was probably much closer to the truth than the much larger prevailing estimates of his time. After Graunt published his estimates in 1662, he learned about an enumeration of London's population that had been made for tax purposes in 1631. This report supported his estimate of less than half a million residents of London.

Graunt was equally meticulous when he made estimates of the total population of England. Again, using three separate approaches, he concluded that the total population of the country in his time was about 6.5 million people.

But Graunt did much more in his small book than make population size estimates. He also theorized about the processes that shaped and influenced population, especially that of London. Graunt made a strong case for the importance of migration from rural parishes as the primary factor adding to the population of London. Since London had more burials than births each year and more and more housing units were added each year, he reasoned that the growth of the population had to come from migration. The rural parishes had the opposite circumstance, since they usually had more births each year than burials. This migration of people from rural to urban places first noted by Graunt is a trend that continues to be true in most Western societies today. In underdeveloped societies today, this rural to urban migration pattern is often very pronounced. In many countries of Latin America and Asia this pattern has led to the rapid population growth of cities.

Graunt was able to show, even without precise data on the ages of those dying, that about one-third of all children born died before age 5. He is also credited with constructing the first crude life table. A *life table* is the statistical method that leads to estimations of life expectancy, which we discussed in Chapter 1 (see Appendix B). Today a life table would be constructed based on data regarding the probability of dying in every single year of life. Graunt had no such information since the ages of those who died were not reported in the bills of mortality. On the basis of the recorded *cause* of death, however, he estimated how many people out of 100 would die in the first 6 years of life, about 36, and how many more would die at 10-year intervals thereafter. While Graunt's life table is not quite the same as a contemporary life table, it did serve as a prototype for the more sophisticated forms that followed.

Graunt also observed that in both the country parishes and in London, more males were born than females. In London, the ratio was 14 males to 13 females, while in the country, it was 16 males to 15 females. This greater number of male births is now a well-known biological fact, but in Graunt's day it was apparently unknown. Despite the substantial amount of data available to him, Graunt displayed a praiseworthy scientific skepticism. He noted that there may be other places where more females are born than males, and he urged those who were curious to look into the matter more carefully (Hull, 1899/1963; Graunt, 1662).

The contributions of John Graunt to the birth of demography are important in several ways, but his major contribution is the way he carefully assessed the quality of his data and cautiously interpreted his statistical calculations. His care in establishing a scientific method for utilizing the observations made by others is a major contribution (Kreager, 1988). Graunt set a standard of scientific rigor and caution that may well have shaped the development of the field. Among the social sciences, demography continues to be noteworthy for the attention its practitioners give to data quality and methodological precision.

Though the principal thrust of Graunt's work was an early form of scientific demography, he also provided some information that was useful for the government. For example, his estimate of the population of England was important because it informed the government on how many men would be available for the army and navy. This illustrates well how the political arithmeticians were motivated to do their work so as to better advise the government concerning its resources in people and things. This task was especially central for Graunt's fellow political arithmetician, Sir William Petty.

Sir William Petty (1623–1687)

Petty was a more political man than Graunt. Both his work and his writing reflect an involvement in the political life of his time. Born the son of a poor clothier, he rose to become one of the most respected and richest men of 17th-century England. He remained in the good graces of all England's governmental heads from Cromwell through a succession of Restoration kings. Petty was so well regarded that he was knighted in 1661 and was granted extensive land holdings in Ireland.

It is not surprising that one of his works on political arithmetic was written to apprise the king about the resources of his country, its trade, and its people. When Petty set out to learn how many residents there were in England, it was for the purpose of informing the king how his realm's population compared with those of Holland and France. The comparison carried out in terms of each country's potential for raising military forces (including seamen) and in terms of the economic value of each country's people.

Petty calculated that English citizens were, on the average, worth 80 pounds of sterling silver. He derived this figure after calculating how much the labor of people added to the total national income and then dividing this sum by the total number in the population (Hull, 1899/1963). Petty pointed out that once such a value had been determined, it would be a simple matter to calculate the monetary losses sustained by the Crown as a result of the kingdom's population losses due to plagues and wars.

Petty was also the author of *The Political Anatomy of Ireland*, in which he analyzed the value the Irish lands and people had for the English Crown. In the case of the Irish people, however, Petty did not find them equal in value to that of English citizens. In noting the numbers of Irish people who had been lost in an earlier rebellion, Petty judged that each one might be given the same value as "Slaves and Negroes," or about 15 pounds of silver on the average (Hull, 1899/1963).

While Petty emphasized in his work the political and economic significance of populations, he also produced some data that were purely demographic. By using the hearth-tax records of Ireland, he made estimates of the total Irish population in the middle of the 17th century. These estimates are judged by today's scholars as too low, because Petty generally accepted at face value the number of houses on the hearth-tax records. He neglected to take into account the number of households that had been either missed or unreported by the tax collectors (Connell, 1950). While his estimates of the size of the Irish population may have been too low, today's demographers can make their more accurate estimates of the 17th- and 18th-century Irish population in part thanks to his groundbreaking efforts.

Other Political Arithmeticians

Among the other prominent political arithmeticians of England and the Continent, several made special contributions. Gregory King (1648–1712), following Graunt, carried out a more extensive study of the total population of England and Wales. He used the hearth-tax returns of England and Wales to establish the number of dwellings in the country and then estimated the average number of persons per home to determine the total population.

A German political arithmetician named Johann P. Sussmilch (1708–1767) was trained as a doctor, but also served as an active minister of religion during most of his life. His work as a political arithmetician was most prominently featured in a book that brought together much of the available statistical data on population for his time. This book had as its theme a fundamental theological argument. Sussmilch believed that many of the regularities and consistencies found in population data were evidences of divine intervention, which he later called the "natural order" (Hecht, 1987). Sussmilch saw the workings of the hand of God in such things as the slightly larger number of boys than girls born each year. He reasoned that since boys lead more dangerous lives than girls, the number of males and females in a population will even out by the age of marriage. God, foreseeing the higher death rate for young males, caused slightly more boys to be born to ensure enough partners for girls of marriageable age. Sussmilch represents an extreme in the nonscientific interpretation of scientific data, though many of the other political arithmeticians allowed their religious views to enter into their interpretations of population facts (Mayer, 1962).

In summary, the political arithmeticians were distinguishable from the earlier writers on population because they relied on empirical data to make their observations. Their writings did not advance any scientific theories, but instead concentrated on facts and data. The political arithmeticians, with their emphasis on fact gathering by scientific methods, were the first scholars to establish one of the two essential components of modern population study. The other component—an adequate theory of population—was still lacking. The last of the major historical landmarks in the development of demography is the introduction of a theory of population that has stimulated work in the field up to the present time: the famous *Essay on the Principle of Population* by Thomas Robert Malthus.

THE CONTRIBUTIONS OF MALTHUS
TO THE STUDY OF POPULATION

Thomas Robert Malthus was born in 1766, just a little more than a century after John Graunt published his population research. But the work for which Malthus gained his fame did not fall primarily into the tradition of scientific, empirical demography that the political arithmeticians had started. Malthus's principal interest in population was more closely connected with the other side of political arithmetic, the side that focused on the relationship between the population and the economy. The political arithmeticians, like the mercantilists, simply assumed that population was an economic asset. A large population was viewed as good for the economy, and a growing population was even better. This view of population is what has been called the *optimistic view* (Hutchinson, 1967).

Malthus did not view the relationship between the population and the economy optimistically. Indeed, his views were so much on the pessimistic side that his very name has come to be synonymous with a pessimistic or negative view of population. The *pessimistic view* is that a large and growing population is a burden for the society and a cause of poverty, misery, and other societal problems.

While Malthus has become the most famous exponent of the pessimistic view about population, it is a matter of historical record that many other scholars had many of the same ideas long before he did. Even Plato and Aristotle, as we have seen, did not favor ever-larger populations. In the 16th century Giovanni Botero, an Italian scholar, expressed many of the ideas for which Malthus later became famous. Botero maintained that the reproductive power of humans is very great and that the means of subsistence (particularly food) are limited. These limitations, Botero thought, would restrict the growth of population numbers, often by increasing levels of mortality (Hutchinson, 1967).

Within England there were also writers who held the pessimistic view long before Malthus wrote. One such scholar, Richard Hakluyt, advised Queen Elizabeth I (in the late 1500s and early 1600s) that England was overpopulated and that an excessive number of people was the cause of many of the realm's social problems. Later, in 1673, the prominent English jurist Matthew Hale wrote: "With us in England . . . the more populous we are, the poorer we are (Hale, 1673, quoted in Hutchinson, 1967, p. 55). He also observed that the poor people in the population would multiply in a "kind of geometrical progression." According to Hale, two people, a father and a mother, by having children and grandchildren, would grow from two to four to eight in about 34 years. This idea of geometrical population growth later turned out to be one of the key elements in Malthus's ideas about population. Hale also saw population size being controlled by such things as plagues, epidemic diseases, famines, wars, revolutions, floods, and fires. This idea too appears in the writing of Malthus. It is almost certain that Malthus was influenced by all of these intellectual predecessors.

But it was not just European writers who influenced Malthus, for one of the most famous early Americans, Benjamin Franklin, also did so. Malthus specifically acknowledged Franklin's contribution in the 1803 edition of his essay (Hodgson, 1991). In 1751 Franklin published *Observations Concerning the Increase of Mankind* in which he observed that the number of children produced by couples could easily exceed the numbers needed for their replacement; that the availibility of subsistence could influence the size and growth of populations; and, at least in Europe, that high population density was responsible for many social ills. These views, taken as a theory of population, are essentially pessimistic (too much population will cause social and economic problems), but paradoxically Franklin also held some contrary values. He favored early marriage, large families, and a growing population, at least *for the American continent*. One interpretation of this contradiction in Franklin's ideas is that he made a distinction between the limited lands of European countries and the vast and rich continent of America that remained to be settled and developed (Hodgson, 1991). Excessive population growth, in his view, would have negative effects for Europe, but on the American continent population growth would be good for the country, at least in the short run. There is also the strong likelihood that Franklin preferred a growing resident population (largely English) over the possible immigrants from other European countries (Hodgson, 1991).

So there is ample evidence that by the time Malthus wrote his essay at the end of the 18th century the pessimistic view of population had been advanced by many different writers. No longer was it automatically assumed that a large and growing population was beneficial for either a country or the lives of its people. Instead, many writers were taking the opposite view, namely, that a rapidly growing population was a cause of societal and human problems. It was in this scholarly tradition that Malthus developed his "principle of population."

Malthus's *An Essay on the Principle of Population*

While Malthus had written some earlier scholarly work, it was *An Essay on the Principle of Population* that gained him instant fame. The work was first published anonymously in 1798. It was quite short, since it only presented the basic idea and a few selected arguments that Malthus wished to make. Five years later, in 1803, Malthus published the second edition of the essay, but this time his authorship and credentials were clearly presented. Furthermore, the second edition of the essay was not simply a presentation of the principle of population but a greatly expanded book (nearly four times as long as the first essay). The second edition, and the five editions that followed, included many chapters devoted to descriptions of the growth, and in some cases the decline, of populations in various parts of the world. Malthus accumulated these population examples and offered them as illustrations and supportive evidence for his theory (Petersen, 1979).

One revealing feature of the first essay can be found in its full title: *An Essay on the Principle of Population, as It Affects the Future Improvement of Society, with Remarks on the Speculations of Mr. Godwin and M. Condorcet, and Other Writers.* The "Mr. Godwin" and "M. Condorcet" referred to in the title were two contemporary utopian writers who had written treatises describing an ideal or utopian society. In the traditions of utopian writers generally, each had attributed the problems of their societies to existing social institutions. Each envisioned a society of the future in which the human and social problems of the world would have been eliminated.

Condorcet was a Frenchman who participated in the Revolution, but later, during the Reign of Terror, was sentenced to death. Before he was captured and imprisoned (he committed suicide during his imprisonment), he wrote *Progress of the Human Mind* (Condorcet, 1795/1970). This essay on the perfectibility of the human mind was soon translated into English. It appeared in England in 1795, where it was read by Malthus and by other scholars and intellectuals of his time.

Condorcet envisioned a time in the future when, through an evolutionary development of humans and their societies, all crime, injustice, ignorance, inequality, disease, and unhappiness would disappear. This would occur in a natural way as the rational minds of humans evolved to a higher level.

Similar ideas were advanced by the other target of Malthus's essay, the English utopian philosopher William Godwin, who published his utopian vision of the future under the title *Enquiry Concerning Political Justice and Its Influence on Morals and Happiness* (Godwin 1793/1946). Godwin's ideas were very much like those of Condorcet, since he too believed that through human reason an ideal world could someday be achieved. Godwin put great emphasis on the individual and opposed any social form that required cooperation between people. This basic position led him to reject the institution of marriage, all forms of organized religion, formal education, and any kind of government. Godwin was a complete anarchist and absolute individualist. The future he saw was one in which all existing institutions would disappear and people would live in independent but harmonious perfection.

After reading the Godwin and Condorcet essays, which offered a vision of perfect human life, Malthus discussed the issues they raised with his father, Daniel Malthus. Apparently, the elder Malthus was much more favorably disposed toward utopian ideas than was his son. In an effort to express his objections to the utopians more clearly, Robert Malthus wrote the first version of his essay. The opening line of his preface says, "The following essay owes its origin to a conversation with a friend, on the subject of Mr. Godwin's essay"; the "friend" was Malthus's father (Petersen, 1979).

The first essay was a refutation of the utopian idea that societal and human problems could be eliminated by the development and application of human reason. As Malthus saw it, the institutions of society were not perfect and could indeed cause some human problems, but there was a more fundamental source for many of the problems that beset humanity. The source of many human problems was what he called "the principle of population." This principle is relatively simple and can be presented in Malthus's own words:

> I think I may fairly make two postulata. First that food is necessary to the existence of man. Secondly, that the passion between the sexes is necessary, and will remain nearly in its present state.
>
> Assuming then, my postulata as granted, I say that the power of population is indefinitely greater than the power in the earth to produce subsistence for man.
>
> Population, when unchecked, increases in a geometrical ratio [1, 2, 4, 8, 16, 32, 64]. Subsistence increases only in an arithmetical ratio [1, 2, 3, 4, 5, 6, 7]. A slight acquaintance with numbers will show the immensity of the first power in comparison of the second.
>
> By that law of our nature which makes food necessary to the life of man, the effects of these two unequal powers must be kept equal.
>
> This implies a strong and constantly operating check on population from the difficulty of subsistence. This difficulty must fall somewhere; and

must necessarily be severely felt by a large portion of mankind. (Malthus, 1798/1960, pp. 11–14)

In a brief elaboration of his principle, Malthus expressed his reason for rejecting the utopian idea that societies and the lives of humans can be perfected. As he saw it, populations have a natural tendency to grow at a faster rate than the means of subsistence. It therefore would follow that at any given time a certain number of people would be suffering in some way. Malthus believed that societal and human problems could not be entirely eradicated. It would be nearly impossible to do so because they have their origins in two fundamental biological "truths." One was the natural tendency of humans to express their sexual passions and thus cause rapid population growth. The second was the limitation of the earth, which allows food production to increase much less rapidly.

The "strong and constantly operating checks on population" that Malthus spoke of in his principle, he later called "the positive checks." *Positive checks* are those things that increase mortality, including wars, famine, and especially disease. Malthus specifically pointed to the way positive checks increase mortality among the children of the poor. The basic reason for high mortality is the lack of nourishing food, but as Malthus noted, except in the cases of actual famine, the causes of death that manifest themselves are not necessarily starvation. The immediate causes are likely to be diseases and scarcities that prematurely "weaken or destroy the human frame" (Malthus, 1872/1973, p. 12).

Malthus also focused attention on what he called *preventive checks*, which could potentially control fertility and thereby limit the growth of populations. While Malthus believed that sexual passions were important driving forces for humans, he also believed that humans could use "moral restraint." Moral restraint would be demonstrated by those who waited to marry and have children until they could adequately provide for them. On this matter, Malthus adhered to a strict moral code. For him, deferring marriage was the only acceptable way to limit childbearing. People had to remain unmarried until they were financially secure. Contraception was not acceptable to Malthus, either within or outside of marriage.

If people were to limit childbearing by any means other than refraining from marriage, it would be, in Malthus's view, a vice. Vice included "promiscuous intercourse, unnatural passions, violations of the marriage bed, and improper arts to conceal the consequences of irregular connections" (Malthus, 1872/1973, p. 14). This list, in more contemporary terms, includes frequenting prostitutes, engaging in homosexual relations, and using contraception ("improper arts" in the quotation above) (Petersen, 1979).

Malthus's view on sexual matters reflects one of the troublesome dimensions of his scholarly writing. He frequently intermingled his political and moral views with his scientific principles and observations. The first essay, for example, included a long attack on the Poor Laws of England which provided economic aid to the poor. While his discussion was couched in terms of economic theory, there can be little doubt about his general political philosophy. Malthus's views were those of the privileged middle and upper-middle classes. From that elevated level, it seemed obvious that the poor were in their wretched condition because of their general improvidence and lack of foresight. While Malthus allowed that moral restraint could be practiced "in some degree through all ranks of society," he clearly did not expect much prudent forethought to be exhibited by the lower classes. He generally presumed that the combined forces of vice and misery (the positive checks) would be much more likely to shape the poor's lot in life. As he put it, "The positive check to population . . . is confined chiefly, though not perhaps solely, to the lowest orders of society" (Malthus, 1798/1960, p. 71).

The Place of Malthus in Population Study

There is no doubt that the place Malthus holds in the history of population study is secure. But the prominent position he holds is certainly not due to a shortage of critics. Many religious leaders of his time attacked Malthus for being sacrilegious. To many of these religious writers, his theory seemed to show a lack of faith in the "beneficence of God." Malthus was also morally suspect because he wrote so directly and openly about sexual passions and sexual vices. For the time in which he was writing, Malthus was very direct about sex.

While fellow religionists (Malthus was ordained in the Anglican church) attacked him as an atheist, one of his most severe critics, Karl Marx, repeatedly emphasized that Malthus was a minister of the Church of England. Even though Malthus is much more accurately described as a professor or scholar, Marx used the term "Parson Malthus" as a verbal brickbat. As late as 1951, in *Pravda*, the official newspaper of the former Soviet Union, there was an attack on Malthus's theory which said: "The Anglo-American imperialists utilize the most disgusting theory of hatred against humanity. . . . What particularly attracts popularity in the reactionary camp is the crazy thought of Malthus, English priest" (Kuzminov, 1951, quoted in Cook, 1967).

Aside from the religious criticism and the Marxist criticism, many other of Malthus's contemporaries found his ideas, for one reason or another, objectionable. Even some of the noted literary figures of that time found reasons to attack him (Petersen, 1979, pp. 68–74). Lord Byron found Malthus's ideas about prudent marriage contrary to his own romantic views of love and marriage. Shelley and Coleridge also found time to attack Malthus, often with vicious personal epithets. But the fact that the literati of his time were numbered among Malthus's critics is little more than a scholarly curiosity. The real issue is what did Malthus say and do that makes him noteworthy to the modern study of demography?

The foremost reason for the continued significance of Malthus is that he took a totally clear stand on an important issue that continues to be debated, namely, does excessive population growth cause increased human suffering? An especially important issue is whether poverty and all of its side effects and ramifications are produced by excessive population growth. Malthus said they were, and many thinkers today believe he was right. While others, as we have seen, said much the same thing before him, Malthus was the scholar who came along when the time was ripe. With his initial essay and the lifetime of work that followed it, his name has become inextricably linked with the idea that excessive population growth has a negative influence on the economic welfare of people.

One may ask if Malthus's theory of population growth is correct. The answer depends on how charitable one wants to be. His idea that populations will have a tendency to grow geometrically, doubling every 26 years if unchecked, is far too mechanistic to be valid. Yet, as we noted in the last chapter, it is true that there are countries in the world today whose populations double in about that length of time. Even though Malthus emphasized rapid population growth, it cannot be said that his principle of population explains the present rapid population growth of the world. Malthus did not anticipate the rapid declines in death rates that have occurred in the 20th century. In this he is not alone, for many demographers, well into the 20th century, did not foresee the cause of rapid population growth that has occurred in the last 50 years.

Some critics have downgraded the significance of Malthus because he did not, in their view, add much to the techniques of demographic analysis. It is true that Malthus did not work with the analysis of demographic data in the same way as the political arithmeticians had done more than a century before him. He was not an innovative demographic technician. However, he did assemble demographic data coming from many countries, and in Petersen's view, "he assayed his evidence judiciously" (1979, p. 148). Petersen further states, "The whole of the *Essay* is replete with analysis indicating that Malthus went far beyond the difficult task of compiling the widely scattered statistics. He applied his high standards first of all to himself" (1979, p. 149). Although Malthus was clearly not an innovator in demographic techniques of analysis, he did substantiate and extend his ideas with demographic data and illustrations.

While the contribution Malthus made to the scientific study of population can be debated, there can be no doubt that he had a profound influence on how population growth has been viewed since the time he wrote his essay. His basic ideas, that population growth is often the cause of societal and human problems and that in one way or another the unbridled growth of population must be curbed, are very much with us today. Contemporaries who subscribe to these basic ideas are often labeled *neo-Malthusians*. These "new" Malthusians believe that populations, human as well as those of other species, do have the potential to grow at a very rapid rate. If something is not done to slow down that growth, these populations will run up against the limits of available resources (food and other necessities for subsistence). For the last 50 years, the neo-Malthusians have watched populations grow in many countries around the world and have viewed this growth as a cause of many social problems and human misery. There is also great concern about how an ever-growing world population may be using the earth's resources and doing damage to the environment.

The neo-Malthusian solution for reducing the problems and misery has been to urge reductions in population growth, primarily through reductions in fertility. While Malthus was morally opposed to fertility control through contraception, the neo-Malthusians are almost always favorably disposed toward a variety of methods for reducing fertility. Neo-Malthusians typically support family planning programs that provide and encourage the use of contraception for the limitation of family size. In addition, they usually support policies that allow voluntary sterilization and abortion (Ehrlich & Ehrlich, 1972, 1990; Ehrlich, 1971; Birdsell, 1980; Brown, 1991; Meadows, Meadows, Randers, & Behrens, 1972).

Not only do neo-Malthusians believe that reducing population growth will diminish social problems and human suffering, but they also see such steps as leading to economic development and improvements in living standards. Rapid population growth is viewed as an impediment to economic development, because resources that could otherwise be used to stimulate development are used to support the young "nonproductive" segment of the population (Brown, 1991).

These neo-Malthusian ideas are not agreed to by all demographers and economists, and a vigorous debate is going on about their merits. The debate that Malthus provoked nearly 200 years ago still affects our thinking about population matters, and there can be no doubt that he deserves recognition as one of the landmark figures of demography.

19TH- AND EARLY 20TH-CENTURY DEMOGRAPHY

While the debates and discussions about Malthus were going on in the 19th and into the 20th centuries, there were other things happening in the development of demography. New ideological concerns, the creation of techniques and measures of demographic analysis, and the emergence of the science of demography were occurring concurrently during this period.

Some governments, notably the United States, began to systematically collect demographic data in the form of the decennial census. Other methods for recording vital events such as births, deaths, marriages, and migration were also being established. Scientists and scholars began to develop the methods and tools of analysis for demographic data. In 1891 Walter Willcox taught his first statistics course (referred to as applied ethics) at Cornell University (Hodgson, 1991). In 1889 R. Herman Hollerith, a Census Bureau employee, developed a card with holes punched in it to represent bits of population information. These cards could then be sorted and counted on a machine that he also developed. Hollerith had created the forerunner of the IBM card that for many years was the primary tool for organizing and analyzing census data (Kaplan & Van Valey, 1980).

The first group of scientists and scholars who identified themselves as demographers in the United States and other countries (notably England, France, and Scandinavian countries) often came out of the ranks of government statisticians. But they were joined by economists, biologists, geographers, sociologists, psychologists, and public health professionals.

As the United States moved into the 20th century, several political issues began to shape the development of demography (Hodgson, 1991). First was the issue of immigration reform. A concern over the origin and nature of recent immigrants led to a series of restrictive immigration laws. Second was the issue of birth control and the rights of women to have access to contraceptives. Entering into the political debates were the eugenicists and neo-Malthusians who were concerned about the "quality" of the human race and who wanted to regulate both immigration and childbearing. The activists representing these issues and population scientists would join together to form the Population Association of America.

The first meeting of the Population Association of America was held in May 1930 at the office of Henry Pratt Fairchild (a sociologist at New York University.) It was attended by 13 people, including birth control activists Margaret Sanger and Eleanor Jones; eugenicist Harry Laughlin; and academics and researchers P. K. Whelpton, D. C. Lowell Reed, and O. E. Baker (Hodgson, 1991). The next year, 70 people were invited to attend. Today, the Population Association of America includes scholars from all over the world who come from diverse academic disciplines. We will see in the next chapter how demographers conceptualize the field today and will present an analytic framework for studying population.

SUMMARY

The beginning of the scientific study of population can be traced to the 17th century and the work of John Graunt. But mankind's interest in population can be traced to the beginning of recorded history and probably before—namely, a concern about the ways in which population affects the well-being or survival of the group, tribe, or society.

Ancient Chinese scholars were especially interested in the ideal proportion between land areas and the number of people. The Greek philosophers Plato and Aristotle focused on the optimum size of the city-state for efficient government and administration. The Romans, through their writing and legislation, were most concerned with increasing the population of their empire.

The mercantilist economic philosophers of the 17th century favored population growth because they believed it would lead to economic and military advantages for the state. A separate group of 17th-century scholars, called the political arithmeticians, also viewed population as a resource of the state, but their distinctive characteristic was an interest in applying scientific research methods to the study of population. Through their empirical studies of population data, they made estimates of population sizes, advanced demographic generalizations, and developed measurement techniques. While they moved the study of population into the scientific realm, they did not generally advance any theories of population.

Thomas Robert Malthus, an English scholar, published in 1798 the first edition of *An Essay on the Principle of Population*, which did offer a theory of population. This theory expressed the pessimistic view that populations have a tendency to grow at a faster rate than the means of subsistence. As a result, there are often positive checks on the population, which are illustrated by famine, poverty, and early death. Malthus also believed that preventive checks on the population could be made if people used "moral restraint" (i.e., had only the children they could support economically). Malthus's theory was criticized from all quarters during his lifetime. It has continued to be influential and controversial up to the present day.

In the 19th and early 20th centuries, techniques of demographic analysis were developed in response to ever-increasing amounts of population data collected by governments. The statisticians and scholars working with these data became the first modern demographers.

3

WAYS OF STUDYING POPULATION

The simplest way to study population is to assemble numerical data on the sizes of populations, such as the number of people in the world as a whole, in various countries, or in states and cities. Population numbers of this sort are called *descriptive demographic statistics*. While they are important, there is much more to demography than learning population sizes. Population numbers are always changing, so even if they are accurate when gathered they are soon out of date and inaccurate. Descriptive statistics are best left to the reference books or other sources where they can be easily found when needed.[1]

Demography is much more interesting when it goes beyond simple description and seeks answers to various questions about population. Why, for example, do some countries have large and fast-growing populations, while others are barely growing at all? Why do some groups within a population have many more children than other groups? Why has the life expectancy of Americans increased greatly over the last 50 years? Why do some groups in the population, such as African Americans and Native Americans, continue to have a shorter life expectancy than whites? What types of people—by age, education and occupation—are most likely to migrate (change residences)? These are the types of questions demographers ask and try to answer. In this chapter we will describe and illustrate the different ways these types of questions are answered. At the end of the chapter we will also describe a relatively new field that is closely associated with demography: *family demography*.

The limited number of demographic processes and characteristics introduced in Chapter 1 will serve as the starting point for understanding how

---

[1]*The Statistical Abstract of the United States* is published yearly and contains comprehensive statistics on the United States and its population. On a worldwide level, the *Demographic Yearbook* of the United Nations provides population information. An organization called the Population Reference Bureau publishes a *World Population Data Sheet* yearly, which conveninetly provides demographic data on every country in the world.

population is studied. The basic demographic processes are fertility, mortality, and migration, and the two most fundamental population characteristics are gender and age. Finally, populations vary in size, growth rate, and density. Using only these eight demographic variables, we can examine the ways population is studied.

As a first distinction, we can say that there are two types of demography: *formal demography* and *population studies*.

FORMAL DEMOGRAPHY

Formal demography is a type of study in which the basic objective is to explain or predict changes or variations in population variables or characteristics. Thus, formal demography aims to explain or predict differences in fertility, mortality, migration, growth rates, size, density and distribution, or the age and sex composition of a population. The distinguishing feature of formal demography is that *only population variables are used to explain variations in other population variables.* In order to provide some examples of formal demography we must first introduce two terms that are widely used in many kinds of scientific inquiry: *independent variables* and *dependent variables*.

In any science it is an important goal to identify the causes of important phenomena through research. Scientists often begin research by hypothesizing that a relationship exists between two variables and that one of these variables influences or affects the other in a causal way. As an example, a demographer might hypothesize that the birthrate of a population would affect the age composition of that population. In this hypothesis birthrate is the *independent variable* influencing or affecting a population's age composition, which is the *dependent variable*.

Thus, if the birthrate of a population is high, it will probably follow that a high percentage of the population will be under 15 years of age. That, in fact, is the case in many countries that are less economically developed. In many African countries the birthrates are very high (40 to 50 births per 1,000 people in a year), and the percentage of the population under age 15 is also very high (40 to 50% of the total). Data from some specific African countries will illustrate how high birthrates are associated with (and probably help to produce) a high percentage of the population under age 15. Ethiopia, for example, has a birthrate of 46 per 1,000, and 49% of its population is under age 15; Nigeria has a birthrate of 44 per 1,000, and 45% of its population is under 15. By contrast, Mauritius, an island in the Indian Ocean off the coast of Africa, has a birthrate of 21 per 1,000, and only 30% of its population is under age 15. Tunisia, in northern Africa, has a birthrate of 25 per 1,000, and only 37% of its population is under age 15.

Were we to examine the birthrates and population percentage under age 15 in all the countries of the world we would find a high correlation between these two measures. Note that this example illustrates formal demography because both the birthrate and the population under age 15 are demographic variables.

One of the paradoxes associated with formal demography is that a researcher could approach these same two general variables, the birthrate and age composition, in the opposite fashion. The age composition of a population could be viewed as an independent variable, which could then be used to explain variations or changes in the birthrate, now viewed as the dependent variable. This might be the case if a population scientist were to examine how the proportion of the population in the childbearing years might explain variations or differences in birthrates.

Suppose, for example, that the years between the ages of 20 and 35 are the years when women are most likely to have children. A demographer might wish to examine how variations in the proportion of women in this age group might account for differences in birthrates. To carry out such an analysis, the demographer might examine each of the states in the United States as a separate population. For each state the percentage of all women who are in the 20–35 age group could be determined, perhaps by using census data collected every 10 years, which would provide the age and gender characteristics of the populations of each state. Treating this percentage as the independent variable, it could then be correlated with the different birthrates of the various states. One would expect that the states with higher percentages of women in their childbearing years would have higher birthrates. Of course, there are probably a number of other factors, such as the economic conditions within particular states, that could also produce differences in birthrates. But if we were to introduce economic conditions as an independent variable this action would take us out of the realm of formal demography, which only examines the relationships between demographic variables.

Population Projection

Formal demography is closely associated with an important tool of demography called population projection. *Population projection* is a procedure that uses certain kinds of demographic information about a population to make estimates, or projections, about the size and characteristics of that population at some future time. The simplest form of projection applies the current yearly rate of growth (or decline) of a population to the present size of a population as a way of estimating how large the population will be at some future date. This method of projection assumes that the rate of growth (or decline) will remain constant during the period in question. For example, the current population of the United States is about 260 million and is growing at a rate of 0.7% per year. At this rate of growth the population will reach 300 million by the year 2010 (Population Reference Bureau, 1994). This example illustrates formal demography because the population growth rate

(0.7%) is an independent variable that produces the population size in the year 2010, a dependent variable. Since we are using one demographic variable to predict the size of another, this is consistent with the concept of formal demography.

A more complicated way of making a population projection would involve making some assumptions about future trends in the rates of fertility, mortality, and net migration (or immigration) and using these to make estimates of the future size of a population (Day, 1993; Haupt & Kane, 1991). The United States Bureau of the Census, as well as many population scientists, uses a sophisticated version of this method to estimate the future size of the U.S. population (Ahlburg & Vaupel, 1990; Day, 1993). The Census Bureau is currently making estimates of the U.S. population going well into the 21st century.

The farther one projects a population into the future the more uncertain and speculative the assumptions about fertility, mortality, and immigration become (Ahlburg & Vaupel, 1990; Keyfitz, 1981). To address this problem, demographers often make multiple projections, based on differing assumptions about what might happen to fertility, mortality, and immigration (Day, 1993). The Census Bureau, for example, makes high, middle, and low estimates of the U.S. population up to the year 2050. These projections are shown in Figure 3.1.

The essential characteristic of formal demography, as we have seen from the above examples, is that all analyses are based solely on demographic variables. Both the independent and the dependent variables, in every case,

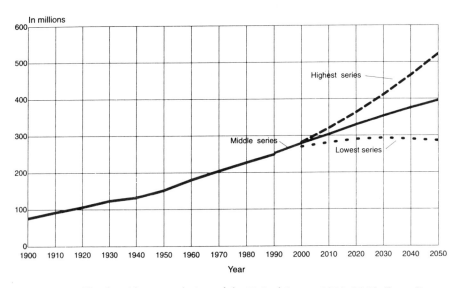

FIGURE 3.1. Total resident population of the United States, 1900–2050. From Day (1993, p. xiii).

have been drawn from the eight basic demographic variables. We have also seen how the same demographic variable can sometimes be treated as an independent variable, and at other times as a dependent variable. While that may seem arbitrary and even somewhat unscientific, it is not, because there is a continuous interplay or interaction among all the demographic variables we have introduced. Whether a particular demographic variable is conceived of as the independent (causal) variable or as the dependent (effect) variable depends upon where the demographer chooses to enter an ongoing process of interaction.

While formal demography is an important part of the study of population, it should already be apparent that there are other influences on demographic events. Earlier we mentioned economic conditions as a nondemographic factor that could influence birthrates. There are many other nondemographic variables that can be used to explain demographic events, and we turn to them now under the heading of population studies.

POPULATION STUDIES, TYPE I

The key to *population studies* is that nondemographic factors are introduced as either independent or dependent variables in relationships with demographic factors. The nondemographic variables can be of many types, but the major ones we will consider are biological, economic, geographic, political, sociological, social psychological, and psychological. We will call studies in which nondemographic factors are used to explain or predict variations and differences in demographic variables *population studies, type I*. In this case, the independent variables are nondemographic while the dependent variables come from the set of demographic variables. *Population studies, type II,* refers to relationships where a demographic factor is used as an independent variable to explain or predict nondemographic phenomena.

In the next few pages we will examine some major systems of nondemographic variables to see how they can help us to understand demographic events. Illustrations coming from these systems of nondemographic variables will familiarize you with population studies, type I. Following this section we will turn our attention to population studies, type II.

Systems of variables refer to the major variables that are of primary interest to particular sciences. Take, for example, biology, a science primarily concerned with living organisms. When biology is related to demography it is concerned with the characteristics of human organisms. Biological variables include such things as genetic dispositions, physiological characteristics, and physical impairments, such as illnesses and disease.

The science of economics also has its particular system of variables, often focusing on economic conditions, such as recessions, depressions, unemployment, and prosperity. These variables from the economic system of variables may affect demographic events such as fertility, migration, and even some forms of mortality. Geography offers another system of variables, those having to do with the natural conditions and characteristics of the earth; as we will see, these variables too influence populations and population processes. All the sciences we will consider in this chapter have what may be viewed as systems of variables. In every case the variables from these systems have the potential for explaining demographic phenomena.

Biological Variables

Biological variables include the genetic, physiological, and anatomical characteristics of humans, as well as the illnesses and diseases that affect the human body. The two demographic processes most likely to be influenced by biological variables are fertility and mortality. Mortality is especially influenced by biological factors. For example, there is evidence that gender is related to one's life expectancy. Females in the United States and many other contemporary societies live longer *on average* than do males. This fact, however, does not necessarily mean that biological factors produce this greater length of life for females. It may be that females live longer than males because of differences in their social roles and life-styles. We will consider this question more fully in Chapter 7, but there is one bit of evidence in favor of a biological explanation. It has been established that during pregnancy, male fetuses have a greater chance of miscarrying than female fetuses (McMillen, 1979). Since the higher fetal death rates of males occur before birth, it is not possible to argue that life-style differences account for this difference.

Biological factors can also influence fertility, a truth illustrated by studies that have found seasonal variations in childbearing. In a study with data on monthly birthrates from 166 different regions of the world, the researchers found that the populations in most countries had their highest levels of fertility in the spring (Roenneberg & Aschoff, 1990a). They concluded that the seasonality of human reproduction must be due to biological factors. However, they also noted that the rhythm of fertility has changed in recent years. They suggest that with industrialization people may be shielded from the effects of daylight hours (by indoor work) and temperature (by air conditioning and heating) such that the old seasonal patterns determined by biology are being modified by environmental factors (Roenneberg & Aschoff, 1990b).

In the United States there is some seasonal variation in fertility, but the high-fertility months are typically August and September, not April and May. In fact, in the United States the lowest fertility months are usually March and April. A recent study conducted in San Antonio, Texas, explains this seasonal pattern, again in biological, not sociological, terms (Levine et al., 1990). Male workers in San Antonio were found to have lower sperm counts in July and August than in other months. While the low sperm count could be attributable to the Texas heat in the summer, these researchers found that even workers in air-conditioned settings had low sperm counts in the summer. This led them to speculate that there is a genetic predisposition for a lower sperm count in the summer, leading to fewer babies being born in late winter and early spring. It may be, they conjecture, that through an evolutionary selection process a higher proportion of babies are born in late summer and early autumn when, after the harvest, food supplies are generally more plentiful. However, since a higher number of August and September births in the United States is not consistent with the vast amounts of data coming from other times and places, we must be skeptical of the genetic interpretations of the Houston researchers. While these studies are not conclusive, they do illustrate a biological approach to the study of fertility.

Economic Variables

Several of the key variables studied by economists have been shown to affect demographic events. For example, the state of the economy may influence rates of migration. Research has shown that during times of society-wide economic recession or depression, the migration of people is likely to decline, while in good times migration is likely to increase (Stahl, 1988). One form of reasoning suggests that economic hard times would cause people to migrate in search of jobs; indeed, that may be the case when there are important regional differences in economic conditions. However, when an entire economy is depressed the rate of migration declines, probably because people are reluctant to leave familiar places and conditions, and take a chance on the uncertainties of some new place. In this case, the causal mechanisms are economic, or the psychological response to economic uncertainty, or a combination of the two. Whatever the reason, there is abundant evidence that migration declines during economic hard times and increases during times of economic prosperity.

Fertility is also influenced by economic factors. This was demonstrated by a recent study that found how changes in tax exemptions allowed for children were related to increases in marital fertility (Whittington, 1992). In the United States, in both 1979 and 1982, the tax-exempt deduction for each child increased, thus effectively reducing the yearly costs of rearing children. This study tested the hypothesis that marital fertility would increase when the costs of rearing children decreased. By following families over the years between 1977 and 1983 this study was able to show that families did respond to the tax exemption by increasing their likelihood of having another child (Whittington, 1992). There is also some evidence that the U.S. birthrate rose again after the Tax Reform Act of 1986 (Whittington, 1992), which reduced the yearly costs of rearing a child by about 8% (Espenshade & Minarik, 1987).

Geographic Variables

Geographers work with many different kinds of variables, but the primary geographic variables include the topography of the earth—mountain ranges, hills, plains, lowlands, and so on—its climate and weather, and its resources, such as fertile soil, water, forests, fossil fuels, and minerals. These basic geographic variables are very useful for explaining variations in the distribution of people and populations over the earth's surface. In mountainous or desert areas, the land is invariably sparsely settled, while fertile valleys with moderate climates usually support large and more densely settled populations. Usually, too, there are population concentrations at the mouths of major rivers.

As an example of how climate can influence population, note that the island continent of Australia has a land area of almost 3 million square miles, but has a population of just 18 million people. The average density is only about six people per square mile, but even that figure is misleading, because in the vast desert that covers the central part of the continent there is almost no population. Most of the population is concentrated in cities, especially along the eastern and western coasts.

The case of Australia illustrates, in the extreme, the way geographic variables can explain variations in population distribution and density. Geographers, using their particular skills and interests, are able to explain how and why populations distribute themselves the way they do, and also why population density and distribution changes over time. When Oklahoma, Arkansas, Texas, Kansas, and other states in the Midwest and West suffered severe droughts in the 1930s (also a period of economic depression), it was necessary for many people in these states to migrate. Even though migration tends to decline in difficult economic times, sometimes people must move in order to survive. When changing physical conditions (weather, resources, etc.) make it impossible to secure the necessities of life, migration must occur. Geographic variables thus can be used to explain both the distribution of populations at any one time or the changes that take place through migration.

Political Variables

Political actions and events may at first glance seem far removed from demographic events. However, governmental and political actions often have significant effects on demographic events. Indeed, as we noted above, when the U.S. income tax laws were changed in the 1970s and 1980s, giving larger deductions for children, marital fertility increased. While we treated that example as an economic influence, it could also have been treated as a political influence since the tax laws were changed by Congress.

For an historical example of political influences we can look to the case of Japan after World War II (Oakley, 1978). Japan had lost the war, and the Japanese had to give up their ambition to expand throughout the western Pacific and onto the Asian mainland. The Japanese population in 1948 already exceeded 72 million people and was rather densely settled on the nonmountainous parts of the Japanese home islands. The birthrate of approximately 32 per 1,000 people in 1948 was twice as high as the death rate of 16 per 1,000. This difference between the two rates produced a growth rate of 1.6% per year, not an especially high growth rate for the time, but one that would have doubled the Japanese population by 1992 if it had persisted. Japan's birthrate, however, did not remain high. It dropped rapidly and now stands at 10 per 1,000, one of the lowest rates in the world. In 1994 the population of Japan stood at about 125 million (Population Reference Bureau, 1994).

The significant slowdown in the population growth rate of Japan was produced largely by political actions in that country after World War II. In 1948, under the prompting of American military occupation forces, the Japanese government passed legislation that both permitted and encouraged sterilization and abortion. Though this legislation was ostensibly passed to safeguard the lives and health of mothers, it enabled Japanese women to seek abortions whenever births would lead to economic hardship for their families. Under this liberal interpretation of the law, the number of abortions in Japan increased rapidly and the birthrate dropped. In the 10 years after the liberalized abortion policy was adopted, the Japanese birthrate dropped from a high of 34 per 1,000 to about 18 per 1,000 (Oakley, 1978); today the birthrate is about 10 per 1,000.

The very low current birthrate in Japan has prompted a new political debate, which may eventually affect the population (Arioka, 1991). Japanese political leaders are getting concerned about the low birthrate because the families of Japan may not be providing enough children for the future labor force. An alternative political view is represented by women's rights advocates who oppose governmental policies that push women into having more children (Arioka, 1991).

The Japanese case is but one of many possible demonstrations of how political policies and political actions can have a profound effect on demographic events. Especially important is the fact that many governmental actions may not be directly aimed at demographic events, but may still have an indirect effect (Johansson, 1991). Sweden, for example, has had an upsurge of fertility in recent years, after many years of low fertility. One Swedish demographer (Hoem, 1990) believes that the upturn in the rate of Swedish childbearing is due to changes made in the law giving employed women (and men) paid maternity leave. When the law was modified in 1986, it became economically advantageous for Swedish couples to have a second child within 30 months of the birth of their first child. This law could account for the increase in childbearing among Swedish women.

The effects of political factors are not, of course, limited to childbearing. Immigration and migration laws and policies can have a significant impact on the movement of people from country to country and from place to place within countries. Public health policies and the control of dangerous substances (chemical wastes, tobacco, alcohol, drugs, etc.) can decrease death rates. Wars and revolutions, which are extreme political events, have often had extraordinary influences on the populations of nations and on particular population groups. The number of ways in which political factors can affect demographic processes could be extended to a lengthy list. While we must be content here to offer a few examples, the major point is that political actions and events do have an impact on population.

Sociological Variables

The field of sociology has had a long-standing claim to demography as one of its subfields. This suggests that sociological variables are particularly important in the explanation of demographic processes and facts. While this is indeed true, it does not reduce in any way the explanatory usefulness of other fields of study. As we have already seen, biology, economics, geography, and political science all have demonstrable contributions to make to the study of population.

Since many parts of this book will focus on sociological explanations of demographic events, one example from sociology will be sufficient for the moment to illustrate both what a sociological variable is and how such a variable can be used to account for demographic events.

Every kind of social system, from the largest society to the smallest informal group, has some kind of structure. A structure is a pattern of relationships among people that persists over time. At the societal level, the most familiar structure is the socioeconomic structure. People fall into different social classes, which have different economic resources, degrees of power, and levels of prestige. Social classes have been shown to be related to various kinds of demographic events, but one of the most persistent relationships is between social-class status and mortality. Put very simply, people in higher social classes have a much better chance of living longer than people in lower social classes (Antonovsky, 1967; Kitagawa & Hauser, 1973; Mare, 1990; Navarro, 1990). This relationship has been found to prevail in society after society (Fox, 1990; Smith, Shipley, & Rose, 1990), both historically and contemporaneously (Antonovsky, 1967).

The relationship between social class and mortality can be raised to the level of a general principle: Whenever social relationships are arranged in a rank order, those people in lower statuses will have higher death rates than people in higher statuses. Minority groups in societies have persistently higher death rates and lower life expectancy than the dominant group.[2] Black

---

[2]"Minority groups" is a term used here to refer to a group's subordinate political and social positions in the society, not to their population numbers. While minority groups are often numerical minorities, that is not necessarily the case (e.g., the black population of South Africa is larger than the white population).

Americans have higher death rates than white Americans; while much of this difference can be accounted for by socioeconomic factors, some of the difference is due to the unequal treatment of black Americans as a minority (Potter, 1991; Rogers, 1992). This latter point is a matter of some debate among researchers, with some concluding that social class, not minority status, is the key influence (Navarro, 1990).

Psychological Variables

In recent years, psychologists have become increasingly interested in certain aspects of population. The psychological reasons people have for bearing children is one such interest. From a psychological perspective, the issue is whether or not there are particular psychological traits or personality characteristics that motivate some people to have more children than others (Miller, 1981).

A study of 401 California couples has shown the importance of three specific personality traits on motivations to have children (Miller, 1992). These three personality traits are: (1) *nurturance*—giving sympathy and comfort, assisting others whenever possible, offering a "helping hand" to those in need; (2) *affiliation*—enjoying being with friends and people in general, accepting people readily, making efforts to win friendships and maintaining associations with people; (3) *Autonomy*—trying to break away from restraints, confinement, or restrictions, enjoying being unattached, free, not tied to people, places, or obligations, sometimes being rebellious (Miller, 1992, pp. 272–273).

Both husbands and wives who were high on nurturance and affiliation were more positively motivated to have children (they were positive about holding and cuddling a baby, viewed a baby as carrying on family traditions, wanted to guide and teach a child, and saw a child as strengthening marriage) (Miller, 1992, p. 271). By contrast, the husbands and wives who scored high on the measure of autonomy were not positively motivated to have children.

This study supports the idea that basic personality traits (in this case, nurturance, affiliation, and autonomy) are related to one's motivations to have children. But it does not tell us how individuals come to have particular personality traits, although the researcher in this case (Miller, 1992) believes that these traits are at least partly genetic in origin (which is again a biological explanation). Other psychologists are more likely to believe that personality traits are produced by one's social environment (especially one's family). But the origin of personality traits is not the issue here; what is important in the study of population is such traits' impact on childbearing.

Psychology also includes the largely unexplored area of subconscious psychology, where motivations may be deeply embedded in the psyche and yet affect demographic events. One study has shown that prominent American historical personalities and people listed in *Who's Who in America* were more likely to die in the month *after* their birthdays than in the month *before* their birthdays (Phillips, 1972). One psychological interpretation of this pattern is that these prominent Americans wanted to postpone their deaths until after their birthdays. Whether conscious or unconscious of such a desire, many were successful in postponing death.

The purely psychological approach to understanding demographic phenomena is supported by only a limited amount of research, but there is certainly some evidence that psychological variables can be useful for the explanation of some demographic variables. A more established research tradition is an approach that is partly psychological and partly sociological. This is the field of social psychology.

Social Psychological Variables

Social psychology, like psychology, focuses attention on the individual. A key idea of social psychology is that individuals have attitudes and orientations toward a wide range of significant social issues and objects. These attitudes and orientations are likely to influence their behavior. A demographic example is found in the attitudes of Americans about abortion. Attitudes about abortion range from those who see it as equivalent to murder to those who see it as one of several choices a person may make about a pregnancy. It is certainly likely that one's attitude about abortion will influence one's childbearing behavior.

Individual attitudes and orientations about all types of social issues and objects can clearly influence childbearing, but they can also influence mortality, and migration. In contemporary societies, people have a great deal of control over the three demographic processes: fertility, mortality, and migration. Individuals today understand how fertility can be controlled and they usually can choose what number of children they wish to have. Most people have children when they have reasons for doing so, or no reasons for not doing so. From a social psychological perspective their attitudes and orientations will play a central part in determining the number of children they have and when they have them.

In the past, having children was often a matter of fate (Back, 1967). When people did not understand the process of reproductive physiology, or even when they did understand it but lacked effective contraceptive methods, having children was "a rather uncontrolled, fateful concomitant of sexual desires" (Back, 1967, p. 92).

Even death, which still sometimes seems to be largely a matter of fate, is increasingly subject to the control of individuals. No longer are infectious diseases the major causes of death. The majority of deaths today, in most modern societies, are the result of failures of the cardiovascular system (via heart disease and stroke) or cancer. Most people readily understand that they can take actions to reduce the chances of death by these causes. They are aware that heart disease is the result of high cholesterol diets, lack of exercise, obesity, smoking, excessive use of alcohol, and stress, but individual attitudes may determine how they respond to these causes. Even cancer, which often seems to be more a matter of fate, is increasingly subject to individual control, and thus reflects personal attitudes. A leading cause of cancer deaths is lung cancer, which has been shown in study after study to be related to cigarette smoking (Ravenholt, 1990). Skin cancer is recognized as being caused in part by excessive exposure to the ultraviolet rays of the sun. Whether or not people avoid excessive sun exposure is determined by attitudes they hold about the beauty of a suntan.

Increasingly, demographers are finding it important to study the attitudes and orientations of individuals as a way of explaining demographic events, including mortality. As this occurs, the social psychological approach will become a more important part of the study of population.

POPULATION STUDIES, TYPE II

Thus far we have focused our attention on the various ways in which population events can be explained and understood. But demographic variables can also be used to explain nondemographic events and variables. This is the case of population studies, type II, where demographic variables are used as independent (or causal) variables to explain variations or changes in dependent, nondemographic variables. As an example, suppose a researcher— perhaps a political scientist—wishes to explain variations in voting behavior on a tax referendum to increase the school budget. To account for differences in the levels of opposition in various parts of the community, this researcher might examine differences in voter age composition. Age composition, as we know, is a population variable, and it could account for variations in voting behavior (the dependent political variable) because young married couples with preschool-age or school-age children might be much more likely to support a tax increase. By contrast, unmarried young adults or elderly people might be less interested in the schools and would oppose a tax increase. Using the age composition of different areas of a community as a way of explaining variations in voting behavior is an example of population studies, type II.

Population studies, type II, often has a very practical or applied character as opposed to the more purely scientific kinds of inquiries. Frequently the future needs of a society must be anticipated and planned for far in advance. Take, for example, the Social Security system of the United States. In order to know what the future financial requirements of the Social Security system are going to be, it is necessary to project the age structure of the population into the future. By knowing how many potential recipients are likely to be in the system some 20, 30, or 40 years from now, one can begin to plan for the future needs of the system. But a complete evaluation of the Social Security system is further complicated by the fact that one must also estimate how many working-age people will be contributing to the system at future dates. This will be determined by the number of people in the younger part of our society's age structure. If we are looking into the 21st century, the youngest age groups will be shaped by current and near-future birthrates. As one can see from this example of population studies type II, the analysis can become quite complicated.

The practical applications of population studies, type II, can also be found in communities where school systems have to be planned on the basis of projected school-age populations. Similarly, colleges and universities must make estimates about the sizes of their future freshmen classes. But it is not just government and educational leaders who must use population information to make decisions about their actions. The managers and executives of businesses and industries must also consider population size as they plan for the future. A recognition of the importance of demographic factors for commercial enterprises has led to the creation of a new word: *demographics*.[3]

Demographics can be defined as management information that helps decision makers identify and then take advantage of knowledge about population size and composition, and anticipated changes in these (Merrick & Tordella, 1988). It is very common to hear the term "demographics" used in connection with the entertainment industry. Radio stations will make programming decisions on the basis of the characteristics of the population in a particular listening area. For example, it would be foolhardy for a new radio station to feature heavy-metal music, popular with teenage males, in an area that has a large number of retirees.

As industries and businesses plan for the future they must anticipate how the population will be changing and adjust their plans accordingly. There is a widely held belief that the large baby-boom generation (born between 1945 and 1963) has already had an inordinate impact on the U.S. economy and will continue to do so in the future. In the 1990s the baby-boom generation will start reaching age 50, and this milestone may see them purchasing differently and changing their participation in the labor force (Samuelson, 1991).

Demographics is also a critical aspect of running successful political campaigns in the United States. Political advisers pay close attention to the age, gender, race, and socioeconomic status characteristics of voting populations. Running a modern political campaign requires "selling" candidates to the voters, just as businesses sell their products. In both cases the target population must be known.

---

[3]Note that *demographics* is a noun and should not be confused with the adjective *demographic*.

Demographics, while it is not, strictly speaking, a part of the science of demography, is nonetheless a part of what we have called population studies, type II. This is the case because population size and population characteristics (acting as independent variables) influence marketing, business, and political decisions (the dependent variables).

This completes our presentation of formal demography and population studies, type I and type II. Table 3.1 summarizes and illustrates these three ways of studying population. Some of the illustrations have already been presented and discussed, while others are new.

TABLE 3.1. Characteristics and Examples of Formal Demography, Population Studies, Type I, and Population Studies, Type II

Type of study	Independent variables	Dependent variables
Formal demography	Demographic variables	Demographic variables
	Birthrate	Age composition
	Age composition	Birthrate
	Population growth rate	Projected population size
Population study, type I	Nondemographic variables	Demographic variables
	Gender (biological variable)	Mortality
	Economic conditions (economic variable)	Migration
	Climate (geographic variable)	Population density
	Abortion policy (political variable)	Fertility
	Social class (sociological variable)	Mortality
	Personality traits (psychological variable)	Childbearing
	Attitudes about health (social psychological variable)	Risk of dying
Population study, type II	Demographic variables	Nondemographic variables
	Age composition	Voting behavior (political variable)
	Population characteristics	Marketing decisions

THE EMERGING FIELD OF FAMILY DEMOGRAPHY

There is a close relationship between marriage and family life and the subject matter of demography. It is therefore quite natural for the two subjects to converge, as they have done in the relatively new field of family demography. Because of the newness of the field there are several different views of what is distinctive about family demography (see Burch, 1979; Glick, 1988; Teachman, Polonko, & Scanzoni, 1987; Sweet, 1977). Drawing on these various views, we can say that *family demography* is the study of family processes (including marriage, childbearing, divorce, remarriage, and others) and family characteristics (household and family size and composition) and their association with demographic processes and population characteristics.

There are a number of familiar illustrations of the closeness between family and demographic processes. Marriage, which is a family process, is often associated with subsequent childbearing. The birth of a child is itself an important family process, but, of course, childbearing is also an aspect of the demographic process of fertility. Entering marriage is also associated with establishing a new place of residence, which in demographic terms is migration. Divorce, another family process, almost invariably leads to a change of residence for one of the spouses, and sometimes for both. Marital status is also related to mortality: Married people live longer than single, divorced, or widowed people (this topic will be discussed further in Chapter 7).

The examples above show how marriage and family variables can influence demographic events, but the relationship can also go in the other direction, with demographic events influencing marriage and family variables. As one example, in populations where one gender outnumbers the other, marriage rates will be influenced. The gender with excessive numbers may not be able to find marriage mates and therefore will have to forego marriage (Guttentag & Secord, 1983; Veevers, 1988). The U.S. population aged 65 and over, in which women greatly outnumber men, illustrates this point. In the 65-and-over age group 41% of the women and 76% of the men are married (U.S. Bureau of the Census, 1993).

As a way of describing family demography systematically, the primary family processes and characteristics most often studied by family demographers are listed below:

1. Entry into marriage
2. Having a child (or children)
3. Children leaving the parental home
4. Separation and divorce
5. Widowhood/widowerhood
6. Remarriage
7. Size and composition of families
8. Size and composition of households

These family processes and characteristics, of course, are only headings for a variety of more specific variables. Entry into marriage, for example, is sometimes measured by the *marriage rates* of a population. Or entry into marriage may be measured by *age of entry into marriage*. Marriage age can be also be measured by the *age differences between brides and grooms* (Vera, Berardo, & Berardo, 1985).

If we take these eight family processes and characteristics as the principal elements of family demography, we can use them as part of an analytic typology that will allow us to see the different types of family demography.

An Analytic Typology of Family Demography

The eight family processes and characteristics listed above, when cross-classified with five of the basic demographic variables identified earlier in this chapter, will define and illustrate different types of family demography. The five demographic variables are fertility, mortality, migration, age composition, and gender composition of the population.

Sometimes the demographic variables will be the independent variables and the family variables will be the dependent ones, and sometimes the opposite will be the case—family variables will be employed as independent variables to explain demographic variables. There is also a part of family demography in which some family variables are used as *both* independent and dependent variables.

Table 3.2 shows the first type of family demography, in which demographic variables are independent and family variables are dependent. Examples are also provided.

Studies of divorce in the United States have found that rates of divorce are higher in those states and regions with higher in-migration rates (Glenn & Supancic, 1984; Glenn & Shelton, 1985). It appears that migration either leads to greater divorce or that those people who are likely to divorce are also more likely to migrate.

A second type of family demography is illustrated by those instances when family variables are the independent variables and demographic variables are the dependent variables. This form of family demography, along with an example, are shown in Table 3.3.

TABLE 3.2. Family Demography, Type I

Independent variables	Dependent variables
(Demographic variables)	(Family variables)
Examples	
Sex ratio	Marriage rate
Migration rates	Divorce rate

TABLE 3.3. Family Demography, Type II

Independent variables	Dependent variables
(Family variables)	(Demographic variables)

Example

Extended versus nuclear families	Fertility

There is a long-standing hypothesis that extended families, because of their structure (three generations in a household) and social roles in the family, will lead to higher fertility (Davis, 1955). The supporting evidence for this hypothesis has not always been as conclusive as one would suppose, but there is some evidence that where the extended family prevails family size will be larger (Bongaarts, 1983).

A third type of family demography examines the relationship between two different family variables.[4] This type of family demography is characterized by research in which one family life-course event is related to another. This type is illustrated in Table 3.4.

There is ample evidence that marriages at younger ages are more likely to end in divorce (Booth & Edwards, 1985; Teachman, 1983). There is also evidence that late marriage (marriages substantially above the average age) increases the chances of divorce, but not as highly as does early marriage (Glick & Norton, 1977).

Some research has shown that having children reduces the likelihood of divorce (Becker, Landes, & Michael, 1977; Cherlin, 1977), though there have been other studies that have not confirmed this relationship (Bumpass & Sweet, 1972). While the results of this research are somewhat contradictory, there is evidence that the gender of children *can* influence divorce—having male children reduces the likelihood of divorce (Morgan, Lye, & Condran, 1988).

Most contemporary family demographers have focused their attention on the kinds of research described in the examples above, which is to say that their primary interest is in studying and explaining family processes. Historical demographers, on the other hand, who are concerned with the family of the past, have usually devoted most of their attention to the composition of families and the numbers of people in households (Burch, 1979).

The field of family demography is one of the most exciting and vigorous in the social sciences today. Research in this field attracts sociologists,

---

[4]It could be argued that since both the independent and dependent variables in this type are family variables, and there are no pure demographic variables involved, this type should not be called family demography. While that is true, the scholars who have tried to define the field have always included this type of research as an important part of family demography (Sweet, 1977; Teachman et al., 1987).

TABLE 3.4. Family Demography, Type III

Independent variables	Dependent variables
(Family variables)	(Family variables)

Examples

Age at marriage	Divorce
Having children	Divorce

demographers, historians, and other family scholars. Family demography covers a wide range of topics, issues, and questions, so the examples presented above offer only a glimpse of this new and growing field.

SUMMARY

Demography is the scientific study of population. Demographic analysis and research is fundamentally concerned with a limited number of demographic variables: Fertility, mortality, migration, and the age and sex structure of populations are the most basic ones. In addition, the three basic demographic processes affect growth rates (or decline rates), size, and density of populations. These too are important variables of demography.

Population research and analysis can be conveniently divided into three different types: formal demography, population studies, type I, and population studies, type II. The distinctions are based on whether demographic variables are viewed as independent (or causal) variables, dependent (or effect) variables, or both.

Formal demography employs only demographic variables. One or more demographic variables are used to account for variations in one of the other demographic variables. The age composition of a population can account for differences in fertility. Population projection, which uses information about a population to make estimates or projections about the size and characteristics of a population at some future time, is an example of formal demography.

Population studies, type I, uses nondemographic independent variables to explain or predict variations or differences in dependent demographic variables. Nondemographic variables come from systems of variables, the major variables that are of primary interest to particular sciences. The major systems of variables include biological (including genetic and medical), economic, geographic, political, sociological, psychological, and social psychological variables. There are research examples representing all of these systems of variables showing their importance for explaining demographic variables.

Population studies, type II, uses demographic independent variables to explain variations or changes in nondemographic dependent variables. "Demographics" is a new term that is commonly used to describe how managers of industries and businesses (or political campaigns) use demographic information to make decisions about marketing goods, services, or candidates.

The emerging field of family demography is the study of family processes, such as marriage, divorce, childbearing, and remarriage, and family and household characteristics (composition and size). In family demography, type I, demographic variables are used as independent variables to explain variations and changes in family dependent variables. An example is the influence that sex ratios may have on marriage. Family studies, type II, uses family variables as independent variables to explain variations and changes in demographic dependent variables. The organization of families, as, for example, either nuclear or extended, may account for differences in fertility. A third type of family demography involves using one family variable to explain variations in another family variable. Age at marriage, for example, may account for differences in the likelihood of divorce.

4

THE DATA OF DEMOGRAPHY

In 1990, the U.S. Bureau of the Census counted 253,000,000 residents of the United States and its territories. Each month the Census Bureau conducts a Current Population Survey that interviews individuals in over 55,000 households. The Demographic and Health Service Program funded by the U.S. Agency for International Development has interviewed over 180,000 women in 32 countries since 1985 and additional surveys are planned (Blanc, 1991). Information from sources such as these provide the raw material of demography.

Millions of demographic events—births, deaths, and migration—are recorded every year. For the people who experience these life-producing, life-ending, and life-changing events, they are of the greatest personal significance. For demographers, they are the data upon which the scientific study of population rests.

To be available for study by demographers, these events must be accurately observed and recorded. Since demographers themselves rarely make these observations, they are dependent upon the observations other people make of births, deaths, and migration. From the first demographers, with their very primitive techniques of analysis, to the most highly skilled, computer-assisted, contemporary demographers, the same basic fact is true: Someone must accurately observe and report demographic events if the demographic analysis of these events is to be meaningful. This truth may be illustrated by returning to one of the earliest demographers, John Graunt, the 17th-century political arithmetician.

Graunt used the "bills of mortality," the weekly summaries of London deaths, to do his analysis. But who actually recorded the deaths? According to Graunt, the parishes of the city often had older women, called "searchers," who were charged with that responsibility. These "ancient matrons," in Graunt's words, would go to the place of the deceased to ascertain who had died and the circumstances of his or her death.

However, Graunt, with his characteristic scientific caution, expressed concern about the accuracy of the searchers' reports, especially the reported causes of death. He feared that sometimes the searchers might have made their observations after drinking too much ale, or perhaps they falsified their reports after having been bribed (Graunt, 1662/1964). Graunt's concern is shared by all demographers because they must deal with events that they themselves have not observed. And to make matters worse, the events (births, deaths, and migration) are generally observed by someone who does it for reasons other than scientific demographic analysis. It is with this problem in mind that we will examine, throughout this chapter, the sources of demographic data, giving special attention to the quality and accuracy of those data.

DEMOGRAPHY AS AN OBSERVATIONAL SCIENCE

The fact that demographic data are generally gathered by people who are not themselves demographers is closely related to a general characterization of demography as an observational science.

Observational sciences are those sciences for which the data are obtained by observing and recording events that occur naturally in the world (Hauser & Duncan, 1959). For example, traditional astronomy is an observational science because astronomers obtain their data by observing the composition and movement of the stars and planets *as they exist in nature*. In much the same way, demographers must generally obtain their data from "naturally occurring" births, deaths, and migration. The nature of observational sciences can be understood more easily if they are compared to an alternative type of science: the experimental sciences.

Experimental sciences are those in which the data are generally produced by experiments conducted in laboratories under conditions that are closely controlled by the scientist. The most familiar examples of experimental sciences are the physical and biological sciences. In a classic experiment, data are produced in the laboratory or in a laboratory-like setting, where the researcher controls and manipulates variable conditions. Biological scientists, for example, may test for the possible cancer-producing effects of a particular food additive by feeding it to, or injecting it into, a random sample of laboratory animals. At the same time, a control group of the same animals will be maintained in exactly the same conditions, except that they will not receive the food additive. The scientific data produced in this type of experiment are the before-and-after observations (or measurements) of the animals. In this example, the data *must* be produced in the laboratory because the animal would probably not have ingested the food additive in its natural habitat.

This example of an experimental science also reveals another reason why demography is not generally experimental. Demographers are, as we know, interested in mortality (including that caused by cancer), but it is not possible to manipulate human populations in experimental ways to study their mortality. There are important practical and ethical limitations to experimenting with humans if one is studying matters of life and death. But that does not mean experimental research in demography is impossible. There have been some experimental studies in demography, and in the following section we will review some of these attempts at experimentation.

Experimental Research in Demography

Basically there are two types of demographic experiments: those conducted with nonhuman organisms, and field experiments with humans. We will illustrate the nonhuman experimental studies first.

During the 1920s and 1930s the biologist/demographer Raymond Pearl (1925, 1939) theorized that human population dynamics were largely shaped by biological factors and that it would be possible to establish a law of population growth that would hold true for all living things. To test his idea, he experimented with population growth among fruit flies (drosophila). In his experiment, Pearl placed a small number of fruit flies in a jar along with an ample amount of food. Then he recorded the increases in the number of fruit flies. After a series of experimental trials, Pearl was able to establish a growth curve for the fruit fly population. The fruit fly population would grow slowly at first, then increase rapidly, experience a slower rate of growth, and finally reach a peak and level off.

Pearl found that when he plotted the human population estimates beginning with the 17th century against the growth curve for the fruit flies, a very similar pattern appeared. Since the curves were similar, he believed that it would be possible to project the curve into the future and predict the ultimate size of the world's population. After some revisions, Pearl estimated that the population would reach a peak of 2.6 billion people by the year 2100. As it turned out, the world's population reached that level about 15 years after he published his estimate! Today, even before the year 2000, the world's population is more than twice as large as Pearl's prediction and it is still growing.

In short, Pearl's efforts to make reasonable estimates of world population growth proved to be dramatically wrong. The curve derived from experimental studies of fruit flies was a hopelessly ineffective tool for estimating human population growth. In general, it is not possible to make statements about human population dynamics by studying insects or animals in the laboratory. To understand human population processes, it is necessary to study the demographic behavior or intentions of people themselves.

Field Experiments in Demography

Since the scientific rigor of experimental research is very desirable, a number of demographic field experiments have been carried out in recent years. *Field experiments* are those studies in which researchers control or manipulate some factors, just as in laboratory experiments, but the subjects are human and they are studied in their natural settings.

The most common demographic field experiments have been associated with family planning programs. One example of a successful field experiment was carried out in Matlab, a rural area of Bangladesh (Simmons, Balk, & Faiz, 1991.) Bangladesh has a very high birthrate (37 per 1,000) and a high infant mortality rate (120 per 1,000). In 1977 the Matlab Family Planning and Health Service Project was established. The goals of the project were to lower birthrates and to provide better health care for women and children in order to lower infant mortality. People living in one area of the region were provided with extensive family planning and maternal health services; this was the experimental group. People living in another area received only the normal and more limited government services: This was the control group. The results of this experiment were positive. The rate of contraceptive use increased substantially for the experimental group, while there was only a modest increase for the control group. In turn, the total fertility rate dropped more rapidly in the experimental area than in the control area. Recent evaluations show that the program is both effective and cost-efficient. The overall success of this program has been widely noted in the international family planning community and has become a model for other developing countries.

It is of course true that this field experiment in Bangladesh probably did not meet the most rigorous standards of laboratory experiments. The researchers could not control exactly who received the family planning and maternal health services (the independent variables in this field experiment). Also some people may have moved from the experimental area to the control area, or vice versa. Such movement would have "contaminated" the experiment to some degree. But these and other similarly uncontrollable factors are some of the practical problems faced by those who conduct field experiments in demography. These practical problems are one reason why demography is generally practiced as an observational science, not an experimental science.

A second major problem associated with experiments is that when human subjects are used, moral or ethical issues often arise. This is especially true when humans are the subjects of experiments in which demographic events are involved. Demographic events are often so significant in people's lives that scientists cannot intrude just for the sake of obtaining scientific knowledge. There is an infamous case, involving the U.S. Public Health Service, that illustrates the kind of harm that can be done when humans are used in experiments where their lives are at stake (Jones, 1981).

In the 1930s, 400 Alabama black men who had syphilis were made the subjects of an "experimental" study to determine the long-term effects of the disease. This study, sponsored by the U.S. Public Health Service and conducted under the auspices of Tuskegee Institute, sought to determine what would happen if the men afflicted with syphilis were left medically untreated. Even after 1953, when penicillin, a simple cure for syphilis, was widely available, these men were not treated for their disease. Not only did this lead to the premature deaths of many of the experimental subjects, but it may have also contributed to the spread of the disease in the African American population. This study was not discontinued until 1972 (Jones, 1981).

While this horrifying case is extreme, it does illustrate how the major areas of demographic interest are often not appropriate for scientific experimentation. Because of these kinds of ethical considerations, as well as some of the practical problems of controlling people's activities over long periods of time, most demographic research is done with data coming from naturally occurring events. This reality returns us to our original point: Demography is primarily an observational science.

DEMOGRAPHIC DATA OF THE PAST

We started this chapter by noting that the events that interest demographers are occurring all the time and all over the world. Often the time we are most interested in is the present, or even the future, but there are also the demographic events of the past, and these too are important and interesting.

This interest in the demographic events and conditions of the past has led to a subfield called *historical demography*. As the name suggests, this is the study of populations in some earlier historical period. For example, if a contemporary demographer were to undertake a study of Ireland's population changes during the 18th and 19th centuries, such a study would be identified as historical demography. While there is no agreed-upon cutoff date before which the study of a population results in historical demography, a rough rule of thumb would be to consider studies focusing on pre-20th-century populations as historical demography.

Historical demographers often have to work with demographic data coming from nongovernmental organizations, since censuses and other forms of government record keeping are relatively recent in origin. Their most common source of data is church records of baptisms, marriages, and funerals. From these records historical demographers have developed a method called *family reconstitution*. Family reconstitution involves bringing together information about the major events in the lives of a family in order to make general demographic statements about the population as a whole (Wrigley, 1966). The major events of a family might begin with the record of a marriage of two people and go on to a listing of all children born of that marriage, the deaths of any children, and the eventual deaths of the original married couple. Family reconstitution is very much like the work done by genealogists who take a particular family line and trace all the offspring in that line through their births, marriages, and deaths. But the objective of historical demographers is not to establish family lineages; it is instead to establish demographic facts and general population trends.

If all events in the lives of most of the people of a parish have been faithfully recorded by a parish priest or his clerk, then there will be enough information to make some demographic statements about the population as a whole. Among other things, the average number of children born to each marriage could be calculated. The number of childless marriages could also be determined. It would also be possible to calculate the level of infant mortality, the average length of life for males and females, the average age at which men and women married, and the amount of mortality resulting from childbirth (called *maternal mortality*).

The kind of demographic data available to historical demographers is illustrated by the case of the city of York, England. York has continuous parish records going back to the year 1538 (Cowgill, 1970). In that year the Anglican church started keeping records of baptisms, marriages, and funerals. The information in these records has been extracted and transformed into a computerized database. The information in this record, from 1538 to 1812, includes 33,000 identifiable births and 11,000 marriages (Cowgill, 1970).

Using these and similar data, historical demographers have contributed greatly to our knowledge of people's lives in earlier times. Often the information derived from the demographic records of the past has been quite surprising because it runs counter to commonly held beliefs. One prominent example of a surprising discovery made by historical demographers is the age at which people married in Europe between the 16th and 19th centuries.[1] Instead of marrying at an early age, as most people suppose, Europeans in those centuries actually married at a relatively late age. For example, the data from York indicates that almost no one married before age 20. Between 1538 and 1812, of the 11,000 recorded marriages, only 50 women married before they were 20 years old. Only 34 women bore children before the age of 20 (Cowgill, 1970).

The same pattern of late marriage occurred in the rural English parish of Colyton (Wrigley, 1969). Between 1560 and 1837, the average age of first marriage for men was between 26 and 28 years of age. For women, the range was greater, in one period going as high as 30 years of age. Furthermore, for more than a century, between 1647 and 1769, the women of rural Colyton were older on average than the men they married (Wrigley, 1969, p. 87). It was also during this period in Colyton that many women remained unmarried, perhaps because there were more women than men in the population (Sharpe, 1991).

---

[1]The interest of historical demography in age at marriage is an example of the overlapping interests of demography and family demography discussed in Chapter 3.

The pattern of late marriage, usually combined with a high proportion of the population never marrying, was not limited to England, but was true of many other countries of northern and western Europe. Studies of the marriage records in the Scandinavian countries, France, Austria, Bavaria, Italy, the Netherlands, and other places in England have all supported the same general conclusion that Europeans from the 16th century to the 19th century delayed marriage until their late 20s, and many did not marry at all. This widespread tendency to delay marriage, or to forgo marriage entirely, is referred to as the *European marriage pattern* (Hajnal, 1965; see also Watkins, 1984).

The European marriage pattern has become one of the most influential and accepted generalizations of historical demography (Alter, 1990). While there can be little doubt about the accuracy of the fact of late marriage in Europe, the explanation for this pattern is less certain. The most widely accepted explanation is that couples had to become economically self-sufficient before they could marry. In rural areas this meant that couples had to wait either to inherit a farm or to save the money to rent or buy one. In either case, their marriages were delayed until they were well into their 20s or 30s. This explanation, while widely accepted, continues to be challenged and modified by researchers who are studying the historical marriage systems of particular European countries, for example, Ireland (Guinnane, 1990) and Italy (Kertzler & Hogan, 1990).

The rural Colyton records can be used to illustrate another important contribution of historical demography. Research into demographic history often forces us to give up linear thinking about the past. "Linear thinking" simply means that any current trend is extended in a straight-line fashion back through history. Since the average length of life has been increasing in our time, it is often assumed that the average length of life decreases steadily as we go back in time. In a general way that may be true, but historical demographic research can show us it is not always or invariably true. When Wrigley (1968, 1969) studied the Colyton data, he found substantial variations in life expectancy in different time periods. Wrigley estimated the expected length of life at birth in the Colyton parish for three different time periods beginning in 1538. These estimates are shown in Table 4.1.

The first notable point in Table 4.1 is that the people of Colyton in the 16th, 17th, and 18th centuries had a relatively short expectation of life at birth, compared to people in contemporary societies. In the United States today, life expectancy at birth is approximately 75 years.

The second thing of significance Table 4.1 shows is how the expectation of life at birth was greater in the earliest time period (1538–1624) than it was in the two later periods. Infant mortality (death in the first year of life) followed a similar pattern during those centuries. The infant mortality rate was highest in the period between 1700 and 1749. In the earliest period, from 1538 to 1599, the infant mortality rate was lower than it would be again

TABLE 4.1. Estimated Expectation of Life at Birth in
Colyton, England, 1538–1774

Period	Number of years
1538–1624	43.2
1625–1699	36.9
1700–1774	41.8

Note. From Wrigley (1968, pp. 546–580). Copyright
1968 by Daedalus. Reprinted by permission.

for 200 years. With simple linear thinking we would have expected the
opposite.

As we go further back in time, population data are less and less likely to
be available. Indeed, the earliest known church records of baptisms, marriages, and burials go back only to the early part of the 15th century in France
(Henry, 1968). For demographic data before that time, historical demographers have very little hard information available to them.

One of the more interesting efforts to say something about the demographic past, despite the absence of church or governmental records, is
Durand's (1960) estimate of life expectancy in the time of the Roman Empire.
Using the birth dates and death dates on Roman funeral tablets (the equivalent of today's tombstones), Durand made estimates of life expectancy at birth
for Romans who lived in the first and second centuries A.D. Actually, Durand
was only able to make estimates for the males, because females were less likely
to have funeral tablets made for them. Durand also had to make adjustments
for class differences in the availability of funeral tablets, since middle-class
Romans were more likely than lower-class Romans to have been commemo-
rated with funeral tablets. Durand also took into account the reality that
children who died in infancy were seldom memorialized with a funeral tab-
let. After making all the necessary adjustments, he estimated that Roman
males in the first and second centuries A.D. had a life expectancy at birth of
between 25 and 30 years. Recent studies have supported Durand's estimate
of life expectancy (Parkin, 1992). No contemporary society, regardless of
how poor the living conditions, has a life expectancy so low. When we recall
that Rome was the richest and most powerful of Western societies at that
time, these estimates of life expectancy give us some insight into how pre-
carious life has been through most of human history.

Demography Deduced from Historical Facts

There is one further element to historical demography that completes this
description. For many historical periods, there are serious gaps in the demographic records. No demographic data on births, deaths, and migration exists,
either because the observations were not made and recorded at the time or

because the records have not survived to the present. When this situation prevails, historical demographers have often turned to an alternative method of research: deductive historical demography.

Deductive historical demography is the estimation of what demographic facts might have been on the basis of known economic, social, medical, or other facts. An illustrative example is estimating a population's death rate on the basis of the prevailing living conditions during some particular historical period. This method would depend on the availability of information on "prevailing living conditions" that would allow a researcher to estimate (or deduce) what the death rate might have been. (Remember that this is the approach used when there is no information on the death rate itself.)

Deductive historical demography has been used many times to make inferences about demographic characteristics of the past, but the conclusions it produces are less secure than those based on actual demographic data. This is not to say that deductive historical demography is useless or without merit. When no demographic data are available, it may be the only way to gain demographic insights into the past. We will consider one example of deductive historical demography to show further how and when it can be useful.

An interesting case of deductive historical demography concerns Irish history between 1780 and 1840. It has been established that during this period the Irish population grew very rapidly. During much of this 60-year period, the population grew between 13 and 20% each decade (Connell, 1950; cf. Clarkson, 1981). While this rapid growth rate has been firmly established by historical demographers, its explanation remains a topic of debate. The problem can be traced to the unavailability of data on births and deaths for that period of Irish history. Thus historical demographers have turned to deductive historical demography (Kammeyer, 1976).

Historical demographers offer several competing answers to the question of why the Irish population started to grow rapidly in the last part of the 18th century. There is general agreement that immigration could not have accounted for the growth of the population. This leaves only one explanation: The birthrate must have exceeded the death rate. But this leads to the further question of whether it was an increase in the birthrate or a decrease in the death rate that produced the difference. Finally, there is the question of *why* either the birthrate or the death rate would have changed.

Historical demographers have theorized three explanations: (1) an increase in the birthrate due to an increase in fecundity, the biological capacity to reproduce, among the Irish people (Drake, 1963); (2) a decrease in mortality, beginning in the 1770s (Drake, 1963); and (3) a tendency toward earlier ages of marriage, which would have led to earlier and more childbearing (Connell, 1950).

The third of these explanations is the most involved and will be treated first because it can then be related to the first two explanations. Why would the Irish have started marrying earlier? According to Connell (1950), the answer lies in the way the Irish, a rural people, passed on their land to their children. In rural societies around the world, marriage for young people generally has to be postponed until the marrying couple can find a farm on which to live. Connell has argued that more land became available to rural Irish youngsters in the late 18th century, thus allowing them to marry young. Not a lot of land was available, but enough so that a young couple could raise enough food (specifically potatoes) to feed themselves and their offspring and enough grain to pay rent to the English landowner.

During the late 18th century, both the Irish farm tenants and their English landlords found it suitable for their own interests to split existing farms into parts, making new farms (though very tiny ones). The landlords found this to their advantage because it allowed them to increase their profits from the sale of grains. Irish farmers were unable to produce grain crops on a large scale because they were limited to hand labor. Since each farmer could produce some grain on his small piece of land, the landlords approved of splitting the farms into smaller and smaller units. The process of splitting the farms into smaller units was augmented by the reclamation of some adjacent lands (bogs, woodland, and hillsides). Through these processes, a farmer who had two or three sons might carve out two or three tiny farms. The sons, under these conditions, could marry young and begin their families. If, as Connell believes, this pattern was widespread through Ireland in the last part of the 18th century, it could have produced increases in childbearing that brought about the rapid population growth.

A key part of this explanation is linked to the importance of potatoes in the diets of the Irish people. On a very small plot of an acre or less an Irish family could raise enough potatoes to feed itself throughout the year. The potato had been introduced into Ireland in about 1600, and by the 1780s it made up the overwhelming part of the Irish diet. It is reliably known that the average Irishman ate between 10 and 12 pounds of potatoes every day when they were in abundant supply, and women and children ate somewhat smaller amounts (Connell, 1950).

Although an unvaried diet of potatoes must have been very monotonous, it did have the virtue of being relatively nutritious. The healthfulness of the potato diet provides the basis of the two alternative arguments for why Ireland's population started to grow rapidly in the 18th century. It is possible, as Drake (1963) argues, that the general health of the Irish population might have improved during this period. Such an improvement in health could have increased Irish fecundity, or the biological capacity for reproduction, and thus produced a higher Irish birthrate.

The final alternative explanation is that the improved health of the Irish people might have lowered their mortality. If the death rate went down, the population might have grown even though the birthrate remained unchanged. This was the case in many European countries where it has been shown that early improvements in diet and living conditions started to bring down the death rate. While we do not know that this happened in Ireland, since mortality statistics for that period are not available, we can deduce that some of the population increase might be attributable to declines in mortality.

The interesting feature of these competing explanations for the growth of Ireland's population is that they have a common base (the central importance of the potato) and they are mutually compatible. There is no reason why all three explanations could not have contributed simultaneously to the same result. We will probably never know which explanation is most accurate, but the analysis done by historical demographers does give us insights into the ways social and biological factors can produce demographic outcomes, especially in combination.

CONTEMPORARY DEMOGRAPHIC DATA

Demographers today do not lack for data on contemporary demographic events. Especially in the economically advanced Western world, there is a tremendous amount of statistical demographic data. These data come largely from censuses (or counts) of populations, the registration of vital events (births, deaths, marriages, and divorces), and special sample surveys. But the widespread availability of demographic data does not mean that all data problems are solved. Demographers must always be concerned with the quality of data they use. We will discuss some of the most prominent contemporary sources of demographic data and the quality of those data in the sections that follow.

Population Registers

A number of countries have systems for registering and assembling data on the personal characteristics and major events in the lives of their citizens (Lunde, 1980). Such systems often provide a continuous record of where all citizens reside and what they are doing. These systems are called by various names, including continuous population registers, central population registers, and person–number systems. For convenience, we will use the term "population register."

The basic idea of a population register is that the government has a continuous, up-to-date file on each person in the population. This is accomplished by assigning each person in the population a unique number in much the same way most people in the United States have a Social Security number. Since computers can easily store and retrieve numbers, it is a simple matter to keep a file, a set of information that is readily accessible for inspection or analysis, for each citizen.

In Sweden, for example, a number is assigned to each new baby (the boys are assigned odd numbers; the girls, even ones). Any immigrant coming to reside in Sweden also receives a number (Lunde, 1980). The standard information found in the Swedish file includes, among other things, information on date of birth, place of birth, current address, marital status, dates of changes in marital status (i.e., dates of marriage, dates of divorce), and whether one is a member of the Church of Sweden (Lutheran).

Israel is another country with a population register for its people. Shortly after Israel became a nation in 1948, the government assigned numbers to each resident aged 16 or over. Over time, that system has expanded, with personal numbers now being assigned at birth. The Israeli system also has a considerable array of information on each individual, including date and place of birth; gender; ethnic identity; religion; marital status (e.g., single, married, divorced, or widowed); names, dates of birth, and gender of children; past and present nationality or nationalities; current address; date of entry into Israel; and date of becoming a resident. Each resident of Israel is bound by law to give notice of any changes in status, such as marriage, divorce, change of address, or other important life events (Lunde, 1980).

Judging by the countries that have them, it appears that population registers are most likely to be implemented where populations are relatively small and stable and where most people are literate. Population registers require the cooperation of almost all citizens if they are to work properly. Since citizens have a good deal of responsibility for reporting new events in their lives, they must be willing to give reports to the government on their activities. Any country with a substantial number of illegal aliens or a minority group that feels threatened by the majority will probably have difficulty in maintaining an accurate population register.

The countries with the most effective systems are those in Scandinavia and northwestern Europe. The countries with the most advanced systems are Denmark, Finland, France, Iceland, Israel, the Netherlands, Norway, and Sweden. Some South American countries have also instituted population registers; Argentina has the most advanced and complete system (Lunde, 1980).

Population registers are extremely useful sources of demographic data when they are comprehensive, universal (covering the entire population) and up to date. The demographic value of a population register can be seen most clearly in the area of migration research. As noted above, in the systems of Sweden and Israel, people must register any change of address. Thus any individual's record will reveal the migration history of that person. From individual records, it would be a simple matter to assemble information on the amount of migration in the total population, the direction and flow of migration, and the characteristics of migrants (e.g., their age, education, marital status, and occupation). Such an abundant and readily available supply of migration information contrasts sharply with migration data available in the United States. There is no systematic and universal record of people who move from one place to another in the United States. A citizen who leaves Des Moines, Iowa, and moves to Sacramento, California, can do so without asking or telling anyone. Certainly there is no requirement that the move be registered with government officials. As a consequence, the United States has less systematic information on migration than countries with population registers. Migration research in the United States is made much more difficult by that limitation in the data.

While the potential for demographic research is greatly enhanced when a population register exists, there is also a negative side to be considered. Population registers by their very nature may pose a threat to the privacy of individual citizens. Even more threatening is the potential that such a system may have for governmental repression of all or of certain categories of its people.

For example, both Sweden and Israel record the religion of each citizen. Both countries have an official state religion, and apparently the people, in general, feel no threat when they are asked to provide religious information to the government. But those citizens who do not subscribe to the official and dominant religion could conceivably be singled out by the government for special treatment. The treatment of Jews in Nazi Germany illustrates what can happen when a repressive government singles out a minority religious group for negative treatment. In the United States, some people are very concerned about the possibility of government obtaining information about their religious affiliation. The doctrine of separation of church and state is firmly embedded in the minds of most Americans. Even the slightest infringement of religious freedom or privacy is viewed as a threat to individual freedom. For this reason, the census of the United States does not ask about an individual's religious affiliation.

The religious question aside, there is a general and perhaps growing concern in the United States about the potential abuses of individual freedom when the government maintains files on its citizens. A population register does give a government the capability of keeping a close watch on the whereabouts and actions of its citizens. Even though the demographic research possible with a population register is not likely to be a violation of individual privacy, the other possible uses may be. This issue of individual freedom probably keeps some countries, including the United States, from adopting a population register system.

The Registration of Vital Events

As an alternative to the population register just described, most countries have some system by which vital events are registered with governmental authorities. This is called a *vital registration system*, or simply a *registration system*. When the data from a registration system are assembled, they are referred to as *vital statistics*. The registration systems in existence today are almost always controlled and managed by civil governments, but their origins can usually be historically linked to parish or church records.

In England, for example, as a result of a civil law passed in 1538, Anglican priests were required to record in a register the weddings, christenings, and burials in their parishes. It was not until 1837 that these events were registered directly by a governmental records office. The first governmental publication of vital statistics appeared in England in 1839, 2 years after the Registration Act went into effect (Shryock & Siegel, 1976).

In the United States, the vital registration system has been greatly influenced by the federal system of government established by the writers of the Constitution. When the legal and political system of the United States was established, there was concern that the central government would become too powerful. To guard against that eventuality and to ensure that individual states would retain power over their own affairs, the Constitution clearly limited the powers of the federal government. The Constitution described the specific responsibilities of the federal government and left all other things to be done by the individual states. As we will see below, the Constitution did mandate a periodic national census (or count) of the population, but it did not require a system of registering vital events. Individual states were given the freedom to register vital events in any way they saw fit or not to register these events at all. As a result of that decision, the vital registration system of the United States is incomplete and inconsistent because the states are not required to collect data in a uniform way. This inadequacy and incompleteness has affected the vital statistics of the United States through most of its history.

It was not until 1900 that the federal government became actively involved in upgrading vital statistics, especially with the aim of increasing the consistency and uniformity of the data. This was done by setting up a Death Registration Area. The Death Registration Area was composed of all states that met a minimal standard of uniformity in recording deaths. States became a part of the Death Registration Area by adopting a standard death certificate that followed the guidelines suggested by the federal government. In 1900, only 10 states and the District of Columbia met the standard. In 1915, the U.S. Census Bureau set up a Birth Registration Area, and again, only 10 states and the District of Columbia met the minimal requirements of completeness and consistency. It was not until 1933 that all states met the requirements of both the Death and Birth Registration Areas. Even today, while there is a U.S. Standard Certificate of Live Birth, which identifies the information that should be on a birth certificate, not all states report all of the information requested (U.S. Department of Health and Human Services, 1992).

Up to the present day, marriage and divorce statistics are not collected uniformly from all states in the United States. A Marriage Registration Area for the United States was established in 1957, and a Divorce Registration Area was established in 1958. As with the Birth and Death Registration Areas, these areas only include the states that have procedures for collecting marriage and divorce statistics that meet the minimal standards established by the National Center for Health Statistics. Currently, the Marriage Registration Area includes 42 states, the District of Columbia, Puerto Rico, and the Virgin Islands (U.S. Department of Health and Human Services, 1991a). The Divorce Registration Area is composed of only 31 states and the District of Columbia (U.S. Department of Health and Human Services, 1991b).

The states that are part of the official registration areas can change at any time. States may be added when their data collection methods conform to the federal standards, or removed if they do not. California, for example, the state with the largest population, allows a marriage procedure that makes its statistics on marriage problematic. Beginning in 1972, California law allowed some couples to have nonlicensed (confidential) marriages (U.S. Department of Health and Human Services, 1991a). Unmarried couples who assert they have been living together may be married confidentially without a marriage license or health certificate. Since 1972, county clerks in California have been required to keep sealed records of these marriages, and these records cannot be opened for inspection without a court order. The county clerks do report the total number of such marriages to the California State Department of Health Services, but the entire system is open to serious questions about accuracy and validity. It is widely believed that many people take advantage of California's confidential marriage procedure for reasons of personal convenience. Simply by declaring that they have been living together, a California couple can avoid the normally required blood test and the regular 5- to 7-day waiting period required for the results of those tests.

The list of inconsistencies and shortcomings in the vital registration system of the United States is extensive. Most American citizens would be amazed to learn just how much the decisions of individual states detract from a uniform system of vital statistics. The following are a few examples of how state actions have weakened the vital registration system.

In 1963, New Jersey stopped recording racial data on birth and death registration forms. Apparently the racial information was removed because it was viewed by many people as a reflection of governmental racism. While that may have been true to some extent, the removal of racial data from the New Jersey records in 1963 made certain kinds of demographic analyses impossible. For example, comparative death rates for the white and black populations, which might have demonstrated the effects of racism, could not be determined. In 1964, race was again included on the New Jersey birth and death statistics, but the data lost in the previous year can never be recovered.

Two states, Arizona and New Mexico, do not have state-level central files on marriage. From these states, data on marriages must be obtained directly from county officials. Although information is requested on the number of *marriages performed*, some counties report only the number of *marriage licenses issued*. The number of licenses issued may be somewhat higher than the actual marriages performed (U.S. Department of Health and Human Services, 1991a).

Divorces and annulments are not reported by all counties in Indiana,

Michigan, Mississippi, New Mexico and California. In Indiana, the number of reported divorces is based on the number of divorce petitions filed, rather than the number of divorce decrees granted (U.S. Department of Health and Human Services, 1991b).

The vital registration system of the United States, which allows states autonomy in the collection of data, clearly has shortcomings. Too often there is a lack of consistency from state to state and even within states. Despite these inconsistencies and other shortcomings, the vital registration system is the most complete record of vital events available to demographers in the United States. Table 4.2 provides an example of the type of statistics that are complied each month by the National Center for Health Statistics.

THE CENSUS

A census of population is a complete count or enumeration of that population. This basic definition can be supplemented with a more formal definition, such as the one arrived at by the United Nations: "A census of population [is] the total process of collecting, compiling, and publishing demographic, economic, and social data pertaining, at a specified time or times, to all persons in a country or delineated territory" (United Nations, 1958, quoted in Kaplan & Van Valey, 1980, p. 4).

Censuses are official undertakings because they are conducted under the auspices of some governmental authority. The results of the census are generally viewed as the official count of the population.

Censuses are also distinguished by the fact that the entire population is counted. However, this statement must be qualified somewhat because many times as a *part of a census* certain kinds of information are collected from only a sample of the total population. The sampling, however, is conducted so that the results will represent the total population.

A sample of the total population is not to be confused with *partial counts* of a population in which only certain types of people are counted. Many of the early historical accounts of population "censuses" were not true censuses because they were only counts of certain types of people. Usually early population counts, as we will see, were for the purposes of counting the taxable population or the men of military age. Modern censuses, which attempt to count the entire population, are much more recent in origin, going back a little over 300 years. Before considering the modern census, we will take a brief look at some early population counts.

Early Population Counts

Historical accounts from many parts of the world report a variety of population counts, some going back to ancient times. These include enumerations

TABLE 4.2. Provisional Vital Statistics for the United States, 1994

| | January 12 | | | | 12 months ending with January | | | | |
| | Number | | Rate | | Number | | Rate | | |
Item	1994	1993	1994	1993	1994	1993	1994	1993	1992
Live births	346,000	325,000	15.7	14.9	4,060,000	4,075,000	15.7	16.0	16.3
Fertility rate	—	—	68.8	64.7	—	—	68.7	69.0	69.7
Deaths	223,000	198,000	10.1	9.1	2,292,000	2,168,000	8.9	8.5	8.6
Infant deaths	2,600	2,800	7.6	8.4	33,000	34,000	8.2	8.4	8.9
Natural increase	123,000	127,000	5.6	5.8	1,768,000	1,907,000	6.8	7.5	7.7
Marriages	109,000	103,000	5.0	4.8	2,339,000	2,353,000	9.1	9.2	9.4
Divorces	97,000	92,000	4.4	4.2	1,192,000	1,203,000	4.6	4.7	4.7
Population base (in millions)	—	—	259.4	256.6	—	—	258.2	255.3	252.4

Note. From U.S. Department of Health and Human Services (1994, p. 1).

Rates for infant deaths are deaths under 1 year per 1,000 live births; fertility rates are live births per 1,000 women aged 15–44 years; all other rates are per 1,000 total population. Data are subject to monthly reporting variation.

Figures include revisions received from the states. Twelve-month figures for the current year reflect revisions received for previous months, and figures for earlier years may differ from those previously published.

of parts of the populations in Babylonia, China, Egypt, Palestine, and Rome. Many of these population counts were undertaken by authorities to determine the size of the taxable population. The very word *census* is a derivative of the Latin word *censers*, which means "to tax" or "to value" (Kaplan & Van Valey, 1980). A second common reason for early population counts was to assess the number of people who would be available for military service or for forced labor.

An Old Testament reference to a "census" taken somewhat before 1000 B.C. illustrates how early counts had a specific but limited purpose or aim:

> David said to Joab and the commanders of the army, "Go, number Israel, from Beersheba to Dan, and bring me a report, that I may know their number." But Joab said, "May the Lord add to his people a hundred times as many as they are! Why then should my lord require this? Why should he bring guilt upon Israel?" But the king's word prevailed against Joab. So Joab departed and went throughout all Israel, and came back to Jerusalem. And Joab gave the sum of the numbering of the people to David. In all Israel there were one million one hundred thousand men who drew the sword, and in Judah four hundred and seventy thousand who drew the sword. (1 Chronicles 21:1–6, Revised Standard Version)

As is evident from this passage, the purpose of this particular count was to learn how many men there were "who drew the sword." The leaders of Israel were obviously interested in knowing the number of able-bodied men who were eligible for military service.

In 1066, William the Conqueror gained control of England. He thereafter assembled the "Domesday Book," which was a list of all the landowners and their holdings. Again the scope of this survey was limited to those who could be taxed. These are but two examples of how most early counts of populations were limited to "heads of households," males of military age, or taxpayers (Shryock & Siegel, 1976).

The Modern Census

Historians of population do not agree on exactly which country should have the credit for conducting the first complete, and therefore modern, census. Some say it should be Canada, since a complete census was carried out in the Province of Quebec in 1666, while others contend that Sicily and various other Italian states were the first to do so in the 15th century. However, these censuses covered only some parts of modern-day countries. The first country to undertake a complete count of the entire nation was Sweden in 1749. Norway and Denmark had complete censuses beginning in 1769.

The United States has an important place in the history of census taking because it was among the earliest countries to count its population on a regular basis. The United States conducted its first census in 1790 (some 11 years before the first census in Great Britain). The results of the first census are presented in Table 4.3. This first census was called for in the Constitution, which specified in Article 1, section 2, that it had to be conducted "within three years after the first meeting of the Congress." Article 1 further provided that a new census should be conducted "within every subsequent term of 10 years." This provision for a population count in the Constitution was required because state populations were to be used to apportion the representatives in the House of Representatives.

This constitutional requirement, which was a direct outgrowth of the federal system of government, had important implications for the nature and quality of American census taking. To begin with, the requirement for a census every 10 years (called a decennial period) has given the United States two centuries of experience in counting its people and processing the resulting statistical data. While we will see that the U.S. census is not without its flaws, it surely has led the way in advancing the techniques of enumerating, analyzing, and reporting population data.

A second implication follows from the use of the census count to apportion the House of Representatives. Because this was to be the primary use of the count, it was necessary for the size of the population to be made public. In fact, names as well as the counts from the first censuses were posted at public places to ensure that no one was omitted. The size of the U. S. population could not be a state secret, and its population counts have always been public information. In many other countries, both in earlier times and in recent years, census counts actually have been regarded as state secrets. In the 19th century especially, the sizes of populations were considered an indication of the military strength of countries. A 20th-century example is found in the former Soviet Union, which conducted a census in 1937 and then treated the results as state secrets. This was done on the order of the Soviet leader Joseph Stalin. It was not until 1991, thanks to Gorbachev and his policy of *glasnost*, that the data from the 1937 census were released and became available for publication (Heleniak, 1991).

Types of Censuses

Before considering U.S. census-taking procedures, a few basic distinctions must be made about types of censuses and methods of census taking. One distinction is related to the place of residence or the location of the people counted. In a *de jure census*, people are counted as being residents of the place where they usually live. The United States conducts a *de jure* census because the count is used to apportion representatives (as well as to allocate various federal funds). It would be unreasonable to count a person as a resident of a state or city just because he or she happened to be visiting there or traveling

TABLE 4.3. First Census of the United States (Population of the United States as Returned at the First Census, by States, 1790)

District	Free white males of 16 years and upward, including heads of families	Free white males under 16 years	Free white females including heads of families	All other free persons	Slaves	Total
Vermont	22,435	22,328	40,505	255	16[a]	85,539[b]
New Hampshire	36,086	34,851	70,160	630	158	141,885
Maine	24,384	24,748	46,780	538	None	96,540
Massachusetts	95,453	87,289	190,582	5,463	None	378,787
Rhode Island	16,019	15,799	32,652	3,407	948	68,825
Connecticut	60,523	54,403	117,448	2,808	2,764	237,946
New York	83,700	78,122	152,320	4,654	21,324	340,120
New Jersey	45,251	41,416	83,287	2,762	11,423	184,139
Pennsylvania	110,788	106,948	206,363	6,537	3,737	434,373
Delaware	11,783	12,143	22,384	3,899	8,887	59,094[c]
Maryland	55,915	51,339	101,395	8,043	103,036	319,728
Virginia	110,936	116,135	215,046	12,866	292,627	747,640
Kentucky	15,154	17,057	28,922	114	12,430	73,677
North Carolina	69,988	77,506	140,710	4,975	100,572	393,751
South Carolina	35,576	37,722	66,880	1,801	107,094	249,073
Georgia	13,103	14,044	25,739	393	29,264	82,548
Total number of inhabitants of the United States exclusive of southwestern and northern territory	807,094	791,850	1,541,173	59,150	694,280	3,893,635

Note. From U.S. Bureau of the Census (1907).

[a]The census of 1790, published in 1791, reports 16 slaves in Vermont. Subsequently, and up to 1860, the number is given as 17. An examination of the original manuscript returns shows that there never were any slaves in Vermont. The original error occurred in preparing the results for publication, when 16 persons returned as "Free colored," were classified as "Slave."

[b]Corrected figures are 85,425, or 114 less than figures published in 1790, due to an error of addition in the returns for each of the towns of Fairfield, Shelburne, and Williston, in the county of Chittenden; Brookfield, Newbury, Randolph, and Strafford, in the county of Orange; Castleton, Claredon, Hubbardton, Poultney, Rutland, Shrewsbury, and Wallingford, in the county of Rutland, Dummerston, Guilford, Halifax, and Westminster, in the county of Windham; and Woodstock, in the county of Windsor.

[c]Corrected figures are 59,096, or 2 more than figures published in 1790, due to error in addition.

through at the time of the census. So the 1990 census form instructs the member of the household who is completing the form to include "persons who usually live here but are temporarily away."

When a *de jure* census is taken, there is always some uncertainty about what should be considered a person's usual or normal place of residence. College students, for example, often live for 9 or 10 months a year in the community where they attend college. Yet during summer months and holiday periods, they may live in the homes of their parents, which may be in different communities or states. Which is their usual place of residence? The Census Bureau has concluded that college students should be counted as residents of the communities where they reside *while* attending college. However, the Census Bureau has not always defined college students in this way. Prior to 1950, college students were counted as residents of their parents' homes regardless of where they lived while attending school. One result of this change in census definition and interpretation is that many towns and small cities with colleges reported tremendous increases in population size between 1940 and 1950 (Shryock & Siegel, 1976). This is but one example of the way in which users of census data must be aware of changes in definitions whenever they use historical census reports.

In contrast to the *de jure* census, there are some countries that conduct *de facto* censuses. A *de facto census* is one that counts all persons as residents of the place they happen to be at the time the census is taken. In practice, *de facto* censuses are conducted in such a way that most people are counted in the places they usually live, but not always. The Polynesian island of Niue conducted a *de facto* census in 1981. To quality for financial assistance from New Zealand, the island had to prove it had at least 3,000 people. Government officials in Niue, who wanted to ensure the highest count possible, instructed the census takers to count all people visiting the island. Thus, four people from a French sailboat anchored in the island harbor were part of the 3,281 people counted in the Niue census (May, 1992).

In countries where a *de facto* census is to be taken, the government typically designates a specific census day, or perhaps a night, during which people are required to stay in their homes. Prior to the day of the count, the people are told where they should be when the counting is done. Servants, for example, who work and sleep in other people's homes most nights, may be instructed to be in their own homes on the night of the census. Or the opposite instruction could be issued, that is, to be in the home where they work, even if they have a home of their own. These decisions may be somewhat arbitrary, but the general guiding principle of a *de facto* census, just as it is with a *de jure* census, is to count the people in such a way that the results reflect the "normal" distribution of the population in the country.

Since a *de facto* census almost literally requires the entire population to stay in one place while the count is being taken, the entire process must be completed in a very short time. As a practical matter, a *de facto* method cannot be used if the population is exceptionally large. Even counting the people in a small country in one day or night would require a large number of canvassers. Often people who are already employed by the government, such as postal delivery people or army personnel, will do the job. This brings us to the second major distinction that can be made about censuses. This second distinction reflects who actually performs the task of collecting and recording the information.

Probably the most long-standing method of census taking is one called the *canvasser* or *direct interviewer* method. This method requires that someone who represents the census-taking agency visits each residence, asks the questions, and records the answers. Often the canvasser method will require the hiring of a vast number of temporary interviewers who carry out the task of taking the census. When China conducted its census in 1990, 7 million people were employed as canvassers or enumerators (Haub, 1991a).

The alternative to having a canvasser who asks the questions is to require some person or persons in the household to read and answer the questions. This method is called the *householder* or *self-enumeration* method. When the householder method is used, the census questions must be delivered to the household in some way. They may be delivered by an employee of the census-taking agency or, as is generally the case in the United States, through the postal service. Once the census form is delivered, the householder follows the written instructions, answers the questions, and returns the forms to the government census office. The forms are usually returned through the mail system, but in some cases they are picked up by a census employee.

The self-enumeration method requires certain characteristics to be widespread in the population. First, most people in the population must be literate, since someone in the household must read and answer the questions. In the United States, it was not until 1960 that some proportion of census forms were completed by householders. Prior to that time, the entire census had been taken by canvassers. Probably the fact that many Americans in the first half of the century were immigrants from other lands made it difficult to use a householder method of census taking. Such residents might not have been able to read English, even when they were literate in their native languages. People not only have to be literate to complete the census form but must also be fairly sophisticated about reading and responding to complicated questions. The 1990 census form is reproduced in Appendix A, and the reader will observe that some questions require a good understanding of legal, medical, and economic terminology. Since, even today, there are many people in the population who have very limited reading skills, it is not difficult to imagine some people having difficulty with some of the questions.

Finally, an important, and perhaps even a necessary, characteristic of the people is a spirit of cooperation and trust regarding the census-taking process. This means that the people must trust their own government. While this characteristic is harder to pin down and measure, it can hardly be denied that if there is widespread hostility toward the government or distrust of governmental objectives, census taking will be ineffective. The citizens must have a spirit of cooperativeness and trust because it does require energy and effort to answer the questions completely and accurately.

When citizens in the Netherlands protested their 1971 census, 2.3% of the population refused to cooperate. By the time of their 1981 census, the protest groups had grown so large and anti-census campaigns were so effective that the census was postponed. At issue was the amount of data that was to be collected and the plan to link the census data with population registers. The people believed that their individual rights were being endangered (Choldin, 1988). Similar protests occurred in Germany in 1983 and in Great Britain in 1971. Since these protests, the governments in these countries have added data-protection laws to ensure the privacy of individuals. In the United States, even though it is illegal for a person to refuse to answer census questions, the system works as well as it does thanks to voluntary cooperation, not to legal coercion.

The 1990 Census of the United States

The most recent decennial census of the United States was carried out in April 1990. Just before April 1, 1990, census questionnaires were delivered by the postal service to about 88 million dwelling units or households. Approximately 83% of these households were instructed to complete the questionnaire and return it by mail. This procedure was used primarily in the more urban areas of the United States. This is, of course, a householder or self-enumeration method and may be simply referred to as a "mail-out/mail-back" system. (See Appendix A for the census questionnaire.)

While most of the population was questioned through the mail-out/mail-back system, a different procedure was followed with 11% of U.S. households typically located in rural areas and concentrated in the South and Midwest. A Census Bureau canvasser visited the household so that an accurate address could be obtained. In rural areas, household addresses are not always the same as the postal address. People living in the household were instructed to complete the questionnaire and return it by mail. In addition, about 6% of all household questionnaires were delivered by the Postal Service but people were instructed to hold it for a Census Bureau canvasser who would collect the completed questionnaire (Robey, 1989).

In the 1990 census a special attempt was made to count the homeless in the U.S. population. During what was referred to as "S-Night," canvassers counted people living in homeless shelters; seeking refuge in such indoor locations as hospital waiting rooms, bus and train stations, or all-night coffee shops; and sleeping on the street.

The Census Bureau based its procedures for counting the homeless on what had been learned from an earlier study conducted in Nashville, Tennessee (Lee, 1989). This study revealed where and how to look for the homeless in a population, in order to reduce the number who would be missed by the enumerators. But the study also found that it was easy to count the homeless more than once. Double counting of the homeless was likely to occur if the count continued over several days or a week because the homeless may move from one shelter to another, or move about on the streets. A count of the homeless population may also vary by the time of month when it occurs. Some homeless people may live in the homes of friends or relatives during some parts of the month, but live on the street or in shelters at other times. The Nashville study found that more people were on the streets or staying in shelters at the end of the month when the friends or family who often took them in were awaiting pension or welfare checks (Lee, 1989).

Many advocates for the homeless were opposed to the Census Bureau's efforts to enumerate the homeless, and in some places made efforts to block or impede the count. They feared that the census count would miss many of the homeless and thus minimize the dimensions of the social problem. Indeed, their fears were realized when the Census Bureau reported its April 1990 count of the homeless in the U.S. population as 228,621 (178,828 people living in shelters and 49,793 people living on the streets) (Haub, 1991b). Many homeless advocates have estimated the number to be at least a million, or maybe more.

It is probably true that homeless people are, in general, likely to be undercounted by a census, because their most salient characteristic—being without a fixed address—makes them more difficult to locate and count. Furthermore, some people who are homeless may be trying to disguise their homelessness by living in their cars or camping in public parks. Although Census Bureau personnel acknowledged from the beginning that it would be impossible to locate and count every single homeless person, they made a serious and systematic effort to find as many as possible. There is no reason to believe that the more casual and unsystematic counts or, as is more often the case, estimates and guesses about the size of the homeless population are any more accurate than the census count. This is especially true when it is recalled that some members of the homeless population are likely to be double counted. In the end, the census count, even with its acknowledged limitations, is the best count of the homeless we have.

Short and Long Forms of the Census

The *first* U.S. census was very brief; only five questions were asked: (1) total number of persons in the household, (2) number of males and females, (3) number of free or slave persons, (4) number of males 16 and over, and (5) the name of the head of the household. Today the census collects much more information, but most of that information is based on a sample of the population, not the entire population. The sample is large—one out of every six households in 1990—but is a sample nonetheless, and it gives us reliable information on many aspects of the population.

The questions asked of the entire population are found on what is called the short form. The short-form questions ask only about age, gender, and racial or ethnic characteristics of persons in the household and about the relationships between household members. There are also six questions about the dwelling unit in which the members of the household live. These questions on the dwelling unit are asked because the U.S. Congress in the 1930s, being concerned about the quality of American housing, started requiring the Census Bureau to ask about the houses in which people lived. These questions include inquiries about the number of rooms, type of unit, and value of unit.

The long form of the census is a much more extensive set of questions and provides great amounts of detailed information about the population. There are 26 questions, including questions about where each person in the household was born, where each lived 5 years earlier (a measure of migration), active military duty, health, work activity, occupations, and income. There are also, on the long form, 19 more questions about the dwelling unit.

The Quality of the U.S. Census Data

While census taking in the United States was originally required for political reasons, it is now obvious that many other purposes are served, including demographic analysis. Demographers who use census data must be aware of the kinds of errors that exist in those data. Basically, there are two types of errors: errors in counting and errors in the characteristics attributed to members of the population. Errors in counting come largely from the fact that some people in the population are not located. This results in an *undercount*. At the same time, some people in the population are counted more than once, which results in a *double count* or *overcount*.

Overcounting may occur, for example, among young college students. The census instructions state that college students living away from their parents' homes should not be listed as members of their parents' households, but parents may not always follow the instructions. Thus, students may be counted as residing with their parents and also be counted as part of their college communities. Interestingly, this is parallel to a form of double counting found in earlier censuses when some children were counted twice. Studying the records of the 1850 and 1860 censuses, Adams and Kasakoff (1991) noted that children, especially young girls, were often counted once in the households of their parents and again in the homes of their grandparents, or of wealthier families where they had been sent to live and work. This "circulation of children" was common in the 19th century, and led to their numbers being overcounted. From these two examples, and the case of the homeless considered earlier, we see that double counting is likely to occur whenever persons live in two different places at different times of the year (or month).

While double counting can be a problem, a more serious type of error in the census is undercounting. When people are missed by the census they are deprived of their fundamental right to representation in the Congress. (Probably more important to many political leaders is the resulting loss of federal funds, which are based on population size.) There has been concern about undercounts of the population from the very earliest censuses. In 1790, when Thomas Jefferson submitted the official count of 3,929,326, he revealed his suspicion that some people had been missed. He recorded the official number in black ink, but entered an adjusted figure of 4 million in red ink to show what he thought was the true population of the country (Robey, 1989).

Since it may not be immediately obvious how census analysts determine that there has been an undercount of the population, a brief description, in general terms, will give a sense of how it is done. An estimate is made of the difference between the number of people enumerated and the number of people *believed* to be in the population at the time of the count. Estimates of the population are made by using the previous census results, augmented by birth registrations, death certificates, and immigration/emigration statistics. Also it is often possible to cross-check the estimates with other independent data. For example, estimates of the number of people over age 65 can be compared with the number of people applying for Medicare (Siegel, 1974).

Using these and other methods, the Census Bureau has estimated that in 1990 there was a net undercount of about 1.8% of the population (Himes & Clogg, 1992). But the people who are likely to be missed by the census are not randomly distributed. Analysis has shown that males are missed more than females; blacks are missed more than whites; and people living in urban inner-city areas are missed more than people living elsewhere. In 1990, estimates are that 5.7% of the African American population was missed by the census count, including 12.5% of black males aged 30–44 (Himes & Clogg, 1992).

For many decades the Bureau of the Census has carefully analyzed the official census count and forthrightly reported its estimates of the extent and nature of the undercount. As a result, more and more people have become aware of the undercount, especially of minority populations and of the residents of the inner cities. This has led to discussion about using *adjusted* or corrected population totals rather than the actual census count. Local officials and political leaders have frequently challenged the accuracy of the count in their areas. Mayors from several cities have filed lawsuits against the federal government, demanding that adjusted figures be used. They argued that the high concentration of people in the inner cities who are most likely missed by the census would deprive them of federal assistance money based on population size. And, as we noted before, this would also affect the allocation of seats in the House of Representatives. None of these lawsuits has been successful.

For the 1990 census the unadjusted census data are being used as the official counts of the U.S. population. The decision about whether or not to use adjusted or actual data had to be decided *before* the census results were compiled. If the decision had been made after the adjusted results were available, the census could have become even more political. States that would gain seats in the House of Representatives or that would have received more federal money would surely have argued that adjusted counts be used. On the other hand, states that would not benefit would be likely to argue for the unadjusted counts.

Errors made in describing the characteristics of a population are referred to as misclassification errors. *Misclassification* occurs when a person is somehow placed in a category that is not correct. A simple illustration is misclassification by age. This occurs in one instance when individuals report their ages erroneously, or even dishonestly. There is ample evidence that many people are not precisely the age they report. There is a noticeable tendency for people to overstate their ages when they are under 21, understate their ages through the middle-adult years, and then again overstate their ages during the advanced years (Shryock & Siegel, 1976).

A more specific age misclassification is *age heaping*, which is a tendency of some people to round their ages to numbers ending in 0 or 5 (30, 35, 40, 45, e.g., or by date of birth, such as 1920, 1925, 1930). Since the ages should normally be distributed evenly among the numbers ending 0 through 9, it is a fairly simple matter to detect how much age heaping has occurred in a census count. Some Asian countries have developed an interesting way of dealing with the problem of age heaping (Fraser, 1982). When a census is conducted, the standard questions on age or date of birth are asked; in addition, each respondent is asked to identify their year of birth according to the zodiac calendar. For instance, the Year of the Horse occurred in 1906, 1918, 1930, 1942, 1954, 1966, 1978, and 1990. If persons during the 1990 census said they were born during the Year of the Horse and yet they also said they were 40 years old, the decision by the census takers would be to record the age as 48 (determined by the calendar Year of the Horse which occurred in 1942).

Misclassification also occurs with respect to gender, marital status, racial and ethnic characteristics, education level, and so on. The extent of misclassification on these and other characteristics is revealed in special postcensus surveys conducted by highly trained interviewers. In the United States, these are called *postenumeration surveys*. The procedure involves interviewers reinterviewing a sample of the population and comparing the results they obtain with the results obtained through the regular census procedure. The assumption is that the postenumeration survey is accurate and the discrepancies are a reflection of misclassification. Even with such seemingly straightforward information as the highest year of education completed, the amount of error can be substantial.

Probably no question on the census is entirely immune from misclassification. There will always be errors, made by someone, in any questioning process. The concern of demographers is "How extensive is the misclassification?" and "Is the misclassification systematic?" The last question is important because a systematic bias in any body of data can lead to erroneous conclusions.

While the preceding discussion has called attention to shortcomings and errors in census data, the purpose has not been to criticize or to diminish the extraordinary work done by the U.S. Census Bureau. The purpose is to emphasize that users of census data must be aware of the types of errors that do occur and take them into account. For demographic analysis or even descriptions of populations, it is often important to identify and, if possible, minimize the impact of errors in the data.

The problem of data quality is persistent in demography, even when the data come from an official census of the population. This basic issue, which must so often concern demographers, stems from the fact that demographic data come from observations of naturally occurring events, and these observations are often made by people who have very little interest in the quality of demographic data.

OTHER SOURCES OF DEMOGRAPHIC DATA

In the last 50 years, more and more demographic data have come from special sample surveys. In the United States, this tradition was given its beginnings by an early study of the social and psychological factors affecting fertility. Conducted in Indianapolis, Indiana, in the early 1940s, this study was initiated by the leading demographers of that era to learn more about the processes that determine family size (Whelpton & Kiser, 1943). This study was groundbreaking because it asked people about their attitudes as they might affect their childbearing, as well as about their use of contraception. While the research findings were generally not conclusive, there were enough suggestive results to give further generations of researchers a sound basis for their special surveys of fertility (Kiser & Whelpton, 1958).

In the 1950s and 1960s, the successors to the Indianapolis fertility study flourished, often under the guidance of some of the original researchers. Two research centers, one at the University of Michigan and the other at Princeton University, carried out national surveys of American childbearing behavior, yielding substantial amounts of data on American patterns of childbearing.

A third generation of fertility studies, the National Fertility Studies, was started in the mid-1960s. These studies were supported or carried out by federal government agencies. The sixth survey in this series, now called the National Study of Family Growth, was conducted in 1988. These fertility studies, covering several decades, have provided demographers with an enormous amount of data on the fertility intentions and behavior of Americans.

For historical demographers, one of the major sources of data is the Princeton European Fertility Project. This project was designed to explain variations in the decline of fertility during the 19th century in countries thoughout Europe. Data on marriage, childbearing, and economic factors have been assembled for countries and provinces throughout Europe. Ansley Coale and other demographers working with these data have developed standard indices so that comparisons may be made. We will cite much of this research in Chapter 9, when we discuss the historical decline of fertility and current issues of world population growth.

At the international level, the demographic and social survey that stands out above all others is the Demographic and Health Survey (DHS) program initiated in 1984 (Blanc, 1991). Through this program, any developing country that wishes to undertake a survey is given financial support and scientific assistance. The typical survey covers a sample of women between the ages of 15 and 49. The survey questionnaire, which is administered by native-language interviewers, asks about personal background, fertility behavior, contraceptive knowledge and use, child mortality, and other health-related issues. Since these surveys ask women about both their attitudes and their actions, the findings have been especially useful for examining how well women achieve their childbearing goals and objectives. The results coming from these survey data are used by countries to establish and evaluate family planning programs and policies. The DHS program provides an invaluable source of demographic, social, and economic information for demographers and other social scientists.

Surveys relating to fertility, both in the United States and internationally, have allowed demographers to go far beyond the data provided by censuses and vital registration systems. Some of the most interesting and theoretically relevant studies are now being done through special surveys, both large and small. While the focus of most surveys has been on childbearing, some address migration or mortality. Throughout the remainder of this book, we will often have occasion to examine the results coming from such surveys as well as from the more traditional sources of demographic data.

SUMMARY

Demography is basically an observational science, not an experimental one. The data of demography come largely from the observations of naturally occurring events. When experimental research with nonhuman subjects has been used, it has usually been ineffective for understanding human population processes. Field experiments have been used, especially in family planning research, and though these have inherent limitations, many have proved valuable.

Historical demography is a subfield of demography concerned with demographic events and conditions of the past. The data for historical demography often come from parish records of christenings, marriages, and burials.

When no direct demographic data exist, inferences about the past must be made from known economic, social, medical, or other historical facts.

Contemporary demographic data come mostly from censuses, the registration of vital events, and special surveys. Population registers are continuous and up-to-date files of the vital events in the lives of every person in the population. They provide a good source of demographic data for those countries that have them. The vital registration system of the United States is less complete and consistent than demographers might wish. This problem arises from the freedom that individual states have to collect vital statistics in whatever way they choose.

A census is a complete count or enumeration of the population, conducted by the government. While partial counts of population go far back in history, modern censuses, which count all of the people, go back only to the 15th century. The first country to conduct a complete census was Sweden in 1749. The U.S. census goes back to the founding of the country.

Censuses are classified as either *de jure*, counting people where they normally reside, or *de facto*, counting people where they are when the census takes place. The major shortcomings of censuses are counting errors (undercounting and double counting) and errors in the characteristics attributed to people (misclassification). While the U.S. census still has errors, strenuous efforts to improve the accuracy of the count continue to be made.

More and more demographic data come from special sample surveys. These are conducted by university-based scholars, governmental agencies, and private organizations. These surveys can be more detailed and focused than censuses and vital registration systems, and thus often can be more effective in answering demographic questions.

5

POPULATION COMPOSITION

Many words have been written about the baby-boom generation, but if you were born between 1970 and 1979, you too are part of an important population age group: the baby bust. While less well publicized than the baby boomers, the baby-bust generation may have just as important an impact on U.S. society. If you are part of the baby-bust generation, this may have implications for your personal life as well as U.S. society. There is a demographic theory that places great emphasis on the size of the birth cohort in which you were born (Easterlin, 1980). If you were born at a time when there were relatively few births per year (called a small birth cohort), you may have some special social and economic advantages in later years.

In 1973 3.1 million babies were born in the United States. That may seem like a large number, but compared to earlier years the number is small. The baby-bust cohort is often hidden from view because of the large number of children born between 1945 and 1964. This last group is part of the baby boom. From 1947, when the baby boom began, until about 1960, when the baby boom ended, 60 million babies were born. These 60 million people cannot hide from view. As a group, whether demanding hula hoops as young children or jeans with a bit more room for expanding waistlines today, they have set the trends and shaped the times in American society. In this chapter, we will examine the effects of the baby-boom and baby-bust generations and in general see how important the age compositions of a population can be.

But age is just one population characteristic. Another important demographic characteristic is gender composition. Differences in the number of men and women in a society affect many demographic rates. For example, if we have data about the ratio of males to females in a society, we can often make statements about the likelihood of marriage. Other characteristics are equally interesting and often important.

We will also consider the measures and tools that demographers use to describe the composition of a population. Demographers usually focus their attention on age and sex composition, but they are also interested in other population characteristics, such as race and ethnicity, place of residence, marital status, and level of education. All of these characteristics have been found to influence demographic behavior in some way. The three basic demographic processes—fertility, mortality, and migration—interact with and are dependent on these population characteristics in a variety of ways. Likewise, it is often necessary to consider the impact of population characteristics when making certain kinds of statements about *social rates*—such things as suicide rates, unemployment rates, and the like. Changes in the age structure of the population are related to the marriage rate and the divorce rate. Before we consider how changes in the age composition may affect these and other social rates, let us first see how the age composition and sex composition come to be in the first place.

MEASURES OF AGE AND SEX STRUCTURE

There are several measures and graphic methods for describing or depicting the age structure, sex structure, or combined age and sex structure of a population. The three most often used measures are the sex ratio, the dependency ratio, and age–sex pyramids. All of these are particularly useful for comparing the populations of different societies or the populations of the same society at various points in time.

Sex Composition and the Sex Ratio

When we study issues related to sex composition, we are often asking the question: What factors influence the sex ratio? The *sex ratio* refers to the number of males in the population per 100 females. The sex ratio is derived by dividing the total number of males in a population by the total number of females and then multiplying the result by the constant 100. This ratio may be expressed by the equation:

$$\text{Sex ratio} = \frac{\text{Total no. of males}}{\text{Total no. of females}} \times 100$$

If the number of males in a population equaled the number of females in a population, then the sex ratio would be exactly 100. However, there are a number of factors that can influence sex composition, and thus the sex ratio rarely equals 100.

An imbalance in the sex ratio begins at birth. Generally about 105 to 107 males are born for every 100 females (Coale, 1991a). In the United States and throughout western Europe the sex ratio *at birth* is usually just over 105. However, because females have better life chances than males after birth, the sex ratio usually moves downward as we move through the childhood and young adult years. This tendency for more males than females to die at every age results in sex ratios that eventually fall below 100. In the United States, the sex ratio is 95; this means that for the total population, there are 95 men for every 100 women.

Cultural factors may influence the balance between males and females. For example, there are many cultures that value males more than females. If males are more highly valued than females, female babies may not be provided with the same levels of nutrition and health care. Such neglect would result in higher than expected death rates among females. This discriminatory treatment toward females seems to occur in India, especially when there is already one female child in the household (Coale, 1991a).

The influence of cultural factors on the sex ratio has been clearly documented with regard to China. China has traditionally placed a greater value on males than females. In 1982, China developed a policy to limit childbearing. The sex ratios reported in China since that policy was implemented show that the sex ratio favors males over females. Female infanticide may account for some of the imbalance. In fact, it was the imbalance in the sex ratio that caused officials to question whether infanticide was taking place. But it is also likely that some female babies are simply not reported to the government officials, while other females are adopted by relatives in the more rural parts of China (Johansson & Nygren, 1991). We will discuss this issue in more detail in Chapter 7, on mortality.

The imbalance of the sex ratio in China is related to social policy. But cultural factors may be more subtle in influencing the sex ratio. In 1966, Japan marked the Year of the Fiery Horse. According to superstition, a girl born during the Year of Hinoe-Uma, or the Year of the Fiery Horse, will be ill-fated. As a consequence of this belief, the motivation not to have a female baby is quite strong. In Japan, during the Year of the Fiery Horse, the *neonatal mortality rate* (the number of infants who die within the first 28 days per 1,000 live births) for females was 7.78, but for males only 6.94. In the years prior to and after 1966, the rates followed the normal pattern of greater neonatal mortality for males (Kaku, 1975).

The sex ratio may also be affected by migration. Generally speaking, whenever an area has been settled by migration that involved movement over great distances, or where the journey was in some way arduous, the sex ratio is very high, indicating more males than females. The settlement of the western part of the United States provides a dramatic illustration of the effect of migration on sex ratios. The sex ratio in California in 1850 was 1,228—1,228 men for every 100 women! As California became settled and more women moved there, the sex ratio dropped rapidly. By 1870, the sex ratio in California had fallen to 150 (Guttentag & Secord, 1983). Currently, the sex ratio in California is 100.2, which means that the number of males and females are approximately the same. Alaska, the last U.S. frontier state, has a sex ratio of 111 (Horner, 1992).

Rural-to-urban migration can also affect the sex ratio. The flow of migration from rural areas to urban areas is often influenced by the job market. In the United States, job opportunities for clerical and service workers, which are typically filled by women workers, tend to lower the sex ratios of cities. The outstanding illustration of this phenomenon in American society is Washington, D.C., which had a sex ratio of 87 in 1992 (Horner, 1992).

Age Composition

The age structure of a society is shaped by the processes of fertility, mortality, and migration. A country may grow "older" or "younger" depending upon changes in the fertility rate, the mortality rate, or the rate of migration. When we refer to a population becoming younger or older, we are actually referring to the age distribution of a population, or the age structure. A population is considered to be young when there are proportionately more young people than people of other ages. A very popular measure of the youthfulness of a population is the percentage of the population under 15 years of age. For example, two countries with almost half of their populations in this young age group are Nicaragua, with 46% under age 15, and Kenya, with 49% under age 15 (Population Reference Bureau, 1994).

It may be equally important to know something about the older part of a population. Currently, in the United States, there is some concern over the fact that the population is growing older. Normally we think of a population growing older when people live longer, but another factor that causes a population to age is declining fertility. The average age of a population will increase when fewer babies are being born, because there will be fewer young people to pull the average down. Conversely, when the fertility rate is high or increases, a population will become younger.

Mortality changes also influence the age distribution of a population. However, under certain conditions, death rate declines may actually cause a population to become younger. Reductions in mortality can occur at any place in the age structure. Typically, when the death rate has been relatively high, the greatest declines in death rates occur among infants and very young children. When there are fewer deaths among children in the first few years of life, more young people are added to the population, and they lower the average age of the population. Furthermore, in about 20 years, these children, who might otherwise have died, will again add to the younger part of the population by having children of their own—and if mortality rates have continued to decline, an even higher proportion of their babies will live, will grow up to marry, have babies at their own, and so on. In summary, if mortality rates decline and the decline is produced primarily by reductions in infant and childhood mortality, the effect will be to produce a younger population.

Migration can also affect the age distribution of a population; however, its effect depends on who the migrants are and what impact they have on fertility and mortality rates. If young adults migrate out of an area (as is often the case), the average age of the population they leave will probably go up. This will occur because the population will not only lose the relatively youthful migrants themselves, but will also lose the babies these migrants will have. By contrast, the receiving area will often become younger because the young migrants will have children in their new communities.

One summary measure used to describe the age structure of a society is the median age. The *median age* is simply an age at which 50% of the population is older and 50% of the population is younger. In the United States, the median age is 33 years. The median age is useful as a summary measure, but it does not tell us much about the *distribution* of a population.

The Dependency Ratio

A detailed and meaningful way to describe the age distribution of a population is called the *dependency ratio*. The idea behind the dependency ratio is that certain age groups in a society are generally productive and other age groups are generally dependent upon the efforts of the productive group. Of course, the label "dependent" will always be somewhat arbitrary, since there will undoubtedly be exceptions. Nevertheless, it has become customary to describe the population under age 15 and the population aged 65 and over as the *dependent population*. The people in the population aged 15 to 64 years are considered to be the *productive population*.

It could be argued that young people in many societies are dependent until they are 18 or 20 years old, since very few people in modern industrial societies enter the labor market before that age. Although it is true that many teenage youths do contribute something to the economic well-being of their families, they usually provide the family with less than they receive. A more compelling reason for using the cutoff point of age 15 is that the dependency ratio is often used for comparisons of different countries. In many countries it is common for young people to enter the labor force during the teenage years. For the purpose of comparing all the societies of the world, age 15 may be used as the best compromise as the upper limit on the dependent youthful years.

The dependency ratio is a comparison of the combined dependent groups (under age 15 and over age 64) and the productive group. Specifically, the dependency ratio is the number of persons under 15 years of age plus the number of persons aged 65 and over per 100 persons between the ages of 15 and 64. The dependency ratio is computed using the following formula:

$$\frac{\text{No. of persons} < 15 \text{ years of age} + \text{No. of persons} \geq 65 \text{ years of age}}{\text{No. of persons } 15\text{–}64 \text{ years of age}} \times 100$$

An interesting comparison of dependency ratios can be found in the contrasts among countries shown in Table 5.1. It is apparent that the less-developed counties have higher dependency ratios (e.g., Iraq, 92.0, and Tunisia, 74.9) than the more-developed countries (e.g., United States, 51.8). This means that in Iraq, for every 100 people in the productive ages, there are 92 people in the dependent ages. By comparison, in the U.S. population, for every 100 people in the productive ages, there are only 52 in the dependent ages. A high dependency ratio means there are fewer productive people supporting the dependent population.

Although the dependency ratio can be used to specify the ratio of the dependent to the productive population, it is equally important for showing how the dependent population breaks down between the young and the old. Less-developed countries typically have higher fertility rates than do developed countries, and this is reflected in the large percentage of the population under age 15. The dependency in less-developed countries is produced largely by infants and children, while older dependents make up only a small part of the dependency ratio. In Iraq, 45% of the population is less than 15 years old; only 3% is aged 65 or older (United Nations, 1992).

By contrast, the older population represents a more substantial part of the dependent populations in the developed countries. In Norway, 16% of the total population is aged 65 or over, while 19% is under 15 years. Overall, this contributes to a dependency ratio of 54.4. In a similar but less extreme way, in the United States, the elderly population represents 13% of the total population, while young people contribute 22%. For the developed countries, the larger percentages of people in the elderly, dependent part of the population will have a strong impact on such programs as Social Security, retirement, and pension plans; for the less-developed countries programs

TABLE 5.1. Dependency Ratio by Country, Selected Years

Country	Year	Age group (percent)			Dependency
		0–14	15–64	65+	
Ethiopia	1990	49	48	3	111.0
Iraq	1988	45	53	3	91.0
Haiti	1990	40	56	4	79.6
Tunisia	1989	38	57	5	74.9
Chile	1991	30	64	6	57.8
Norway	1990	19	65	16	54.4
France	1991	20	66	14	52.1
United States	1990	22	65	13	51.8
Canada	1991	21	67	12	48.3
Cuba	1989	23	68	9	46.5
Japan	1990	18	70	12	43.3

Note. From United Nations (1992, Table 7, pp. 152–187).

directed toward lowering fertility and child health and education receive more attention.

The Age–Sex Pyramid

A graphic method used to show the age and sex composition of a population simultaneously is called the *age–sex pyramid*, or the *population pyramid*. It is called a pyramid because in many cases it has a triangular shape, although as we will shortly see that is not always true. An age–sex pyramid is a variation of a graphic presentation technique called a bar chart, in which the length of a bar represents a proportion of the total. A vertical line divides the males from the females in the population. It is customary for the males to be represented on the left side of the pyramid and the females on the right.

Basically, there are three typical shapes for a population pyramid: a "true" or *expansive* pyramid, a *constrictive* type, and a *stationary* type. These are presented graphically in Figure 5.1. The first pyramid takes the shape of an actual pyramid—triangular with long, sloping sides and a large base. This type of pyramid is typically associated with rapidly growing countries with a young age structure. This pyramid shows that each birth cohort is larger than the one that preceded it, and thus the population is growing. The country is "young" because the largest proportion of the population consists of young people. The age–sex pyramids in less-developed countries are often of this type.

The second pyramid is a constrictive one. The shape of the age–sex structure is constrictive at the bottom. This type of population is one that has a declining growth rate. The constricted base reveals that birth cohorts are getting smaller. Thus each succeeding generation begins with fewer people than the previous generation. If this trend continues, the population will grow progressively older and smaller. Western Europe provides an illustration of this constrictive type of pyramid (Ashford, 1995).

Finally, it is possible that a country may stabilize its vital rates (births and deaths) and become stationary. A stationary population pyramid has a somewhat rectangular shape because the proportions of the population in each age group are equal: the population is neither increasing nor decreasing. In any one year, it would be possible to have *zero population growth* (any additions to the population, births plus immigration, equal losses, deaths plus emigration). A *stationary population* can only be attained where the age-specific birthrates and death rates are at low levels, remain equal and constant for a long time, and immigration remains constant. Austria, Belgium, Denmark, Luxembourg, and the United Kingdom are a few of the countries currently facing the possibility of becoming stationary in the near future.

Most populations, rather than having an exact fit to one of these ideal-type pyramids, approximate one type or another but have their own special irregularities. Population pyramids for the United States in 1990 and pro-

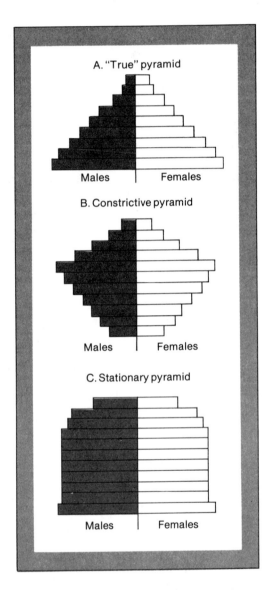

FIGURE 5.1. Types of age–sex pyramids.

jected pyramids for the years 2000, 2025, and 2050 are presented in Figure 5.2(a–d). These particular population pyramids are divided into 5-year age intervals, with the percent of the population along the bottom.

These age–sex pyramids reveal a variety of things about the U.S. population. Beginning with the 1990 pyramid, there are four unique birth cohorts we can follow. One is the Great Depression cohort, or those people who were born in the 1930s and are represented by the bars for 50- to 59-year-olds in 1990; the baby-boom cohort, or those who in 1990 were aged 25 to 44; the baby-bust cohort, or babies born between 1970 and 1979; and the baby-boom echo cohort, the current birth cohort. Notice that as we approach the year 2000, these four cohorts are easily distinguishable even though they are all 10 years older. The cohort sizes of the population below age 10 in the year 2000 are largely based on estimations. The population pyramids for 2025 and 2050 take a completely different shape from the previous two. First of all, the survivors of the baby boom are represented at the top. Notice, too, that the greater longevity of females can be seen quite clearly in the upper ages where the male side of the pyramid diminishes faster than the female side. The continuing projections for the population after the baby bust indicate a hypothetical stationary population for the United States. Keep in mind that these projections are based on assumptions government officials make about trends in age-specific fertility, mortality, and international migration.

Age Standardization

We explained in Chapter 1 that when rates are calculated for demographic events, they are most useful for making comparisons between populations (or for the same population over time). Even when the populations are very different in size their rates can be meaningfully compared. While this is generally true, there is sometimes a demographic circumstance that makes comparisons between populations less meaningful, and, in extreme cases, misleading. An illustration will show how this can come about.

Suppose we are comparing the death rates of two countries. Country A might have a crude death rate of 9 deaths per 1,000 people, and country B might have exactly the same death rate, 9 per 1,000. On the surface, it would seem that the two countries have mortality conditions that are identical. But what if we were to learn that country A has a much higher proportion of older people—perhaps it is a country where people go to retire. Country B, by comparison, is made up of a much larger proportion of 20- and 30-year-olds, with relatively few people over age 60. If this situation were to prevail, we would certainly expect the death rate in country A, with many more old people, to be higher than the death rate in country B with proportionately more young adults (who generally have lower death rates). So, if these two hypothetical countries have the same crude death rates, their mortality conditions must be very different. Something must be causing far more 20- and 30-year-olds in country B to die than we would normally expect.

The point of this illustration is that when populations have age-structure differences, comparisons of demographic rates are less meaningful, and they may even be misleading. This is because age is related to the factor under

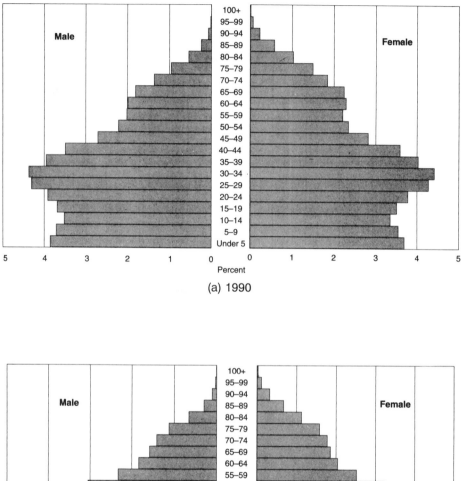

(a) 1990

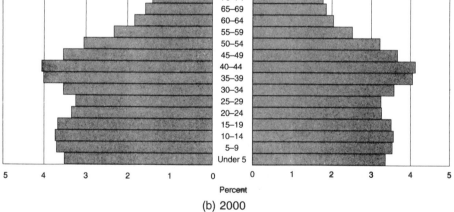

(b) 2000

FIGURE 5.2. Age distribution of the U.S. population, by sex: for the years 1990 (a), 2000 (b), 2025 (c), and 2050 (d). From U.S. Bureau of the Census (1992i, p. 25).

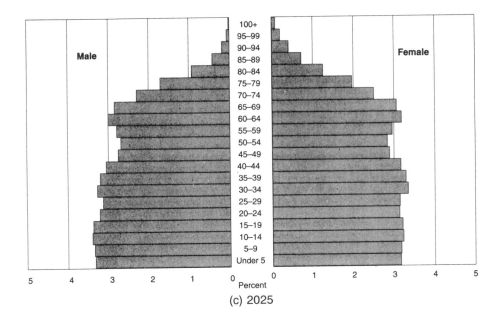

(c) 2025

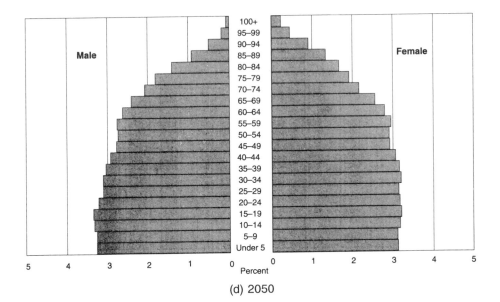

(d) 2050

consideration, in this case death, and the age differences of the two populations are confounding the comparisons of death rates.

There are different solutions for coping with the problem of dissimilar population characteristics.[1] One solution is to obtain the age-specific rates for the different age groups in the two populations. Then one could compare the death rates of the 20-year-olds, the 30-year-olds, the 40-year-olds, and so on, in the two countries. After making a series of comparisons between various age groups in the two populations, a clearer picture of mortality conditions in the two countries would emerge.

While making a series of comparisons of age-specific death rates for two populations might be done, suppose it were possible to remove the effect of the age differences between the populations and thus make only one comparison. *Standardization* does exactly that for us; it is a demographic technique that allows us to compare rates, for example, death rates, between two populations while eliminating the effects of their age (or gender, or other) differences. If we apply this technique to our original hypothetical case, standardization can provide the answer to this question: What would be the death rate of country B if it had the same age structure as country A? To standardize the rate of one population on another means that through statistical adjustments it is possible to compare the rates of two populations *as if they were the same on other relevant characteristics.*

In order to see how age standardization is done, we will consider a realistic and familiar demographic example. We often wish to compare rates over time in a particular country (we will use death rates again). We will use the United States, comparing the death rates for each 10th year between 1940 and 1990. A quick glance at Table 5.2 reveals that the *number* of deaths in these selected years has increased from decade to decade. This is, of course, to be expected since there were more people living in the country each decade. But we also know that the age structure of the United States has changed. The population of the United States is getting older. To make meaningful statements about the historical trend of mortality rates, we must control for the changes that have taken place in the age structure.

To standardize, we must first choose a standard. This is an arbitrary decision based on the specifics of the issue under consideration. In our example, we will use the population of 1940 as the standard. This will be the base year for making comparisons. Once we have our standard, the next step is to multiply the age-specific death rates for each new population, for example, the 1950 population, by the number of people in each age group in the standard population. The result is the number of deaths, and death rates, we

---

[1]Age is not the only population characteristic that can affect comparisons of crude rates. Gender, for example, is also related to mortality, so comparing the death rates for two populations with very different sex ratios might also be misleading. It is possible to make adjustments for any population characteristics, such as race or ethnicity, employment status, marital status, and others.

would expect in 1950 and so on, if the age structure in 1950 had remained the same as the age structure in 1940. This is known as the *direct method of standardization*. The indirect method of standardization uses age-specific death rates as the standard and then applies those rates to different population structures. Using the example above, the age-specific death rates for 1940 might be the standard, and then the actual populations in each age group for each decade would be compared. In each method, we arrive at the expected number of deaths and expected death rates after making adjustments to control for apparent differences related to age.

The direct method of standardization was used in Table 5.2 to compute age-adjusted death rates for the United States population as a whole and separately for males and females. The total population of the United States for 1940 was used as the standard. Thus, the rates reflect what we would expect the death rates to be if there were no changes in the age structure since 1940. Note in Table 5.2 that the age-adjusted death rate for the United States has declined from 1076.1 in 1940 to 520.2 in 1990. In using these standardized rates, we attribute the change to actual mortality differences. Keep in mind that we are applying the age structure of 1940 to the age-specific death rates of 1990. By controlling for the changes in the age structure in this way, we can compare age-adjusted death rates and have confidence that the differences in the rates over time are due to changes in mortality and not artifacts of the age structure.

TABLE 5.2. Deaths, Death Rates, and Age-Adjusted Death Rates, United States, 1940–1990

	Year	Total	Male	Female
Number	1990	2,148,463	1,113,417	1,035,046
	1980	1,989,841	1,075,078	914,763
	1970	1,921,031	1,078,478	842,553
	1960	1,711,982	975,648	736,334
	1950	1,452,454	827,749	624,705
	1940	1,417,269	791,003	626,266
Death rate, per 100,000	1990	863.8	918.4	812.0
	1980	878.3	976.9	785.3
	1970	945.3	1,090.3	807.8
	1960	954.7	104.5	809.2
	1950	963.8	1,106.1	823.5
	1940	1,076.4	1,197.4	954.6
Age-adjusted death rate, per 100,000	1990	520.2	680.2	390.6
	1980	585.8	777.2	432.6
	1970	714.3	931.6	532.5
	1960	760.9	949.3	590.6
	1950	841.5	1,001.6	688.4
	1940	1,076.1	1,213.0	938.9

Note. From U.S. Department of Health and Human Services (1993).

THE NEEDS OF THE ELDERLY AND CHILDREN
IN AMERICAN SOCIETY

Currently, the median age of the U.S. population is 33—half of the population is older and half is younger (O'Hare et al., 1991). Unless fertility increases, we may expect the average age of the population to continue upward through the remainder of the decade, and for several decades thereafter.

The population over 65 years of age increased 21% between 1980 and 1990. But the age group between 25 and 44 made the greatest percentage change, with a 28% increase (O'Hare et al., 1991). In each decade that follows, this baby-boom group will add to the older segment of the population. While their children (the under-15 age group) are also in an increasing category, this cohort will not be nearly as influential as the baby-boom generation itself. At this time, the U.S. population is growing older, and indications are that it will continue to do so into the 21st century.

The growing number and greater proportion of people now found in the 65+ age group is an important demographic feature of America's future. In 1990, there were 31 million people, or about 12% of the total population, aged 65 and over (U.S. Bureau of the Census, 1991a). The older people in the population are not a socially homogeneous group. They are rich and poor, black and white, married and unmarried, and "young-old" (aged 65–74) and "old-old" (aged 75+). There are disproportionately more women than men. This is a diverse group. However, the older members of the population are becoming increasingly influential in regard to governmental policies that affect them, especially in the area of Social Security and medical care. The first Social Security Act in the United States was passed in 1935; this act officially conferred the status of "elderly" on persons aged 65 and older. Today, when persons reach age 65 they become eligible to receive full retirement benefits from Social Security and Medicare. They also receive a number of other services, ranging from housing assistance to reduced bus fares. But because the elderly population is growing, policies dealing with retirement and health care are receiving special attention.

The Elderly and Retirement

When the first Social Security Act was passed, life expectancy for males and females was less than 65 years. Today it is about 75 years, and there are more and more people who live well beyond age 65. Social Security coverage has increased over time and has reached a point where 92% of the elderly population receive Social Security benefits. It accounts for at least 50% of total income for most elderly (Taeuber, 1992). There is fear that the economic demands of the program cannot be met in the future. Given the dynamics of the age structure, this fear may be justified.

A closer look at the issue of retirement shows that retirement has become institutionalized and is expected by most of the working public. American workers are retiring earlier than they did in the past. The labor force participation rate for men over age 65 has declined steadily. In 1950, 45.8% of men 65 and over were employed. By 1990, this figure had fallen to 16.4%. For women 65 and over, the rate has remained relatively constant at about 9%. This is a significant decline in labor force participation rates. But the greatest change has been for men who are 55 years old and older. In 1950, 68.6% of males aged 55 and over were employed, but by 1990 only 39.3% were in the labor force (Taeuber, 1992).

How much support should the government provide for old age, survivor's benefits, disability benefits, and health insurance for this growing group of retirees? The answer to that question cannot be provided by demographers. However, there is a demographic aspect to this problem. It centers around the relationship between the productive population, which supports the Social Security program through taxes and helps to establish a reserve fund, and the elderly, dependent population, which draws on reserve funds.

Currently, the Social Security program is a transfer program. Workers are taxed and the money from that tax is transferred to pay current Social Security benefits. If the proportion of the elderly who are not working increases relative to the proportion of individuals who are working, the relative amount of money available to provide Social Security benefits decreases. As we noted earlier, the proportion of the elderly who are not employed has been increasing. That alone should have reduced the money available for Social Security payments. However, today's dependent elderly are supported by the larger baby-boom cohort, who are currently in the labor force.

It may be apparent at this point that if trends in retirement continue, problems with the Social Security program will become more intense as the baby-boom cohort reaches retirement age. The changing age structure has been a major reason for concern with the solvency of the Social Security program. Whether or not the Social Security program remains solvent and will be available to the baby-boom and the baby-bust cohorts in the future will depend on legislative action and voters' choice.

Children and School Enrollment

Each year, in communities all over the United States, school boards and community leaders must make decisions about the allocation of classroom space and the hiring of teachers. Knowing the absolute number of people in each age group is only the beginning of their planning. Other factors affecting the projections of school enrollments include childhood mortality, migration, and such social factors as the choice between public or private schools.

At the elementary school level, we are concerned with the number of children between the ages of 6 and 13. In 1989 there were 28.7 million children enrolled in elementary school (U.S. Bureau of the Census, 1991c). Many communities must provide new classrooms for these students because they represent a group that is twice as large as the elementary group that preceded them. Not all communities will have to add new classrooms. Some areas of the country have experienced less population growth than others. Economic conditions and migration patterns will determine which communities will have to expand to meet the needs of elementary students.

At the secondary school level, we are concerned with the number of children between the ages of 14 and 17. In grades 9 through 12 there were 12.9 million students enrolled (U.S. Bureau of the Census, 1991c). Knowing the absolute number of young people and the impact of residential mobility is not enough. Administrators must also take into account the number of students who drop out of school and the number of students who do not attend public school. Between 1988 and 1989, 4.5% of students in grades 10 to 12 dropped out of school, and 6.6% of students attended private schools.

For colleges and universities, the potential applicant pool for students is much more complicated. Currently only 21.4% of persons over age 25 have completed 4 or more years of college. Traditional college students, those aged 18 to 24, declined by 2.7 million people over the last 10 years, even though the total number of college students enrolled increased by 1.8 million (U.S. Bureau of the Census, 1991c). This is because many more nontraditional, or older, people are entering or returning to the college classroom. In 1989, 39.4% of all college students were aged 25 and older.

Conversely, we may be interested in how the presence of a college or university affects the age distribution of a community. In Figure 5.3, a population pyramid for Blacksburg, Virginia, shows the effect the univeristy (Virginia Polytechnical Institute and State University) has on the local community. Notice the large and disproportionate number of people between the ages of 20 and 24.

THE EFFECTS OF AGE AND SEX COMPOSITION ON SOCIAL RATES

Knowledge of the age and sex composition of a population can be very important for understanding social issues. Social rates, such as marriage rates, suicide rates, crime rates, and voting rates, change over time. The demographic question is whether or not changes in the age and sex structure can account for the changing social rates. The answer is almost always, "Yes," but sometimes in surprising ways. In the next sections, we will examine in greater detail the way that age and sex composition influences other social rates.

Age and Sex Composition and Marriage

The cultural norms regarding marriage may create an apparent sex imbalance where in reality none exists. An example of this can be found in what

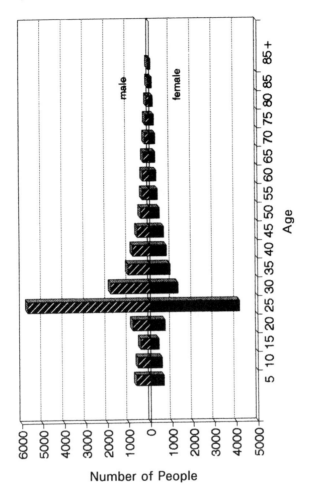

FIGURE 5.3. Population pyramid for Blacksburg, Virginia, 1990. From U.S. Bureau of the Census (1992h, Table 79).

has been labeled the "marriage squeeze." A *marriage squeeze* refers to an imbalance between the number of males and females at the prime marriage ages of the two sexes. Generally, no serious imbalance would occur if the prime marriage ages were the same. However, in most societies females marry males older than themselves. In recent decades in the United States, women have been marrying men who on average are about 2½ years older than themselves. For females the median age at first marriage is 23.9, and for males it is 26.1 (U.S. Bureau of the Census, 1991a). As a result of this pattern, there will be an imbalance of the sexes at the prime marriage age about 20 to 25 years after a pronounced increase or decrease in the number of births. An increase in the number of births will lead to a shortage of marriage-aged males. But a decrease in the number of births will result in the opposite: a shortage of marriage-aged females.

In the late 1960s, the United States began to experience a marriage squeeze. The number of females aged 18–25 exceeded the number of males aged 20–27 by 10%. This imbalance between the sexes at the prime ages of marriage was the result of increasing fertility 20 years earlier. Beginning in the late 1940s, the number of births each year exceeded the number of births of the previous year. As a consequence of rising fertility, there were more females born during 1947 and 1948 than there were males born during 1945 and 1946.

There can also be a marriage squeeze favoring females. If the birthrates fall, there are likely to be fewer total births over successive years. When this occurs, there will be more males than females who are 2 years younger. Since 1980, there has been a "favorable" marriage squeeze for women under 25 years old (Goldman, Westoff, & Hammerslough, 1984). There were estimates that by 1990 there would be 5–20% more men than women in the peak ages where first marriages occur (Guttentag & Secord, 1983). These estimates proved true. Once again, this is because of the drop in the birthrates during the 1970s. Of course, this is a culturally created sex imbalance, produced by the preference that males be older than the females they marry. It could be resolved if men and women were to marry at the same age.

This example shows us how important it is to know something about the age and sex composition of a population. But there are other demographic and social characteristics that add complexity to explaining social rates. Marriage opportunities for men and women also depend on the distribution of the population by race, education, and area of residence (South & Lloyd, 1992).

Age Composition, Marital Status, and Suicide

In 1897, the French sociologist Durkheim published his classic study of suicide (Durkheim, 1897/1951). Durkheim advanced a sociological explanation of suicide (as opposed to a psychological or other explanation), and a key element in his explanation was that persons who are socially integrated are less likely to commit suicide than persons who are not socially integrated. For example, social integration may prevail when a person belongs to a highly cohesive religious community. Durkheim found that Jews and Catholics had lower rates of suicide than did Protestants and concluded that this was so because the two former religious groups were more integrated and more supportive of collective life. Following this line of thought, Durkheim examined the relationship between marital status and suicide rates. Using the data on suicides from all of France for the years 1873–1878, Durkheim found that 16,264 suicides were married persons and 11,709 were unmarried. Durkheim noted in his analysis that if these figures were taken at face value, a person would conclude that being married (with all of its attendant burdens and responsibilities) increased the tendency toward suicide. But he observed that a very large number of unmarried persons were less than 16 years old, while almost all married persons were older. Furthermore—and this was the important point—up to the age of 16, the tendency to commit suicide was very slight. In other words, suicide was and still is related to age, so that in a comparison of the suicide rates of the married and unmarried segments of the population, the different age composition of the two categories has to be taken into account. That is exactly what Durkheim did. According to his calculations, unmarried persons above age 16 had a suicide rate of 173 per million people, while the married persons above 16 had a rate of only 154.4 per million. By comparing the rates for married and unmarried people who were over 16 years of age, Durkheim had at least partially taken age into account. The figures supported his theory that married people would be more socially integrated and have a lower suicide rate.

But Durkheim noted that even this analysis assumed that all unmarried and married persons over age 16 were of the same average age, that is, that the two populations had the same age structure. Durkheim guessed that this assumption was not true, and indeed it was not. The average age of all unmarried men was about 27 and of all unmarried women about 28, while the average age of married persons was between 40 and 45. Since, as Durkheim noted, suicide was not only rare among the young but increased progressively with age, one would expect the married people in the population to have a higher rate of suicide simply because they were, on the average, about 15 years older than the unmarried people. By taking the difference in age structure into account, Durkheim concluded that marriage reduced the danger of suicide by about one half. It was the great care that Durkheim took to consider the age structure of the married and unmarried populations that allowed him to substantiate this part of his theory of suicide.

Age Composition versus Period Effects

As we have already seen in our consideration of suicide and marital status, the age composition of a population can go a long way toward explaining change or differences in social rates. Whenever phenomena such as suicide or marriage occur more frequently at some ages, there is always a possibility that a changing age structure may be producing changing social rates. This is known as an *age effect*.

However, changes can occur in rates and yet that change may not be completely attributable to changes in the age structure. It is possible that societal conditions associated with a particular period of time are also affecting social rates. Perhaps marriage rates go down because unemployment is especially high and the economy poor. These are called *period effects*.

Period effects may be best illustrated by reference to the case of fertility changes over time. Fertility in the United States was low during the 1930s and high during the 1950s. These two very different levels of fertility cannot be accounted for by differences in the numbers of women of childbearing ages. More likely, the effects of economic depression in the 1930s and of economic prosperity in the 1950s explain the differences in fertility and are thus period effects.

Some demographers put a great deal of emphasis on the way the size of a population age cohort can shape a period. Easterlin (1980) and Berger (1989) suggest that the size of a birth cohort (all of the people born during a given year or set of years) will, in part, determine a person's social and economic well-being in later years. Individuals who are members of small birth cohorts, such as those born during the 1930s or 1970s, will find less competition in schools, employment, and housing during their lifetimes. Individuals from large birth cohorts, such as those born during the baby boom, can expect to have more difficulty competing for available job opportunities and housing. This implies that there is an inverse relationship between the size of a birth cohort and well-being. As the number of people in a cohort increases, chances for economic and social success decrease. We can refer to this phenomena as a *cohort effect*.

RECENT CHANGES IN POPULATION COMPOSITION

In addition to the age and sex structure, many other population characteristics are of interest to demographers. These include race and ethnicity, place of residence, and household composition. Each of these has some impact on demographic behavior. In the sections that follow, we will look at recent changes in the population composition of the United States.

Race and Ethnic Composition

There may be societies where the factors of race and ethnicity can be ignored in demographic analysis, either because the composition of the population is homogeneous (e.g., Iceland) or because these are not important social facts. However, in the United States, the racial and ethnic characteristics of people are deemed to be socially important and have clear social consequences. The Census Bureau uses three sets of questions in order to compile a race and ethnic profile of the American population. These include questions on race, Hispanic origin, and ancestry. Developing ways to measure race or ethnicity is not easy. First, the concepts are not based on scientific or objective standards. Second, many people have multiple ethnic identities. And third, some people do not find any meaning in the categories, preferring to define themselves as Americans without reference to any other heritage.

Defining and Classifying Race

Race is technically a biological concept, at least insofar as it has been used by physical anthropologists and geneticists. Nevertheless, one of the basic sources of demographic data, the census, treats race nonbiologically. Not even minimum biological standards for the classification of racial groups are met when the Census Bureau reports the racial composition of American society.

Going back to the census of 1890 in the United States, the race of a person was determined by census takers on the basis of their own observations. Enumerators were instructed to record the race of a person based on what they saw. Consequently, a person with a small fraction of black ancestry would probably have been classified as black, especially if he or she lived in a black community. Since even a trace of black ancestry could be used to classify a person as black, this procedure has been labeled the "one drop rule" (Davis, 1991). The person could have been classified as white, but census takers followed the social custom of determining race according to the social environment.

Beginning in 1960, the U.S. census method of determining race changed. Instead of enumerators making the decision, the individual being enumerated was asked to designate his or her race. This is referred to as the *self-designation method*. In 1960, the census question asked, "Is the person white, Negro, American Indian, Japanese, Chinese, Filipino, Hawaiian, part Hawaiian, Aleut, etc." Note that the choices extend beyond race to include national origin, regions, and ethnicity. No attempt is made to set an objective standard by using skin color, hair texture, or other physical attributes.

By the 1990 census, the question on race asked the respondent to circle the race that the person considers himself/herself to be. The following categories were then offered: White, Black or Negro, Asian or Pacific Islander (with instructions to specify Chinese, Filipino, Hawaiian, Japanese, Laotian, Asian Indian, Korean, Somoan, Vietnamese, etc.), American Indian (with instructions to specify the name of the enrolled or principal tribe), Eskimo, Aleut, or Other (with instructions to specify other race). These categories may be terms commonly accepted by the public, but they do not represent scientifically-defined categories. Individuals may choose which race they belong to, but they are not given the choice in this question to identify multiple identities.

Defining and Classifying Hispanic Origin

Recently the Census Bureau has attempted to differentiate the Hispanic population from other groups. In the past, attempts were made to count the Hispanic population by asking questions on language and surnames. In the 1990 census, individuals were asked to indicate if they were non-Hispanic or to indicate among several Spanish/Hispanic groups. Included in the choices of self-identification were Cuban, Puerto Rican, Mexican, and Other.

There is a problem with classification of the population based on Hispanic origin when used in combination with race. "Hispanic" is a broad term, and includes people from many diverse racial and ethnic groups. When demographers, and others, want to make statements about the composition of these groups, they often combine the race and Hispanic data. Thus, as categories, they are not mutually exclusive. This means that there is overlap in the categories. For example, Hispanics are often grouped with whites, even though Hispanics may be of any race. Del Pinal (1992) suggests that Hispanics should not be grouped with whites in statistical analyses. In his review of socioeconomic status, he found that including Hispanics with whites underestimated the differences found between these groups. The socioeconomic characteristics of Hisanics are lower than for the white population, and to include Hispanics in this category masks important differences.

Defining and Classifying Ethnicity and Ancestry

The census of 1850 was the first to include questions about the foreign-born population. The large number of immigrants from Europe in the late 1800s and early 1900s meant that many people in the population were foreign-born or had at least one parent who was born outside of the United States. Today, the foreign-born population is still counted, but the focus of the census question is on ancestry and a person's sense of ethnic identity, rather than on length of residence in the United States (Farley, 1991).

In the 1990 census, 16% of the population was asked an additional open-ended question about ancestry or ethnic origin. The question was open-ended because individuals could write in any ancestry name or combination of names with which they identified. Thus, people could respond that they were of English-German, African American, or any other ethnic origins. The instructions indicated that religious affiliation should not be reported, but place of birth of parents, "roots," or other decent identities were appropriate. Over 630 different ethnic codes were eventually used for this question (Kalish, 1992).

Profile of Race and Ethnicity in the United States

The people of the United States have always stressed that this is a nation made up of people with diverse origins. Results from the recent census show that this continues to be true. Table 5.3 shows the distribution of the population by race and national origin for 1990. Blacks comprise the largest group after whites, followed by those who responded "Other." As noted earlier, this category is largely made up of persons of Hispanic origin.

TABLE 5.3. Racial and National Origins, U.S. Population, 1990

Race	Number (1,000s)	Percent
White	199,686	80.3
Black	29,986	12.1
American Indian	1,878	0.8
Chinese	1,645	0.7
Filipino	1,407	0.6
Japanese	848	0.3
Asian Indian	815	0.3
Korean	799	0.3
Vietnamese	615	0.2
All other	11,031	4.4
Total	248,710	100%

Note. From U.S. Bureau of the Census (1993a, p. 18).

In addition to the distribution of the population by race and national origin, the census has also revealed that the foreign-born population represents 7.9% of the total population. Recent changes in immigration patterns show that the largest immigrant groups come from Latin America (42.4%) and Asia (25.2%). Europeans have traditionally been the largest group, but today comprise only 22% of all immigrants (U.S. Bureau of the Census, 1993a).

One of the most striking features concerning the American population is that so many U.S. citizens, even those whose families have lived in the United States for several generations or more, still maintain an identity with their family's European country of origin. The recent census showed that most Americans describe themselves as members of a single European ancestry group. Germans are the largest such group (58 million), followed by the Irish, the English, and the Italians. These third- or later-generation Americans still retain a European national identity. However, 13 million persons defined themselves simply as "American."

The fastest growing ethnic subgroup in the United States is the Hispanic population. This represents a diverse group of people of Spanish heritage, but the greatest number trace their ancestry to Mexico (60%), Puerto Rico (12%), and Cuba (5%) (Horner, 1992). Estimates are that by 2020 the Hispanic population will double in size and will account for one third of U.S. population growth between 1992 and 2000 (U.S. Bureau of the Census, 1993a). Hispanics will replace blacks as the largest minority group. This large increase is expected because of the higher-than-average fertility of Mexican Americans and high rates of immigration from many Spanish-speaking nations.

Residential Distribution

The geographical distribution of the population refers to the type of community in which people live, ranging from a simple classification of rural versus urban areas to the more complex classification of metropolitan versus nonmetropolitan areas. The Census Bureau defines an *urban* area as any community of 2,500 or more residents. Metropolitan areas include central cities, suburbs, and their counties.

Several criteria are used to identify metropolitan areas. The Census Bureau uses the designation *Metropolitan Statistical Area* (called MSA) to classify central cities and their surrounding areas. An MSA includes a central city with a total population of at least 50,000 people or an urbanized area of 50,000 people with a total metropolitan population of 100,000 people. Surrounding areas are considered to be part of the metropolitan area when a certain percentage of their population commutes into the central city for work or if a significant number of people who live in the central city are employed in the outlying area (U.S. Bureau of the Census, 1991b). For example, the Metropolitan Washington, D.C.–Baltimore MSA added several new counties after the 1990 census, as a result of new residential development and employment opportunities in those counties. When changes such as these occur, the whole county is added to the MSA even though parts of the county remain rural. The new Washington, D.C.–Baltimore MSA added part of Maryland's Eastern Shore of the Chesapeake and part of the Blue Ridge mountains in Virginia and West Virginia. Today, people can watch black bears on the Appalachian Trail without stepping outside of an officially designated MSA (Garreau, 1992).

The shift in the United States from an essentially rural population to a predominantly urban one has been the biggest demographic change in American history. When the census of 1790 was taken, about 5% of the population lived in urban areas. By the year 1900, about 40% of the population lived in urban areas, and by 1960, the percentage had risen to 70%. Since 1960, the shift to urban areas has progressed more slowly; by 1990, approximately 73% of the population lived in urban areas. Among the 27% of the population who lived in rural areas, only about 7% lived on farms (U.S. Bureau of the Census, 1992c).

Until 1950, most of the population residing on farms lived in the South. In the last 20 years, as the metropolitan areas of the South have grown, the Midwest has become the center for farm activity. Approximately 50% of all farmers live in the Midwest, followed by the South (30%), the West (15%), and the Northeast (5%). The total population in rural areas is still growing. But most of the rural population growth has occurred in areas at the boundaries of metropolitan areas. Indeed, it is likely that as these rural settlements grow, they too will be reclassified as urban areas. This reclassification of rural to urban areas may account for much of the change noted in the South.

The predominant place of residence in the United States is a metropolitan area with a population of one million or more people. The five largest metropolitan areas in the United States are New York, Los Angeles, Chicago, San Francisco, and Philadelphia. However, the fastest *growing* metropolitan areas are all located in Florida. There has been a shift of the population out of the North and Midwest to metropolitan areas in the South. The development of retirement and recreation communities, along with service and retail employment opportunities, has contributed to the rapid growth in this region.

Household and Family Composition

There are many different household structures: married couples with or without children, unmarried couples, single-parent households, and persons living alone. The predominant household structure in the United States is the married-couple household. That may come as a surprise since so much media attention has focused on the breakdown of the family. But, if we look at the population aged 18 and older, 63.7% of all whites, 61.1% of all Hispanics, and 43.6% of all blacks are currently married (U.S. Bureau of the Census, 1992a).

Although the married-couple household predominates, there have been significant changes in household structure. Among married couples, there are now more households without children than with children. The number of persons living alone and the number of single-parent households have increased. Overall, this has caused the average household size to decrease. In 1991, the average household size was 2.6 persons; 57% of all households have one or two people, while 26% have four or more persons (U.S. Bureau of the Census, 1992b). Most of the decline in household size occurred between 1970 and 1980 when the average fell from 3.14 to 2.76 persons per household. The change in household structure between 1970 and 1991 is graphically presented in Figure 5.4. Note that the greatest changes have been the decrease in the percent of married couples with children, the increase in single-parent households, and the increase in the percent of women who live alone (U.S. Bureau of the Census, 1992b).

There are several explanations for the changes in household structure. Perhaps the single most important demographic factor that has influenced the decline in household size has been the change in the age structure. During the last 20 years the baby-boom babies reached young adulthood and the peak years of family formation. An unprecedented number of people reached the age when people customarily begin their first jobs, get married, and begin having children. But the baby-boom cohort has been different. The baby-boom cohort did not follow the mid-century pattern. They have delayed getting married; once married have had fewer children; and have been more likely to divorce than the previous generation.

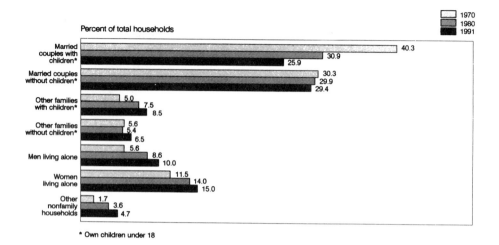

* Own children under 18

FIGURE 5.4. Household composition, 1970–1991. From U.S. Bureau of the Census (1992b, p. 2).

Household structure in the future is difficult to predict. The divorce rate has stabilized since the early 1980s and the birth rate has slightly increased. The baby-boom cohort will continue to have great impact on household structure because of its large numbers. But the marital, childbearing, and divorce patterns of the cohorts following the baby-boom generation will determine the future of household size and composition.

SUMMARY

The age and sex composition of a population is shaped by the processes of fertility, mortality, and migration. In turn, the age and sex structure of a population influences social trends. The sex ratio and dependency ratio are two measures of composition. Age–Sex pyramids provide graphic representations of the population structure. Age standardization is a statistical procedure that adjusts differences in the age structures of populations so that comparisons may be made.

An analysis of the dynamics of the age and sex structure is important for making social policy decisions. Projections of school enrollments and retirement needs are two examples in which knowledge of the age composition is important. Age and sex composition also affect social rates, such as marriage and suicide rates.

Demographers study other population characteristics. Race and ethnic composition, residential distribution, and household and family composition influence a wide range of both demographic and nondemographic factors.

6

MIGRATION

resident Franklin Roosevelt once started an address to the Daughters of the American Revolution (D.A.R.) by saying, "My dear fellow immigrants." Roosevelt knew that the members of the D.A.R. were particularly proud of their American pedigrees, but he wanted to remind them that both he and they had descended from people who had immigrated to the United States. They came long after it had been inhabited by even earlier immigrants: the Native Americans. It can be safely said that all Americans are descendants of people who came to this continent from some other place—some relatively recently, some many, many years ago—but we are all descendants of immigrants nonetheless.

Most people in the United States will change their residence during the course of their lifetimes. Some people change residences many times, while a few will never move from where they are born. It is to this phenomenon that we direct our attention in this chapter.

There are three important issues in the study of migration:

1. *The nature of migration.* What is migration? That may seem obvious, but several important distinctions can be made between the different types of migration. If these differences are clarified, it will be easier to understand the causes and effects of migration.

2. *The causes of migration.* The next important issue is why people migrate. What is it that makes some people move many times and a few not at all? The principal task of this chapter will be to explain why migration occurs or fails to occur.

3. *The effects of migration.* What are the consequences of migration? When people migrate, it affects the place they leave and the place they enter. Migration is also likely to affect the personal lives of the migrants. As we identify the effects of migration on individuals, on communities, and on societies, we will know the full importance of this major demographic phenomenon.

107

THE NATURE OF MIGRATION

The word *migration*, and the related words *migrant* and *immigrant*, tend to stimulate oversimplified and erroneous pictures in the minds of many people. One way to illustrate the misunderstanding that exists about these words is for the reader to try a little self-experiment. Take the word *immigrant*. Construct a mental picture of an immigrant or, better yet, of a group of immigrants. What image comes to mind? For many Americans, the word evokes a picture of refugees or "boat people" coming to the United States without the required visas. Or the word *immigrant* might evoke a stereotyped picture of a cluster of people who have just arrived at some East Coast port of entry at the turn of the century. The cluster is a family in which the wife is wearing a kerchief and the husband is wearing a flat, European-style cap or a broad-brimmed hat. Both are dressed in shabby, rumpled clothing, and they have slightly dazed or bewildered expressions on their faces. Huddled around them are several small children with equally puzzled or frightened expressions. This stereotypic image of the late 19th- and early 20th-century immigrant to the United States probably still shapes the thinking of many Americans when they hear the word.

The same experiment may be tried with the word *migrant*. Americans are likely to think of a migrant as a migratory farmworker or as someone who has entered our country illegally.

The images evoked by the words *immigrant* and *migrant* are not totally erroneous, but they are far from accurate descriptions of contemporary immigration and migration. As a case in point, the characterization of the European immigrant family above may have been accurate for many southern and eastern European immigrants at the turn of the century. But even then, many of the immigrants were not married couples with young children; instead, they were often single young adult males who intended to return to their families and homelands. A more realistic portrayal of recent immigration to the United States would include individuals who are concentrated in professional and skilled occupations. In general, U.S. immigration laws today are not designed to admit unskilled and uneducated people. American immigration laws favor those who have special occupational skills and jobs awaiting them. The new immigrants are no longer likely to be the "weary and downtrodden" coming to American shores for refuge, although a few special cases of political refugees in recent years have come closer to that description.

Similarly, while the word *migrant* suggests to most people a migrant farmworker (either legal or illegal), such a characterization does not describe most of the internal migrants in the United States today, for two reasons. First, migrant farmworkers are not appropriate, even as illustrations of migrants in the United States. This is the case because in terms of the formal definition of migrants, they do not qualify. From a demographer's perspective, a migrant is defined as a person who changes residences, more or less permanently. Yet many migrant farmworkers have a home base, which they leave only *temporarily* during certain parts of the year to work during various harvests. When the season for harvesting is over, they return to their permanent homes. Those who follow this pattern are not, technically speaking, migrants at all.

A second and more important reason for migrant farmworkers being misleading illustrations of migrants is that they fall at the opposite end of the economic scale from the most frequent migrants. Farm laborers are among the most poorly paid members of U.S. society, while migrants in the United States are likely to be well educated, have high-prestige jobs, and receive relatively high pay.

To understand migration properly, we must first give up the stereotyped and erroneous ideas we often hold. Then it will be possible to turn to a more systematic analysis of the definition and various forms of migration.

A DEFINITION OF MIGRATION

Migration is one of the three processes of population change, (fertility and mortality are the other two). It has an impact on population growth and population composition. It can produce social integration as well as social conflict. It is a familiar and important part of the working of many societies, and yet it remains one of the most difficult concepts to define and measure. One reason for this difficulty is that the definition of migration can change depending on the issue being considered.

A *migrant* is a person who makes a permanent change in his or her regular place of residence. Similarly, *migration* is the movement of individuals or groups from one place or residence to another when they have the intention of remaining in the new place for some substantial period of time.

At first glance, the definitions of migrants and migration seem straightforward, yet questions quickly arise. If we were actually going to try to classify people as migrants or nonmigrants, would a change of permanent residence include an individual who moves from one apartment to another in the same building? Would a person have to move across a political boundary, for example, by moving into or out of a city or a county or a state in order to be categorized as a migrant? How long is the "substantial period of time" that someone must intend to remain in a new residence before that person's movement is considered migration? How long does a person remain classified as a migrant after a move? All of these questions need to be answered when demographers set out to study migration.

The first question raised above is concerned with defining *migration boundaries*. Many times demographic researchers will limit migration to moves from one political unit to another. Demographers may distinguish between a mover (or residential mobility) and a migrant (or migration): A *mover* changes residence but remains in the same community, city, or county, while a *migrant* changes residence and crosses a political boundary.

In addition, when an individual crosses a national boundary, the terms *emigrant* and *immigrant* are used. In 1990, 30,667 people moved from India to the United States (U.S. Immigration and Naturalization Service, 1992). To the Indian government, these people are classified as emigrants, people who left their country of origin. The U.S. government classifies these same individuals as immigrants, people who are entering the United States.

The remaining questions, those having to do with the time period, are concerned with specifying a *migration interval*. The U.S. Census Bureau uses a 5-year migration interval for the census, and a 1-year interval for its Current Population Report on geographical mobility. The residential mobility status of the U.S. population is presented in Table 6.1. All people who were residing in a different county or state in March 1990 from where they were living in March 1991 were classified as migrants. Using this time interval, over 15 million people moved to a different county or state between 1990 and 1991 (U.S. Bureau of the Census, 1992f). This method does have some limitations. Even if people moved more than one time during the interval, only the last move is recorded. It is also possible for someone to have moved from one county to another and back to the original county during that time. These are *return migrants*, but the Census Bureau would have no way of knowing that any migration had occurred. In countries with population registers (e.g., Norway), it would be possible to count all of the migrants (Nicholson, 1990). When population registration data are compared to census data, the results show that the census will miss a significant number of migrants.

In general, the longer the migration interval, the greater the number of individuals who are likely to be classified as movers or migrants. The actual migration interval used depends upon the issue being explored. In some cases, the interval may be quite short. For example, a researcher might be inter-

TABLE 6.1. Residential Mobility in the United States, 1990–1991

Mobility status	Number	Percent
Nonmovers, same house	203,345,000	83.0
Movers, within same county	25,151,000	10.3
Migrants, different county but within same state	7,881,000	3.2
Migrants, different state	7,122,000	2.9
Immigrants	1,385,000	0.6
Total	244,884,000	100.0

Note. From U.S. Bureau of the Census (1992f, p. 1).

ested in the number of migrants who moved into an area during the last 6 months. In other cases, a demographer might be interested in making a statement about the percentage of people who migrate at least once during a lifetime (or the percentage or people who do not migrate in a lifetime). In this case, the interval would have to be the life span. Or if a demographer wishes to describe the number of people living in the United States who have migrated from some other country, the time interval would have to be as long as the oldest living immigrant.

MEASURING MIGRATION

The measure of migration used most often is the *net migration rate*. The net migration rate reflects the effect of migration on the total population of an area. It is calculated by subtracting the number of people who have moved out of a specified area from those who have moved into the area, dividing by the total population for that area, and multiplying that number by a constant:

$$\frac{\text{In-migrants} - \text{Out-migrants}}{\text{Total pop.}} \times 1{,}000 = \text{Net migration rate}$$

A positive net migration rate means that the number of people moving into an area exceeds the number who have moved out of that area. Similarly, a negative net migration rate means that an area has lost more residents than it has gained through migration.

When we know not only the number of people who are moving but also their origins and destinations, an analysis of net migration can be used to document migration streams. *Migration streams* refer to the flow of people from one city, state, or region to another designated area. For example, in the 1-year migration interval of 1990 to 1991, the Northeast region of the United States had a net migration loss of 585,000 people: 346,000 people moved into the Northeast, but 932,000 people moved out of the area. During this same period, the South gained 433,000 people through net migration even though almost 1 million people moved out of the southern region. An examination of the migration stream between these two regions shows that 195,000 people moved from the South to the Northeast, while 565,000 moved from the Northeast to the South (U.S. Bureau of the Census, 1992f).

THEORIES ABOUT THE CAUSES OF MIGRATION

There is no one theory of migration to explain the cause of migration; rather, there are several different yet complementary theories. These deal most directly with voluntary migration. Voluntary migration means that individuals make decisions about whether or not to move and about where to move. Most of the theories will have only limited application to the migration of refugees and other types of migration where the individual's ability to make decisions is constrained. In this section, the causes of migration will be examined from three perspectives: Ravenstein's "Laws of Migration," ecological theory, and individual decision-making theory.[1]

Ravenstein's "Laws of Migration"

Ravenstein made one of the first attempts at theorizing about migration in "The Laws of Migration" (1885, 1889). These "laws" were statements in the form of propositions about the nature of migration trends, streams of migration, and migration differences among categories of people. For example, Ravenstein noted that migration tends to flow from rural places toward urban places (an empirical observation that was also made in the 17th century by John Graunt). He also observed that migration occurs in streams, that is, in distinct flows of people from a particular place of origin to some specific destination. When migration occurs in a stream, often there is also a counterstream of migrants moving in the opposite direction.

Migration also differs in terms of the characteristics of the migrants. Ravenstein noted that migration streams tend to include more females than males if the distance of migration is short. He also found that migration is related to age, with the peak years for migration being associated with young adulthood.

Ravenstein's theory of migration offered a number of important propositions, most of which were fairly well grounded in fact at the time he made his observations. It is noteworthy that many of the propositions (or "laws") advanced by Ravenstein in the 1880s still hold true in contemporary life.

A theory should have some underlying mechanism of explanation. In this respect, Ravenstein held that the "desire of most men to better themselves in material respects" was the most influential factor causing migration (1889, p. 286). For him, economic motives provide the primary motive for migration, and in this regard, he set the tone for almost all subsequent migration theories.

Ecological Theories

An ecological theory of migration attempts to explain migration at a societal level. The underlying assumption of ecological theory is that an equilib-

---

[1]For a review of theories on international migration, see Massey et al. (1993). The theories reviewed are general theories such as macro- and microeconomic theory, world systems theory, and network theory as applied to international labor movements.

rium exists between population size, social organization, technology, and the environment (Hawley, 1950). If a society undergoes substantial changes in its social organization, technological innovations, or significant environmental changes, then population will either grow or decline in response to those changes. For example, political unrest may bring about social organizational changes. In the former Yugoslavia, for example, where ethnic groups are in conflict over regional control, the number of migrants is quite high, with many people seeking residence in Germany, Switzerland, Sweden, Austria, and Hungary (Hershey, 1993). The many people who are leaving the area are doing so in response to disruptions in the social organization of the region. The three demographic processes—fertility, mortality, and migration—may all help to restore equilibrium, but migration seems to be the most immediate and efficient response. Specifically, when a change occurs, the distribution of the population will shift, through the process of migration, until there is a balance between areas where opportunities are expanding and areas where opportunities are frozen or contracting.

The ecological model has been widely used as an explanation for migration from rural to urban areas in the United States. In rural areas, farming and mining became more mechanized (a change in technology), and opportunities for employment declined (a change in social organization). Meanwhile, new industries and service occupations expanded in urban areas of the country. Without a shift in the distribution of the population, there would have been an imbalance between population sizes and opportunities in rural and urban areas. Since many people did move from rural to urban areas, the equilibrium between population size and social organization was restored. This type of analysis has been used, for example, to explain the migration of blacks out of the Cotton Belt from 1940 through 1960 (Sly, 1972; Stinner & DeJong, 1969).

Recently, an ecological model has been used to explain, at the societal level, the "migration turnaround": the shift in population from metropolitan areas to nonmetropolitan areas that occurred in the United States between 1970 and 1980, and the return to the more usual migration pattern of nonmetropolitan to metropolitan areas between 1980 and 1990 (Frey & Speare, 1992; Fuguitt, Brown, & Beale, 1989; Sly & Tayman, 1980; Poston & White, 1978; Frisbie & Poston, 1975, 1976). The general trend in migration has been from nonmetropolitan areas to metropolitan areas. That pattern was reversed in the 1970s and is often referred to as a "rural renaissance." Changes in U.S. occupational structures and technological advances created new opportunities for individuals in nonmetropolitan areas. For example, retirement communities in nonmetropolitan areas expanded job opportunities in service occupations. The result was net in-migration to these areas. Even today, much of the current migration from metropolitan to nonmetropolitan areas can be seen as a response to shifting occupational opportunities in these areas. Results from the 1990 census show that metropolitan growth is once again greater than nonmetropolitan growth, especially along the coasts (Frey & Speare, 1992).

The ecological model is useful for describing the societal conditions that prompt migration. It is also useful for suggesting areas that are likely to experience out-migration and potential areas of in-migration. However, in the final analysis, migration is the result of individuals making independent decisions. The rest of this section will discuss decision-making theories and migration.

Decision-Making Theories

An early decision-making theory of migration, one that builds on Ravenstein's propositions, was advanced by Everett Lee (1966). Lee posits four very general factors from which he derives a series of migration hypotheses. For every decision to migrate there will be (1) positive and negative factors associated with the place of origin, (2) positive and negative factors associated with the place of destination, (3) intervening obstacles, and (4) personal factors. The positive and negative factors at the area of origin and the area of destination may include evaluations of employment opportunities, living conditions, climate, the availability of cultural and leisure facilities, the presence or absence of discriminatory treatment, as well as cost factors. Intervening obstacles are factors that make the actual migration from one place to another difficult. Obstacles include physical barriers and political barriers. Physical barriers include mountain ranges, oceans, and deserts; political barriers are immigration laws that keep people out of a country or national laws that prohibit migration within or out of a country.

The concept of personal factors has two different meanings. First, personal factors are simply characteristics of a person or a family, such as family size or stage of the family life cycle. But the term also refers to "personal sensitivities, intelligence, and awareness of conditions elsewhere" (Lee, 1966, p. 50). This second meaning adds a substantial new dimension to the theory, for if personal factors are to include the knowledge, perceptions, and awareness of an individual, then the first three factors in Lee's scheme must always be filtered through this fourth factor. The positive and negative values attached to the origin and destination, as well as the perception of intervening obstacles, can only have meaning as they exist in the minds of the potential migrants. The core of the theory identifies three general sets of factors—origin values, destination values, and intervening obstacles—that enter into the decision making of individuals through their perceptions and knowledge. Lee states, "Clearly the set of +'s and −'s at both the origin and destination is differently defined for every migrant or prospective migrant" (1966, p. 50). Lee's theory is diagrammed in Figure 6.1.

While Lee introduces the idea of personal factors into his scheme for analyzing how migration decisions occur, his discussion is quite limited and

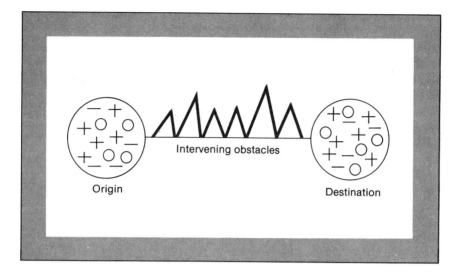

FIGURE 6.1. Origin and destination factors and intervening obstacles in migration—Lee's Theory of Migration. From Lee (1966, pp. 47–59). Copyright 1966 by the Population Association of America. Reprinted by permission.

only hints at how decisions are made by individuals or family units. More detailed analyses of how individuals come to make the decision to change their place of residence have subsequently emerged. This approach to migration has come to be known as the *decision-making model of migration*. In fact, there is not one single decision-making model but a number of different ones that in one way or another deal with the question of what kind of mental processes people go through when they decide to move (DeJong & Gardner, 1981; Mueller, 1982; Speare, 1974). These theories try to answer the following questions: What factors influence people to move? How do people make the decision to move to a particular place? Why do some people decide *not* to move? These questions are deceptively simple, but the answers reveal just how complex the decision-making process can be.

Two assumptions must be made clear about decision-making theories. Both have been touched on above, but they are important enough to call attention to them again. First, decision-making theories are only relevant when individuals are free to make choices. By "free," we mean primarily that individuals are not reacting to coercive government policies that force populations to migrate or keep them from doing so. The extreme example of slavery, and to a more limited extent the displacement of groups of people through war, are instances in which individuals are not free to make choices about migration. Decision-making theories are only partially appropriate under these conditions.

The second assumption is that there is some degree of rationality involved in making decisions to migrate. By "rationality," we mean that individuals can think about objectives or goals that they have for themselves. Next, they can make some kind of choice as to which goal or goals they wish to achieve. Finally, they can examine the different ways of achieving these goals and select the one that will most effectively help them to reach their goal(s). Most individuals in contemporary societies behave in a rational manner most of the time in their daily lives, although no one is perfectly rational all the time. One does not have to assume perfect rational behavior in order to use a decision-making model.

Sell and DeJong (1978) have argued that there are four different components to what they call a motivational theory of migration decision making.[2] The four components are called: (1) *availability*, (2) *motive*, (3) *expectancy*, and (4) *incentive*. In a sense, these four components can be thought of as steps, where each component in turn leads to the next. For each of the four components we will first describe what it means and then offer some illustrations from hypothetical or actual migration cases.

Availability

In order for anyone to make a decision to migrate, it is first necessary for migration to be a possibility—that is, to be an available option. There are two distinct senses in which migration can be available to an individual: physical and psychological.

Physical availability in its simplest sense simply means that it must be physically possible for a person to move. If, for example, a person is in prison, it might be very advantageous for that person to migrate (to obtain greater income, greater personal freedom, etc.), but he or she faces physical constraints that make it impossible. Frequently, nations or countries will have laws or regulations that make migrating a physical impossibility (e.g., Jews were often prevented from leaving the former Soviet Union even though they wished to do so very much).

Physical availability also refers to the constraints that keep people from *moving into* an area. Many countries, including the United States, have immigration policies that place restrictions on migrants' origins and generally specify the characteristics of those who may migrate. (The U.S. immigration laws will be discussed in Chapter 10.) In the United States, there have been many instances in which African American families have wanted to move into a particular neighborhood, but *de facto* segregation policies have kept them from doing so. In these cases, physical availability was denied black Americans.

---

[2]We will draw upon the Sell–Dejong theory, though we will continue to refer to it by the more general term *decision-making model*.

Physical availability is very closely related to the concept of "intervening obstacles" in Everett Lee's theory. When an intervening obstacle is so great that it prohibits migration, then the decision to migrate is made physically unavailable.

The second sense in which migration is either available or unavailable is in the mind of the potential migrant. This is referred to as "cognitive unavailability." A migration decision will be cognitively unavailable when it does not enter the mind of a person to migrate. This is often a matter of degree, but there are many people for whom migration is literally unthinkable. These are often people who are so emotionally attached to their communities or countries that no matter how bad things get or how good the alternatives might be, they cannot think of leaving. Emotional attachment, or a sense of belonging, may not always make migration unthinkable, but it has been shown in a number of studies to be an impediment to migration (Stinner, Van Loon, & Byun, 1992; Speare, Kobrin, & Kingkade, 1982; DeJong, 1980; Heaton et al., 1979; Back & Smith, 1977).

While either physical or cognitive availability may not be possible for some people, for others it clearly is possible. If we now assume that for many people moving is both physically and cognitively available, then we can go to the next component, or step, in migration decisions.

Motive

The term "motive" refers in migration decision making to the things people value in life. Values are embedded in the cultures of every society, and most people learn the most prominent cultural values of their society through the process of socialization. Thus in American society and its culture great value is placed upon success, which for all practical purposes means economic success. It is not surprising, then, that one of the prominent motives for moving in the United States is economic. As Table 6.2 shows, employment-related factors are a major reason for moving from one place to another.

There are, of course, other motives for moving besides purely economic ones. These are also noted in Table 6.2. Many people value their relations with family members. Some people value the welfare of family members or their goodwill more than economic considerations. For them, a decision to migrate or not to migrate will be greatly influenced by how it will affect family members and family relations. For example, throughout the 1980s many young adults moved out of the Appalachian region in search of jobs. Now, several years later, some of the parents of these migrants are retiring in areas near to their children (Watkins, 1990). In fact, the elderly are more likely to migrate when they have several adult children "living in their destination area" (Sommers & Rowell, 1992). Clearly this is a case where the motive for moving is not economic, but instead a yearning for the social support that families provide to each other.

TABLE 6.2. Reasons for Moving among All Respondents and the Elderly, 1991

	Total (16,753)[a]	Elderly (65+) (865)[a]
Displacement/disaster	7.2%	9.7%
Employment-related	28.1	8.4
New job/job transfer	11.3	0.6
Closer to work/school/other	10.9	2.5
Other employment-related	5.9	5.3
Family-related	32.9	44.9
To establish own household	13.8	4.7
Married	3.1	0.7
Widowed/divorced/separated	5.2	8.3
Other family-related	10.8	31.0
Housing-related	47.1	46.3
Needed larger house or apartment	12.9	5.7
Wanted better home	12.7	11.2
Change in rent/ownership	6.7	5.0
Wanted lower rent or maintenance	7.3	12.3
Other housing-related	7.5	12.1
Other/not reported	16.8	28.8

Note. From U.S. Bureau of the Census (1993d, p. 60). Percentages may not total 100 because more than one category may apply.

[a]Number in parentheses represents total number of respondents in thousands.

Other motives that could influence migration decisions are more abstract. Values such as freedom or patriotism have sometimes figured greatly in the decisions of people to migrate. Many people have so strongly valued their freedom to practice particular religious beliefs that they have moved to retain or achieve them. The movement of 19th-century members of the Church of Jesus Christ of Latter-Day Saints (Mormons) provides a good illustration. Founded in New York in 1830, and immediately faced with opposition and violent attacks, the group moved westward to Illinois. Later, when they were attacked in the Midwest, they migrated again, and finally settled in Utah (Kephart & Zellner, 1991).

The list of motives that influence migration decisions could be extended almost without end. Some of the recent migration into nonmetropolitan areas is credited to people over age 55 who are not moving for economic reasons. Many of the migrants in this group are retired and are thus not economically bound to any particular area for employment or income. For these people, climate and recreational opportunities are predominant motives for their moving (Morrison, 1990; Long, 1988; Heaton et al., 1981). Whenever people place great value on their health, happiness, independence, or some special activity (surfing, skiing, year-round tennis, dance, theater), this may be enough to shape their migration decisions. While we cannot make a definitive list of motives, it is possible in most cultures to identify a small number that play an important part in migration decisions.

Expectancy

The third component of migration decision making is expectancy, which means individuals have to believe that migration will help them to achieve their goals. People often know that high-paying jobs exist in some other part of the country (cognitive availability). They may know that it would be physically possible to go there (physical availability). They may also value the money those jobs would pay (motive). However, if such people do not believe it will be possible to get those jobs, they will not migrate. Expectancy would be missing, and without it people will not move.

On the other side of the coin, when people do expect they can get high-paying jobs or some other valued thing, they may migrate. Junior executives are often given the opportunity to manage a company branch plant in an isolated and otherwise unattractive place in the country, and they often take these positions. They migrate because they expect to do well in these positions and ultimately achieve the success they are seeking with the company.

The expectations that people have about the new place of residence may or may not be realistic. People may have unrealistically high expectations that encourage them to move. In John Steinbeck's novel *The Grapes of Wrath*, the people who left the Dust Bowl in Oklahoma had expectations about California that proved to be much too high. It turned out that California was not the land of milk and honey they had expected. Nonetheless, the expectations they had were just as important in their decision to go to California as if they had been totally realistic. It is important to keep in mind that if migration decisions are to be understood, the perceptions of the decision makers are as influential as the objective conditions of different geographical regions and communities.

Geographer Julian Wolpert (1965) was one of the first researchers to recognize the significance of people's perceptions of different places. Wolpert's ideas may be illustrated by using the case of Denver, Colorado. Many people perceive Denver as a city of crystal blue skies with the snow-capped Rocky Mountains as a scenic backdrop. Given this perception, Denver is often viewed as a very attractive migration destination. What many people do not know is that the air quality index of Denver is poor; the city residents endure numerous days when carbon monoxide and other pollutants exceed healthly standards. But even though Denver's climate ranked 200th out of 333 American metropolitan areas, it nonetheless retains its positive image as an attractive migration destination (Boyer & Savageau, 1989).

We noted earlier that Ravenstein observed in the 19th century that streams of migration to particular places are almost always associated with counterstreams. One possible reason for counterstreams is the movement of some disappointed migrants going back to their place of origin. DaVanzo and Morrison (1982) assert that return migrants are not as careful planners and are more likely to be misinformed about migration destinations than are other migrants. There are likely to be other reasons for counterstreams, but the return of some disappointed migrants is likely to be an important component.

Incentives

The final component of migration decision making is incentive. In Lee's theory of migration, shown diagrammatically in Figure 6.1, the pluses and minuses at the places of origin and of destination are quite similar to the meaning of incentives. Incentives are the specific characteristics at the present place of residence and at a potential residence ("origin" and "destination" in Lee's terms) that are positively or negatively related to a person's goals. For example, if a married couple has children, the quality of the educational system in the places of origin and destination may be a very critical factor for their decision making. If they judge the school system in a new community to be better than the one in their present community, then the school system will be a plus factor for moving.

Just as with the motives (or values) discussed earlier, the range of incentives can be extremely wide and varied. However, again we can reflect upon the values of American culture and make some good estimates of the kinds of specific characteristics people will evaluate in their decision making. We know that economic well-being is an important goal in American society; therefore, most potential migrants will check out the occupational opportunities in a possible place of destination. Or, if the potential migrant is retired, the cost of living will be considered. Of course, even though different areas may have similar economic opportunities, it may be important to examine the general levels of income in new areas. For example, the median income for all households in the United States was $29,943 in 1990. However, the median income varies by region. Earnings of $30,000 a year in one community may be relatively high, but in another community such earnings may be at the low end of the scale. The data presented in Table 6.3 show the median incomes of households by regions and for several states in 1990. Clearly an annual income of $30,000 would be more than adequate in Mississippi or Missouri, but perhaps not in Connecticut or Maryland.

In addition to economic characteristics, there are the many possible noneconomic incentives people may wish to consider. Among the obvious ones are community size, climate, crime rate, recreational opportunities, closeness or distance from extended family members, and so on. To know what incentives (or disincentives) any particular individuals may consider important, we can only turn to the individuals themselves for answers.

TABLE 6.3. Median Income of Households by Region
and Selected States, 1990

Region	Median Income, 1990
All households	29,943
Northeast	32,676
Connecticut	38,870
Maryland	38,857
Midwest	29,897
Kansas	29,917
Missouri	27,332
South	26,942
Florida	26,685
Mississippi	20,178
West	31,761
California	33,290
Washington	32,112

Note. From U.S. Bureau of Census (1991e, p. 11).

However, both in terms of general motives (or values) and particular incentives, there is a substantial amount of evidence that two major societal institutions are extremely significant in decisions to migrate (or alternatively, decisions not to migrate). These are the *economic* and the *familial* factors. A substantial amount of empirical evidence points to the conclusion that economic and family considerations are almost always involved in migration decisions. This is not to say that other factors are not important (climate, health, freedom, etc.) but only that the importance of economic and family factors must be dealt with first. In the sections that follow, we will review the research that has shown the importance of economic and family factors on migration.

LEVELS OF ANALYSIS AND THE STUDY OF MIGRATION

In the last few pages, we have focused on the basic components of migration decision making. We concluded by pointing out that the economic and family institutions are usually the most important determinants of migration decisions. While the emphasis in the previous section has been on the personal decision making of individuals or families, we must also recognize that the effects of economic and familial factors can be demonstrated at other levels of analysis. Demographers have traditionally chosen to study demographic phenomena at higher levels of abstraction. Instead of studying individuals and their motives, expectations, incentives, and so on, demographers have been inclined to study demographic rates for various population groups (Back, 1967). Migration has often been studied at a societal level, or what is sometimes called the *macroscopic* level. (The ecological theory of migration is at the macroscopic level.)

The emphasis in societal or macroscopic studies is on the way certain societal characteristics can have an impact on demographic factors. For example, if economic factors influence migration decisions at an individual level, then economic conditions at a societal level should also have an impact on the migration rates in that society. In the section that follows, we will examine the way economic and familial factors at the societal level influence the rate of migration. Later we will turn to the individual, or *microscopic,* level, where the impact of economic and familial factors will be assessed.

Societal Level of Analysis of Migration

At the societal level, demographers have given much more attention to the effect of economic factors on migration than they have to familial factors. The result is that a substantial body of research findings has been amassed. As more and more research is generated, especially in the area of international migration, more refined propositions, hypotheses, and theories are being tested.[3] For example, neoclassical economic theories and dual labor market theory are being applied to migration research (Massey et al., 1993). We will review the economic research at the societal level before turning to the family.

Economic

Everett Lee (1966) based part of his theory of migration on the hypothesis that the volume of migration in a society will vary with fluctuations in the economy. But more importantly, he analyzed how and why migration would vary. Does migration increase or decrease during times of economic growth and prosperity? Alternately, does migration increase or decline during times of economic recession and depression? The answers to these two interconnected questions are not altogether obvious. Lee deduced from his theory what might happen, but when he did so, he had the advantage of knowing what a substantial body of research literature had already revealed. Migration increases during times of economic prosperity and decreases during economic hard times.

Lee reasoned that during times of economic prosperity, new businesses and industries would be created. These new economic enterprises would require more new workers than might be available locally. Furthermore, it is likely that there would be some unevenness in prosperity throughout the

---

[3]For a review of these theories, see Massey et al. (1993). Massey et al. note that there are many competing and complementary theories to explain international migration. Each carries its own set of assumptions and more empirical work is needed.

country. Some areas might not be enjoying economic growth even as other areas were flourishing. Under these circumstances, the places of origin of many people would have economic minuses compared to the economic pluses of the areas with new businesses and industries. This point may be partially illustrated by what occurs when the federal government consolidates and closes military installations. For example, the navy facility in Warminster, Pennsylvania, is closing and the work will be transferred to the navy base located in St. Mary's County, Maryland. Many people are expected to move to St. Mary's County to take advantage of the expanding economic opportunities provided by the expansion of the navy base and the supporting defense industries.

Returning to Everett Lee's consideration of economic factors and migration, he argued that during times of economic depression, migration would go down. First, during times of depression, new businesses and industries might fail; others would not expand. This condition would reduce the number of positive factors at potential places of destination. Second, during economic hard times, people might prefer the security of their current place of residence to the uncertainties of other places. Even though economic conditions at home might be difficult, they are known and familiar difficulties.

A number of research studies over the years have supported the hypothesized relationships between economic prosperity and migration. The generalization applies to both immigration and internal migration (Fassmann & Munz, 1992; Castaño, 1988; Stahl, 1988; Jerome, 1926; Thomas & Kuznets, 1957, 1960, 1964; Gober-Myers, 1978; Lewis, 1982).

Another example of how a country's economic situation can influence in-migration is the case of international labor migration throughout western Europe. The migration streams between many western European countries clearly document the relationship between economic prosperity and migration (Fassmann & Munz, 1992; Salt, 1981). From World War II to the present, there has been a flow of immigrants into the northern industrialized European countries. The receiving countries, led by Switzerland, experienced short labor supplies beginning in the late 1940s because of low population growth rates and expanding economies. The sending countries, led by Italy, experienced an opposite set of conditions: high rates of unemployment and underemployment. To encourage the flow of immigration, Switzerland adopted liberal immigration policies and offered extensive benefits to immigrants to help them adapt to their new environment. However, most of the labor contracts the immigrants received were short term so that during recessions or other adverse economic conditions, when less labor was needed, the immigrant workers could be released from their jobs. When that occurred, as it did in the mid-1970s, these immigrants often returned home.

Studies that have focused on return migration also support the hypothesis that migration flows decline during times of economic recession. The impact of declining oil prices and a worldwide recession in the early 1980s had a great impact on the migration patterns in Latin America and the Caribbean (Castaño, 1988) and in the Middle East (Feiler, 1991). Beginning around 1970, there was a large immigrant flow from Colombia to Venezuela, an oil-rich nation. At the onset of the recession in Venezuela caused by the oil glut, many Colombian immigrants returned home. They did so because they no longer enjoyed high wages in Venezuela which they could send back to relatives in Colombia. At the recession deepened and unemployment rose in Venezuela, the government started to deport Colombian immigrants.

While the impact of economic conditions on migration has received the most attention from demographers, it should not go unnoticed that the very structure and character of an economic system can greatly influence migration. In the extreme case of a totally agrarian society where people are tied to the land and agriculture is the only economic activity, migration is going to be very limited.

On the other hand, if the economy is an urban industrial one made up of large bureaucratic organizations, migration is likely to occur at a very high level. Bureaucratic organizations of all types—in the public and private sectors, producing goods or providing services—require a vast array of occupational specialists. Everett Lee (1970) has argued that this need for special skills in the bureaucratic organizations of the United States has made this country one characterized by especially high migration rates. Lee argues that specialists of all types, from cardiologists to college professors, are needed all over the country. Most cardiologists are trained in major medical centers, but every reasonably sized hospital or clinic needs at least one. The result: migration. Specialists may often be seen as interchangeable parts who are able to make the move from one system to another with ease. Engineers, architects, teachers, computer programmers, welders, and hundreds of other occupational groups have the ability to leave one job and step directly into a similar job anywhere in the country.

Family

It is now time to look at the family systems of societies to see what effect the family has on migration. There are two basic forms of the family: the extended family and the nuclear family. The *extended family* is one in which three or more generations live together, either in the same household or in close proximity. Most commonly, in societies around the world, the household is headed by a family patriarch, with the support of his wife. The other members of the household include their sons, daughters-in-law, and grandchildren.

In the extended family system, sons not only stay in their parents' households but are also financially dependent upon the economy of the household. In short, in a patriarchal, extended family system, sons rarely migrate. On the other hand, daughters are likely to move from the family home when they marry, although they usually do not go too far away.

At the other extreme, societies with a *nuclear family* form have what is called *neolocal residence*: The newly married couple sets up a residence separate from that of both their parents. There has been much discussion about the nuclear family system being particularly suited to the urban–industrial society. The nuclear/neolocal family allows young people to move away from their parents and to those places where job opportunities exist. To some degree, the nuclear family system may be seen as a facilitator that allows the specialized bureaucratic economic system to work. Nonetheless, the general theoretical proposition is clear: A society that has a nuclear family structure will have more migration than a society organized along extended family lines.

The empirical research for the above generalization is limited; thus its validity is largely a matter of definition. However, there is an intermediate family form that has been studied and about which the conclusions are clearer. This intermediate family form is called the stem family. The *stem family* is one in which there is a family home, which remains the family home for generation after generation. In every generation it is passed on to only one of the children. Other children usually go out into the larger world, typically to urban–industrial occupations, but they may at any time, and as the need arises, return to the family home. This is called the stem family because the family home is a kind of stem or tree trunk from which the branches (children) extend outward (LePlay, 1872).

One of the vivid examples of the stem family and its operations may be seen in the Irish family during the last half of the 19th century (Arensberg & Kimball, 1968; Connell, 1962; Kennedy, 1973). After the severe Irish famine of the 1840s, the Irish family took on the character of a stem family. Irish farm families had an average of six children, but only one was able to inherit the farm. This favored child, typically one of the sons, was given control of the farm at the time he married. The remaining sons had only a few options. They could remain in the family home, but they could not marry (because they had no farm on which to live) and they would always be subordinate to the brother who had inherited the family farm. If they did not choose this limited and subordinate role, they could leave home, which typically meant that they emigrated either to England or the United States. The priesthood was another choice, but that, too, entailed migration. The stem family in this instance not only allowed migration but seemed to stimulate it. It is probably the case that many male Irish immigrants to the United States were the sons who did not inherit their fathers' farms.

As for the daughters, they too often emigrated because to marry a young man with a farm a woman had to have a dowry (money, furniture, jewelry, etc.). Those who could not accumulate a dowry sometimes migrated to another country in the hope of working and accumulating money. Some perhaps did eventually return to their native land, but many others probably remained as permanent residents of their new home.

The stem family, or a variation of it, has also been found to exist in and to affect migration in the United States (White, 1983; Harbison, 1981; Davis & Donaldson, 1975; Brown, Schwarzweller, & Mangalam, 1963). Brown's studies of Kentucky mountain communities and Davis and Donaldson's study of black migration revealed how family members, one after another, left their home communities for particular northern cities. Their movement was not random but followed a pattern in which various family members went to the same location in the North. This pattern indicates the importance of communication in migration, communication that often follows family lines. When one family member follows another to the same destination, this phenomena is referred to as *chain migration*. Brown and his colleagues found both chain migration and return migration to the home communities of Kentucky families. Davis and Donaldson documented specific migration chains from 21 states to different metropolitan areas. In White's (1983) study of eastern Kentucky, he found that migration into that isolated area could not be explained by the revitalization of the coal industry or other economic factors. Instead, it was the stem family that had the greatest influence on returning migrants. The home community and the family home did seem to be a "haven of safety" for many migrants. This case supports LePlay's idea that the stem family is highly suited to migration in an urban–industrial economy. When jobs are available, people can migrate to them; when the jobs end, they can return to the safety of the stem family home or community.

Personal Status Characteristics and Migration

The economy and the family, at the societal level, have both been shown to be related to migration patterns. However, we can examine the effects of these institutions more closely if we move to the level of individuals. This is called the *microscopic level* because it deals primarily with the smallest units in society: individuals, their characteristics, and their behavior.

Individuals as they go through a lifetime in any society will have a number of social statuses or positions (e.g., child, adolescent, young adult, etc.). With respect to the family institution and the economy, there are a number of significant statuses, and many are closely related to migration behavior. We will refer to these as *personal status characteristics*. Personal status characteristics in the discussions that follow will refer primarily to the positions or statuses that individuals may have in the economic system and the family and which influence migration. For example, in the economic realm, a personal status characteristic might be the position one has in the occupational world, such as seeking a first job, being unemployed, being promoted, being demoted, being transferred, or retiring. It is easy to see how any of these could be in some way connected with migration.

To illustrate the level of personal status characteristics further, consider the various positions (or statuses) that one may have with respect to the family. The most prominent statuses include being unmarried or married, being a parent, being divorced, and being widowed. In the discussion that follows, we will see how personal status characteristics in the economic and family realms influence migration.

Economic

In contemporary urban–industrial societies, the key economic status that influences migration is occupation. Migration occurs most commonly in young adulthood. This generalization is as true today as it was when John Graunt was assessing migration to the city of London in the 1600s. Between the ages of 20 and 30, migration is at a higher level than any other age in the life cycle (Fuguitt, Brown, & Beale, 1989). The relationship between age and other socioeconomic characteristics is shown in Table 6.4. The principal reason for the high level of migration at this age is that people move to prepare themselves for and to pursue occupations. Preparing for an occupation includes educational preparation, which in itself often entails migration. But more importantly, as people complete (or leave) their educations, they move to the jobs for which they have prepared themselves.

Educational characteristics are themselves closely related to migration. At all adult ages, the more education one has the more likely one is to migrate. The only exception to that rule concerns people who do postgraduate work: Those with postgraduate degrees are slightly less likely to migrate than those who graduate with baccalaureate degrees (see Table 6.4). Even with this minor exception to the pattern, it can be said that college graduates (including those who have gone on to postgraduate work) are more likely to migrate (move across a county line) than someone who did not finish high school (U.S. Bureau of the Census, 1991a).

Educational status characteristics of individuals reflect occupational and professional training. People often migrate in order to use or maximize their training and skills. But do all occupations and professions influence migration equally? The answer is no. Indeed, particular occupations have very different influences on the likelihood of migration (Sell, 1983). Roughly, occupations and professions may be divided into three categories: (1) migration-inherent, (2) migration-probable, and (3) migration-inhibited.

Migration-Inherent Occupations. There are some occupations where migration is more than expected: It is an inherent condition that cannot be avoided. Probably the most familiar example is a military career. In a 20-year career, it would not be uncommon for a military person to have 10 or more permanent assignments that require relocation. And one may add to these major moves a number of temporary-duty assignments that require temporary relocations.

A somewhat similar profession for which migration is inherent is the

TABLE 6.4. Selected Characteristics of Movers and Migrants, United States, 1990–1991

Age	Total population	% Movers	% Migrants to different county
1–4 years	15,297	22.7	7.7
5–9 years	18,466	17.6	5.7
10–14 years	17,601	14.1	4.3
15–19 years	16,839	17.8	6.1
20–24 years	17,986	35.3	12.9
25–29 years	20,767	32.6	12.5
30–34 years	22,138	21.6	7.3
35–39 years	20,379	16.1	5.9
40–44 years	18,286	13.3	5.1
45–49 years	14,129	11.1	4.7
50–54 years	11,557	9.5	3.6
55–59 years	10,692	7.9	3.3
60–64 years	10,654	6.8	3.0
65–69 years	10,123	5.6	2.1
70–74 years	8,114	4.3	1.9
75–79 years	5,835	4.6	1.5
80–84 years	3,629	5.0	1.8
85 and over	2,391	5.0	1.9
Completed education (age 25 or over)			
Total 25+	58,694	14.5	5.5
Elementary (0–8 years)	16,849	11.5	3.0
High school (9–12 years)	78,651	14.1	5.0
College (1–4 years)	49,270	16.2	6.8
College (5+ years)	13,925	14.2	6.5
Median years of school completed	12.7	12.8	13.0
Labor force status Civilian labor force			
Not in labor force	65,164	12.3	4.7
Employed	115,187	18.2	6.6
Unemployed	8,887	28.7	11.4
Armed Forces	997	42.8	17.9

Note. From U.S. Bureau of the Census (1992f, pp. xiv–xv).

foreign service. Anyone joining the foreign service can expect to be reassigned to a series of foreign consulates and embassies during a lifetime career.

A very different occupation, but one that also has migration built into the career pattern, is professional baseball (and most other professional sports). Once drafted by a major league baseball team, a player is usually required to play for regional minor league teams. Moreover, the player might well have to change residence seasonally in order to play "winter ball," "spring ball," and so on. It is not uncommon for baseball players to move 10 to 20 times during an active career because of reassignments, trades, and seasonal relocations.

These few examples indicate that certain occupational statuses have migration probabilities of nearly 100%. Most occupations, however, are not as certain to lead to migration as those we have just considered. Instead, migration for a much greater array of occupations is just probable.

Migration-Probable Occupations. The occupations or professions that are likely to lead to migration are those in which migration is a likely requisite for advancement. It is commonplace in the world of corporate management that movement from one office or plant to another within the same company often indicates success in moving up the corporate ladder. After professionals prove themselves within one organization, they often move to some other organization to gain a higher salary and a more responsible position. Many management and staff positions in the corporate world have an interchangeable quality, such that movement from one company to another is relatively easy and is one route to upward mobility.

Many other occupations and professions have a similar built-in tendency for migration. College professors, the clergy, radio and television personnel, scientists, engineers, and social workers are just a few of the many illustrative examples. In general, it appears that occupations where migration is valued are those that are white collar and are either professional or quasi-professional. There are, of course, examples of occupations in the blue-collar ranks where mobility is also possible—consider plumbers, machinists, and electricians—but in general it appears that geographical mobility in these occupations is not connected quite as closely with advancement or promotion.

The case of plumbers and electricians reveals another dimension of occupations that relates closely to migration. If a plumber or electrician has a private business, success in that business is closely related to developing a regular set of customers or clients. Whenever an occupation or profession depends on a clientele, migration is greatly diminished. This leads us to the third category of occupations, those for which migration is inhibited.

Migration-Inhibited Occupations. Doctors, dentists, and lawyers with private practices and any other business owners or professionals who depend upon a regular clientele cannot easily migrate. It is commonly the case that anyone who establishes a private professional practice, or begins a retail business, has an early period where profits are small (or there is an actual economic loss). Often it takes a considerable period of time before an enterprise becomes successful. Whenever a practice or a clientele has been established, migration is inhibited. This is not to say that a successful professional or businessperson cannot sell the practice or business and move to a new area, only that making such a move is more difficult.

Migration is also inhibited in occupations where there is a fixed resource that provides a person with a livelihood. The most prominent examples are farms or ranches. Farmers and ranchers do not move from one place to another because they are tied to their land. There are other occupations which would have similar limitations on moving, including mine owners, fishing-boat owners, and resort owners.

Family

Families typically go through a series of life stages. There are many exceptions, but marriage is usually followed after a few years by children, who then grow up and eventually leave home. Divorce is another family status change. As individuals go through these various changes in family status, their likelihood of migrating also changes.

Entering marriage, as we have seen, causes couples to establish new residences. Later, as children are born and family size increases, there is pressure to move locally to a larger, more suitable home (Rossi, 1955/1980). However, as children grow older, enter schools, and develop friendships, they usually become reluctant to move away from friends, familiar neighborhoods, and schools; and to the degree that their wishes are considered, they are likely to inhibit migration. Families often make decisions to move at times when the change for their children is seen as least disruptive (e.g., between middle school and high school, or after high school graduation).

Since children act as impediments to migration, it is not surprising that as children grow older (and are still in the home), migration decreases. The relationship between life-cycle stages (e.g., preschool children, oldest child in elementary school, oldest child a teenager) and migration is strong and has been affirmed in a number of empirical studies (McAuley & Nutty, 1982; Sandefur & Scott, 1981; Miller, 1976; Long, 1972; Speare, 1970).

Social Psychological Characteristics

The third level or analysis for studying migration is the social psychological level. Basically, the term "social psychological" refers to the attitudes, orientations, and ideas that individuals have about significant things. The societal level of analysis and personal status characteristics provide the objective criteria on the selectivity of migrants—who moves and under what conditions. However, there are people in similar economic situations with the same personal statuses, and yet some move and others do not. The social psychological level of analysis may help to explain these differences. Again, we may illustrate this level of analysis with respect to both economic and family matters.

Economic Attitudes and Orientations

A preeminent economic attitude is the way people feel about achievement, success, and money, often referred to as "economic aspirations." There is a great deal of variation in the economic aspirations of individuals. Some individuals are extremely concerned with economic success, in some cases to the point of being obsessed with making money or accumulating wealth. Other people are much less concerned with money or economic success. It is apparent that economic aspirations should have an important effect upon migration. If, for example, a person is in an occupation or profession where migration indicates advancement (the probable migration category above), then those people with high economic aspirations will be likely to migrate. A young junior executive may not relish going to the branch plant in some out-of-the-way place in the country, but will accept his or her assignment to ensure further advancement (and more moves).

In addition to illustrative cases, some research evidence shows that economic aspirations are important in migration decisions. Miller (1976) surveyed 280 couples who had children ranging from preschoolers through high-school-age. The economic aspirations of these couples were measured by asking the husbands and wives independently to respond to statements such as "Moving up economically is an important objective to me." For both husbands and wives, those with higher economic aspirations had a high propensity to migrate. (Subjects were asked questions about their willingness and eagerness to leave their present place of residence.)

Family Attitudes and Orientations

Social psychological factors with respect to the family refer to any attitudes or orientations that an individual has toward family members, family relationships, or family welfare. In the previous discussion of families with children, we noted that children are often reluctant to move. We also saw how in many occupations migration is a way to secure advancement and achievement. Many times parents must make difficult decisions about moving when they feel the move may be economically beneficial for them but harmful for their children. In this situation, the orientation toward the welfare of the children may be in direct opposition to economic aspirations. What a person will do in that situation is influenced by the intensity of their attitudes and orientations. Increasingly, studies of migration must focus on this level in order to understand migration decisions.

In addition to attitudes about children, adults will also have attitudes about parents and other kin. For example, young adults, when selecting a place to live, may feel strongly that they want to have close contact with their parents, siblings, grandparents, aunts, uncles, and so on. Other young adults may be equally eager to move away from these extended family members because they prefer more limited contact. Orientations about extended kin may also influence a person's migration decisions. In general, individuals who have positive attitudes toward their extended family members also have less propensity to migrate (Stinner et al., 1992; Harbison, 1981; Miller, 1976).

While attitudes and orientations are important in migration decision making, it is clear from our earlier discussion that migration can be analyzed at several different levels. There is often an interesting interplay between the personal status characteristics of individuals and their attitudes and orientations (both economic and familial). Furthermore, when we recall that migration decisions take place against a backdrop of societal conditions, it is apparent that migration patterns are shaped by an intricate complex of factors.

THE EFFECTS OF MIGRATION

When people make a permanent change in their place of residence their lives are certain to be affected in some ways. In addition, the communities they leave and the communities they enter will also be affected. And when there is extensive mobility in a society, it is going to have some effect on the nature of that society. In this final section, the effects of migration are considered.

The character and distribution of the U.S. population has been greatly influenced by migration; first, by the migration from other countries, and, second, by the continuous movement of people across the continent toward the frontier. It has been argued that this continuing movement of the American people was a factor of the utmost importance in shaping American society and the character of the American people. Lee (1961) has stated that migration "has been a force of greatest moment in American civilization, and that from the magnitude and character of migration within this country certain consequences logically follow" (p. 82). While migration cannot explain everything about Americans, "it was and is a major force in the development of American civilization and the shaping of American character" (Lee, 1961, p. 82). American historian George Pierson (1962) has stressed that the most prevalent characteristic of Americans is their migration and "excessive mobility."

What, then, does the migratory nature of Americans explain? The list of ideas that has been suggested is fairly long and varied, and most are hypotheses and do not rest on a solid base of fully substantiated facts. For example, the migration of Americans may make them more individualistic and self-sufficient, since those who move from their home communities cannot so easily rely on their families or kin groups for support. Paradoxically, however, this lack of familial support may cause internal U.S. migrants to exhibit patterns of conformity as they make adjustments to new places, new situations, and new people. Migration may be an important factor in producing a social conformist who is sensitive to the signals coming from the new environment. But migration may be an important component in the American desire for change, and migration can act as an "equalizer" or "democratizer," producing a less rigid social structure. One recent study suggests that migration promotes greater tolerance. Using two scales, one reflecting the ideas of leftist groups and another the ideas of rightist groups, Wilson (1992) found that migrants show greater tolerance toward both groups than nonmigrants.

The list of American characteristics possibly attributable to migration can be extended to include the practicality and nationalism of Americans as well as their tolerance for disorder and corruption. It has also been suggested that migration has been the great stimulus for American optimism.

All of these ideas and hypotheses have at one time or another been seriously advanced by scholars and social observers. In essence, they argue that migration has shaped the social structure and culture of American society. Such hypotheses are difficult to test because they are difficult to state with any specificity. Empirical evidence is hard to collect, and the criteria for rejecting or accepting any one theory are nearly impossible to establish. But through these macroscopic hypotheses, one is alerted to the possibility of more limited and specific hypotheses that are much more amenable to testing, and thus might in the long run be more fruitful.

The Social Adjustment of Migrants

The problems people have in adjusting to a new community environment are of considerable importance in the study of population migration. It may be offered as a general proposition that when migration occurs, there will be some disruptive effects, both for the migrating individuals and their families and for the receiving community. Just how disruptive a migration experience is depends not only on the selectivity of the migrants but also on the acceptance of the receiving community.

Let us consider an example. Suppose two individuals of the same age, sex, educational attainment, and occupational status emigrate to the United States. Given their similarities, should not, the two individuals adjust to their new environment at the same rate? But suppose that only one of them can speak the English language. Obviously, the one who does not speak English will have a more difficult time adjusting to life in the United States. Over time most immigrants learn English. The immigrant groups that are most likely to retain their native language are those who are highly educated professionals and entrepreneurs, such as Asian immigrants. It is not that they do not learn English, but that they tend to be fluent bilinguals (Portes & Rumbaut, 1990). The general perception that immigrants from the lower socioeconomic classes retain the use of their native language is probably the result of the influx of new immigrants, not those who have resided in the United States for more than one generation. Cultural factors, such as language, play an important part in the adjustment of migrants.

Other research on social adjustment has focused on the relationship between mental illness, stress, and migration. Portes and Rumbaut (1990) found that refugees to the United States experience an initial period of happiness, followed by a period of stress and depression, and then settle into a pattern of well-being. Among other types of immigrants, the pattern of psychological adjustment seems to become more positive as length of residence increases.

Problems of adjusting to cultural differences may be offset by other factors. You will recall that we earlier discussed how people often move in a pattern called chain migration. When migrants have a support group made up of family members and friends in their new areas of residence another dimension is added to the adjustment process. For example, one study (Moskos, 1980) shows that many Greeks who came to the United States arrived with the idea that they would return home. But because they found so many Greek establishments and associations in their new home, many Greek immigrants made a quick transition to the American way of life. These Greek institutions served as such an effective bridge between the two cultures that many Greeks who intended to return to their homeland never did so.

Another factor that may affect the social adjustment of migrants is prior residence in or prior contact with the area of destination. Research has shown that second-time migrants are more likely to be satisfied with their residence than are first-time migrants. In addition, individuals who have had even a minimal amount of contact with an area are more likely to have realistic expectations and adjust more easily to the new environment than are first-time migrants without such contact (Dailey & Campbell, 1980).

Adjustment and the Family

The family or kin group appears to play an important role in the adjustment of migrants to their new environments. Studies of migrants coming from the Kentucky Appalachian area showed that having family members already living in the urban destination reduced some of the problems of moving (e.g., Watkins, 1990). However, the degree to which family or kin-group members aid in the adjustment appears to be affected by the status of the migrant group. If the migrants have a relatively low social status, then the presence of kin or family in the new community takes on special significance. This was found to be the case with a number of migrant groups who arrived at their new communities as the poorest and lowest status members. Examples are the European immigrants in the 19th and early 20th centuries, rural migrants from Appalachia, southern rural blacks, Americans of Mexican descent, and Puerto Ricans. On the other hand, when migrants are not members of a lower status ethnic or racial group, the presence of family members has much less influence upon adjustment.

One illustration of this point is found in a study of migrants to Utah. Stinner and Toney (1980) found that when the sociocultural background of migrants is different from that of the local residents, the migrants tended to be dissatisfied with their new community. However, among the migrants, those who had high levels of education tended to be more satisfied with the community and with interpersonal relationships than were the other migrants. They were much more likely than migrants with low educational attainment to participate in community activities. This tendency suggests that adjustment to a new place is considerably influenced by the status position of the new migrant. The higher this status, the easier and quicker the adjustment will be.

The overall effect of migration on individuals have been well documented. Whether immigrants or migrants, the longer the period of residence, the stronger one's ties become to the community.

SUMMARY

Migration occurs when individuals, families, or groups make a permanent change in their place of residence. The definition of migration and the application of migration rates must specify geographic or political boundaries, length of residence, and many other factors.

Many theories have been presented to explain the causes of migration, beginning with those of Ravenstein in the 1880s. Increasingly, in most societies, migration is the result of decision making by individuals in the nuclear family. The decision-making model takes into account the availability, motives, expectancy, and incentives for migration. The decision to move or not move can be influenced by many factors, but economic and family considerations seem to be the most important.

The effects of economic and family factors can be studied at three separate levels of analysis: societal, personal status characteristics, and social psychological. Under certain societal economic conditions, migration will increase, while under others it will decline. Career and occupational considerations may keep people from migrating or force them to move. Family life-cycle stages are strongly related to migration, with children in the home being inhibitors of migration. At the social psychological level, economic aspirations and attitudes toward family members and extended kin all enter into and affect migration decisions.

Of equal importance are the effects of migration on the social systems the migrants leave and enter and on the migrants themselves. Families, neighborhoods, communities, and societies are all influenced by the movement of people.

7

MORTALITY

E very living organism will eventually die, and humans are no exception. Humans *are* exceptional in that they, apparently in contrast to all other organisms, are aware of the fact that they will die. Our awareness of death raises many questions about the subject of mortality, some of which demographers can answer, others of which they cannot.

Demographers are best at answering certain factual questions about death, such as what the major causes of death are and how they have changed over time. Demographers can also answer questions about how the causes of death differ from one part of the population to another (e.g., what variations can be found at different educational, occupational, or socioeconomic status levels). Demographers can also answer questions about how long an average person can expect to live, and how life expectancy varies for different parts of the population (e.g., the life expectancy of females vs. males, or of blacks vs. whites). A much more difficult question, and one that engages biologists and health professionals as well as demographers, is how much can life expectancy be extended?

By starting with this question, we can explore some important demographic questions about mortality. We will begin by defining two terms that sound very similar and that in everyday language are often used interchangeably, but which demographers use in very distinct ways. The two terms are *life span* and *life expectancy*.

LIFE SPAN AND LIFE EXPECTANCY

Life span is the biological maximum number of years human beings are capable of living. If we ask what the life span of humans is, we are really asking how many years has the longest living person lived? While various claims have been made for the oldest recorded living person, the authenticated record for a human is 120 years (Waldrop, 1992). According to United States Medicare records, the oldest recorded deaths were at 124 years, but these cases may well be based on erroneous date-of-birth claims (Coale,

1991b). The countries of France, Italy, Netherlands, Sweden, Switzerland, and Japan, which have very high-quality population data—often from population registers—have much lower records. Sweden, for example, carefully authenticates the ages of every person who is claimed to be over 100 at the time of death. Since Swedish data are very reliable, it is noteworthy that the oldest living Swede was a woman who lived 113 years, 214 days (Garson, 1991).

The most extreme claims for longevity have come from the Caucasus region of the former Soviet Union (Garson, 1991). After the 1970 census of the Soviet Union, Soviet officials claimed that the oldest living resident of the country was 165 years old, with many others over 120. An entire group of people, the Abkhasians, who live in the Caucusus Mountains, are widely known for their alleged longevity. Many Abkhasians, it is claimed, live healthy and vigorous lives until well over 100 years old. (Some readers may remember the Abkhasians from a television commercial created by a name-brand yogurt; the yogurt diet allegedly contributed to Abkhasian longevity.) While there may be some truth to the claim that these are especially long-lived people (due to biology, diet, hard work, or whatever), it is also true that the birth dates of many very old people in this region cannot be authenticated. Further, since longevity in the Caucusus region is culturally valued, there is a high likelihood that many people of the Caucusus exaggerate their ages to add to their social status. Claims regarding extremely old people (over 120 years) or populations with large numbers of very old people must be viewed with skepticism (Garson, 1991). As a general rule, when the quality of demographic data improves, the number of alleged very old people declines.

The issue of how long human beings are biologically capable of living is intriguing. There will no doubt be recurring claims for the new world records for the person who has lived the longest. But from a demographic perspective, the life span of humans is not as important as their life expectancy.

Life expectancy, as we noted in Chapter 1, is the average number of years newborns can be expected to live under the death rates that prevail at different ages (age 1, age 2, age 3, and etc.). This concept focuses our attention on how long the people of an entire population, or some significant subset of those people (e.g., females and males), are likely to live. The concept of life expectancy directs our attention to *populations*, not just to individuals, and serves as a measure of how successful the people in that population are at avoiding death.

The method used to calculate life expectancy is dealt with in more detail in Appendix B, the Abridged Life Table. In Appendix B we will describe the data used in life tables and how life expectancy is calculated. Even without such knowledge, however, most people intuitively understand that the concept of life expectancy reflects prevailing death rates in a population. When we hear that one society has a life expectancy of 55 years and another has a life expectancy of 75 years, it is obvious that death tends to strike at earlier ages in the former.

What Are the Limits of Life Expectancy?

There is a question relating to life expectancy that is parallel to the question asked earlier about life span: What are the limits of life expectancy for whole populations? (Baringa, 1991; Manton, Stallard, & Tolley, 1991; Olshansky & Carnes, 1994). At present, in the United States, the life expectancy of newborns is approximately 76 years. (But, as is well known, the life expectancy for males is only 72, while the life expectancy for females is nearly 79.) How high can the life expectancy of the entire population go? To 85? To 95? To 100? Or higher? (See Waldrop, 1992.)

There are three perspectives on how high the life expectancy of a population can go (Manton et al., 1991). There are "traditionalists" who believe that life expectancy cannot go much higher than it is at present (Fries, 1980; Olshansky, 1992; Olshansky, Carnes, & Cassel, 1990). The traditionalists base their conclusion largely on the idea of biological senescence, by which they mean that the human body simply wears out, or deteriorates, after a certain age. Senescence, they argue, occurs because as the human species evolved there was no particular biological advantage gained by keeping humans alive after their reproductive period passed (Carnes & Olshansky, 1993; Olshansky & Carnes, 1994). These scholars argue that although the human species is not exactly "genetically programmed" to expire at a certain age, still the "basic biology of our species . . . places inherent limits on human longevity" (Olshansky & Carnes, 1994, p. 76). The traditionalists do not see life expectancy going much beyond 85, because, they claim, the human body generally wears out at about that age. Advances in medical care and healthful living conditions may keep more people alive until about age 85, but thereafter death will soon follow for most people.

A second view of the limits of life expectancy has been called the "visionary" perspective (Manton et al., 1991). While the visionaries also accept the importance of biological senescence, they envision advances in biomedical research that will slow biological deterioration and give people many more years of life. For example, osteoporosis, a deterioration and weakening of bones, can perhaps be forestalled by dietary supplements or treated with new therapies. Growth hormones, which diminish normally with old age, can be injected into older people and perhaps slow down the aging process. Visionaries put their faith in new preventive and innovative therapies for averting and treating the degenerative diseases of old age. The visionaries often foresee a life expectancy of over 100 years; some speculate that life expectancy could reach 125 years or more (Manton et al., 1991).

The third view of the limits of life expectancy has been called the "empiricist" perspective. The empiricists look at recent declines in mortality, which have been about 2% per year, and project them into the future. This view also leads to the conclusion that life expectancy for entire populations could be raised considerably over the present level—perhaps eventually to age 100 and over (Manton et al., 1991).

Researchers have explored the limits of life expectancy by finding groups in the United States that have exceptionally favorable living environments, access to good medical care, and healthy lifestyles (Manton et al., 1991). In some small populations with these characteristics, life expectancy is as much as 98 years. These actual life expectancies for select populations suggest that in advanced countries like the United States life expectancy could approach 100 years (Manton et al., 1991; Waldrop, 1992).

At present, the scientific debate about the limits of life expectancy is provocative, but far from resolved. It is not easy to choose one view over the other, but the next two or three decades will go a long way toward answering the question of how long humans can ultimately expect to live.

Another way to look at the biological aspects of mortality is to see how biological factors are associated with variations in death or dying.

BIOLOGICAL INFLUENCES ON MORTALITY

Death is an eminently biological fact. In an ultimate sense, every death has a biological cause, but our interest here is in those instances where the biological characteristics of certain categories of people are associated with their death rates or longevity. There may be purely biological factors (e.g., genetic or physiological) that give some individuals or groups of people a better chance of living longer. Very often, as we will see, it is difficult to sort out the relative importance of purely biological factors and other influences, especially social factors. In this section we will consider the possible significance of biological ancestry, gender, and, perhaps surprisingly, left-handedness.

Ancestry and Longevity

It is part of the folklore of some families that they live particularly long lives. Or, in the opposite case, some families believe their members die young. Are these just fanciful family myths, or is there any scientific justification for believing that some families have greater or lesser longevity?

The principal scientific evidence on this issue comes from research on long-lived families rather than on families that are reputed to have short lives. One early study focused on individuals who had lived into their 90s—nonagenarians (Pearl & Pearl, 1934). The Pearls found that the immediate ancestors (parents and grandparents) of the nonagenarians had lived longer, on average, than the ancestors of a sample drawn from the family case history files at Johns Hopkins University. However, this research was flawed because the cases coming from the Johns Hopkins's files were from families with a high incidence of tuberculosis (Hawkins, Murphy, & Abbey, 1965).

In an alternative analysis, the longevity of *only the spouses and the off-spring* of the original nonagenarians in the Pearls's research were studied (Abbott et al., 1974, 1978). The researchers compared the ages of death of the *spouses* of the nonagenarians with the ages of death of their children. When the parents lived until age 81 or older, their children were likely to live 5 to 7 years longer than the children of parents who had died at age 60 or younger. The researchers concluded that a genetic component produced this difference, since the relationship held up regardless of whether there were similarities or differences in the environmental factors experienced by the parents and their children (Abbott et al., 1978). This line of research does seem to support the conclusion that people who have close kin, especially parents, who live to an old age will themselves have a better chance of living longer. This apparent genetic influence probably cannot override life-style choices, but in general it may contribute to somewhat greater longevity.

Gender and Longevity

It is well known that in most populations women have a greater life expectancy than men. This demographic fact certainly suggests that females have a biological advantage over males. However, since many societies have gender roles that might contribute to the higher mortality of males (e.g., more dangerous activities and occupations), the differences may be produced by social factors rather than biological ones.

Several pieces of evidence lend support to the biological argument. We can be sure that social factors will not generally influence fetal deaths since the gender of the fetus is not usually known (though with amniocentesis and ultrasound tests that is now changing). It is therefore significant that research persistently shows that more males than females die during gestation. McMillen (1979) examined the data on fetal deaths in the United States for the years 1922–1936 and again for the years 1950–1972 and found that male deaths exceeded female deaths by 6–10%. These findings are supported by the research of other investigators who have also found higher rates of miscarriage and stillbirths by male fetuses (Hassold, Quillen, & Yamane, 1983; Waldron, 1983).

Similar evidence can be deduced from infant mortality, and especially mortality in the first 28 days of life, which is called *neonatal mortality*. Mortality during the early weeks of life is almost always related to biological problems. The neonatal mortality rate for males is higher than for females; in 1991, the neonatal mortality rate for males was 6.2 per 1,000, but for females was only 5.0 (National Center for Health Statistics, 1993, Table E,

p. 11). Mortality in the rest of the first year of life (called *postneonatal mortality*) continues to be higher for males than females. In 1991, the postneonatal mortality rate for males was 3.8 per 1,000, while for females it was 2.9 (National Center for Health Statistics, 1993).

The higher death rates of male fetuses and of males during the first month and year of life do suggest that females are more viable than males. Females seem to gain a slight mortality advantage over males from their biological makeup, but, as we will see below, in many societies they may lose that advantage because they are socially and culturally less valued than males.

Handedness and Longevity

Newspapers and newsmagazines in 1991 reported the startling results of a study that seemed to have found left-handed people living 9 years less than right-handed people (Gladwell, 1991; Halpern & Coren, 1991). The researchers who conducted this study were candidly surprised by the results. They had previously studied 2,271 major league baseball players who were deceased and found that right-handers had lived an average of about 9 *months* longer than left-handers (Gladwell, 1991; Halpern & Coren, 1988).

The more recent research, which showed a 9-year difference in length of life between right-handers and left-handers, used data derived from 1,000 randomly selected death certificates from two California counties. The researchers contacted the next of kin of the deceased and determined whether the person had been right- or left-handed. The mean age at death for the right-handers was 75 years and the mean age at death for the left-handers was 66 years.

The methodology of this research has been criticized by a number of scholars ("Correspondence," 1991). One critical point is that a larger proportion of elderly are right-handed because earlier in this century many left-handed people were forced by parents and schools to become right-handed. As a consequence, the combined natural and forced right-handed population is older on average than the left-handed population. Thus, when deaths occur in the older right-handed population, the average age at death will be greater than the average age at death among the (somewhat younger) left-handed population. One skeptical critic of the Halpern and Coren (1991) study has observed that if there is a 9-year difference in the life expectancies of left- and right-handers, it is incredible that this great difference had never before been noticed (Gladwell, 1991).

Even if the 9-year difference between right- and left-handers is not realistic, it is interesting that left-handed people are much more likely to die in accidents (7.9% vs. 1.5%) (Halpern & Coren, 1988). This highlights the point that left-handed people live in a world created for right-handed people, and thereby are placed in greater danger.

Left-handed people may also have certain disadvantages that are more clearly biological in origin. Studies have shown that left-handedness is related to birth stress. A higher percentage of premature and low-birth-weight babies are left-handed. Left-handedness is sometimes viewed as "marker for a disruption in the typical neurological development of infants" (Gladwell, 1991, p. A18).

Left-handedness may be one of those physiological or even genetically inherited characteristics that influences the probability of dying. It is highly unlikely that the influence of handedness produces *years* of difference in life expectancy, but it may have some limited impact on mortality.

AMERICAN MORTALITY IN THE LAST HALF CENTURY

Mortality patterns in the United States during the last 50 years can be divided into three periods (Crimmins, 1981). Beginning in 1940, and lasting until nearly the mid-50s (1954), the first period was one of rapid mortality decline.[1] In the next period, between 1954 and 1968, the death rate leveled off. Indeed, for young to middle-age males (between ages 15 and 44) the death rate went up slightly during the middle period (Crimmins, 1981). Since 1968, the third period, the death rate has declined gradually (National Center for Health Statistics, 1993). The three periods are labeled in Figure 7.1, below the age-adjusted death rates between 1940 and 1990.

The two periods of mortality decline during the last 50 years were produced by changes in the major causes of death. The causes of death can be roughly grouped into four different types: *cardiovascular diseases* (diseases of the heart, cerebrovascular diseases, and arteriosclerosis), *cancer* (malignant neoplasms), *violence* (accidents, homicide, and suicide), and *infectious diseases* (these are primarily viral, bacterial, and parasitic diseases, with the two most prominent being influenza and pneumonia). During the period between 1940 and 1954, over 80% of the decline in the death rate was produced by declines in deaths due to infectious diseases. This decline in infectious disease deaths was, in fact, a continuation of declines that started at the beginning of the century. The leveling off of the death rate in 1954 can be viewed as a significant triumph in the long battle against infectious diseases in the United States.

Success in the fight against infectious diseases does not mean that bacterial and viral diseases no longer cause death today. Tuberculosis, which was a major killer through the first decades of this century and then was controlled by antibiotics, has recently been on the rise again (Brown, 1992b; Specter, 1992). In 1991 there were 1,713 deaths caused by tuberculosis in the United States (National Center for Health Statistics, 1993). The rise in

---

[1]Mortality in the United States had been declining for most of the 20th century, except for several years between 1916 and 1920 when an influenza epidemic struck the nation.

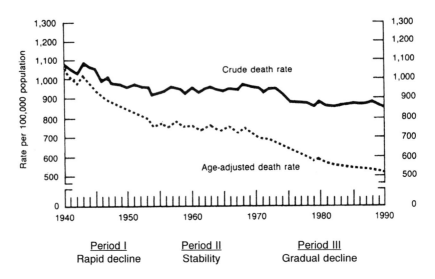

FIGURE 7.1. Three mortality periods and age-adjusted death rates for the United States, 1940–1990. From National Center for Health Statistics (1993, p. 3).

tuberculosis is linked, in part, to the emergence of AIDS in the United States, since those who have AIDS are more likely to become infected. AIDS itself is an infectious disease, since it is transmitted by a virus. We will discuss AIDS later in this chapter.

The decline of the death rate in the United States since 1968 has been produced primarily by reductions in the various cardiovascular diseases. Diseases of the heart, for example, have decreased from an age-adjusted rate of 286.2 per 100,000 population in 1960 to 166.3 per 100,000 in 1988, a 41.8% decrease. The age-adjusted death rate for cerebrovascular diseases declined from a 79.7 per 100,000 rate in 1960 to a 29.7 per 100,000 rate in 1988, a 62.7% decrease. Since these two causes of death still account for over 40% of all U.S. deaths, these rate declines had a considerable impact on the overall death rate (National Center for Health Statistics, 1991).

INFANT MORTALITY:
U.S. AND INTERNATIONAL COMPARISONS

The human infant is a totally dependent and highly vulnerable organism at birth. Unless someone cares for the infant for an extended period of time, its chances of survival are nil. Even with care, humans during their first year of life have a greater risk of dying than at any other age except the very advanced years. When living conditions are harsh (e.g., due to poverty, food shortages, or social disorganization), the mortality of children, especially during the first year of life, is likely to be especially high. In the words of one analyst, "No cold statistic expresses more eloquently the differences between a society of sufficiency and a society of deprivation than the infant mortality rate" (Newland, 1981, p. 5).

Infant mortality rates do reflect the economic and social well-being of societies, but there are many exceptions to this rule. The most economically developed countries do have the lowest infant mortality rates. Among the major populated continents, North America (Canada and the United States) and Europe have the lowest rates. North America has eight deaths per year per 1,000 live births, while Europe has 11 infant deaths per 1,000 births. In all of Asia the infant mortality rate is 63 per 1,000; in Latin American countries (South and Central America, plus the Caribbean) the rate is approximately 50 per 1,000; and on the African continent the rate is 92 per 1,000 (Population Reference Bureau, 1994). In a rough way, these different infant mortality rates reflect the level of economic development on these continents.

At the level of individual countries, especially among the most economically developed countries of Europe, North America, and the Pacific Rim (Australia, New Zealand, and Japan), the relationship between infant mortality and societal wealth is not as consistent (Feinleib, 1992). The United States, itself, is a notable exception, since it has a highly developed economic system and the greatest amount of total wealth in the world, and yet it still has a higher infant mortality rate than many other countries.

Table 7.1 ranks the major countries of the world with infant mortality rates lower than the United States, along with their 1992 gross national product (GNP) per capita (in U.S. dollars). On the basis of this measure of economic well-being, there are six countries with higher per person wealth than the United States (Switzerland, Japan, Sweden, Norway, Denmark, and Iceland), but the remaining 16 countries have lower levels of per capita wealth. Yet, all 22 of these countries have lower infant mortality rates than the United States. Greece, which has an infant mortality rate that is just slightly lower than that of the United States (8.2 vs. 8.3), has a per capita income that is less than one-third that of the United States ($7,180 vs. $23,120).

It is also noteworthy that the United States has higher per person health expenditures than other major industrialized countries, and yet has a higher infant mortality rate (Feinleib, 1992). The United States also spends a higher percentage of its gross domestic product on health care and yet its infant mortality rate is higher (Feinleib, 1992).

Some demographers have argued that too much is made of the infant mortality rankings of the developed countries (Haub & Yanagishita, 1991; Shryock, 1991). The differences among the countries ranked in the top 20 may be relatively insignificant. For example, in Japan about 996 babies out of every 1,000 live through the first year, while in the United States about

TABLE 7.1. Infant Mortality Rates and Per Capita Gross National Product (in U.S. 1991 Dollars) for Major Countries with Lower Infant Mortality Rates than the United States

	Infant mortality rate	Gross national product
1. Iceland	3.9	23,670
2. *Finland*	4.4	22,980
3. Japan	4.4	28,220
4. Sweden	4.8	26,780
5. *Singapore*	5.0	15,750
6. Taiwan	5.7	Not available
7. Norway	5.8	25,800
8. *Germany*	5.9	23,030
9. Switzerland	6.2	36,230
10. *Netherlands*	6.3	20,590
11. *Hong Kong*	6.4	15,380
12. *Australia*	6.6	17,070
13. Denmark	6.6	25,930
14. *Slovenia*	6.6	6,330
15. *United Kingdom*	6.6	17,760
16. Austria	6.7	22,110
17. *France*	6.7	22,300
18. *Canada*	6.8	20,320
19. *New Zealand*	7.3	12,060
20. *Spain*	7.9	14,020
21. *Israel*	8.1	13,230
22. *Greece*	8.2	7,180
23. United States	8.3	23,120

Note. From Population Reference Bureau (1994). Countries with lower per capita gross national product are shown in italic.

992 babies out of 1,000 live through the first year. From this perspective the difference in the infant mortality of Japan (ranked number 1) and that of the United States (ranked number 23) is relatively minor.

Another factor to be considered is the comparability (and validity) of infant death statistics, as collected and reported by various countries. In France, for example, many infant deaths have customarily been classified as stillbirths if the death occurred before the birth was registered. The former Soviet Union did not count certain infant deaths if they occurred during the first week of life. Japan's statistics on infant deaths are suspect, because the ratio of stillbirths to infant deaths is very high. Traditional Japanese view an infant death as a family stigma, and thus there may be pressure to misclassify some infant deaths as stillbirths (Haub & Yanagishita, 1991). Some of these cultural and social practices could have the effect of lowering the measurement of infant mortality of specific countries.

One partial explanation for the relatively high infant mortality rates of the United States comes from the fact that minority populations have much higher infant death rates. African Americans in particular have persistently higher infant mortality rates than white Americans. Over the last 40 years the African American infant mortality rate has usually been double that of the white population. In 1991, the rate for white infants was about 7.3 per 1,000, while the rate for black infants was 17.6 per 1,000 (National Center for Health Statistics, 1993).

A study of infant mortality rates for the Hispanic and white populations in Texas has shown that Spanish-surname populations have had higher infant death rates for over 50 years. There has been a convergence of the rates for Hispanics and whites in recent years, but a difference still persists (Forbes & Frisbie, 1991).

There are also differences within the Hispanic population, at least in Florida, where Puerto Ricans and Mexicans have higher infant mortality rates than Cubans and other Hispanics (Hummer, Eberstein, & Nam, 1992). The generally lower socioeconomic status of Puerto Ricans and Mexicans, compared to Cubans, reveals again how lower status groups have higher mortality.

SOCIAL FACTORS AND MORTALITY

While some biological factors can influence mortality, there is ample evidence that social factors are even more influential in contemporary societies. Social factors can be divided into two types: *position in a social structure* and *lifestyle*. These two are often closely related, but it is fruitful to treat them separately. We will begin by introducing each type and then offering an illustrative example.

Position in the Social Structure

All societies have hierarchically ranked social structures. This means that some groups or categories of people have positions that are subordinate to some other groups or categories. The lower status groups are always disadvantaged in some way. The disadvantages will typically include less power, fewer economic resources, or fewer privileges. These disadvantages seem to translate, inevitably, into diminished chances of living. It may be stated as a principle that *whenever groups or categories of people have lower status in a society they are very likely to have higher death rates and shorter life expectancies than would be expected.*

We may illustrate the effect of socioeconomic status by reference to African Americans in the United States population. There is a preponderance of evidence, both historical and contemporaneous, that the average economic status of black Americans in the United States is lower than that of white Americans (U.S. Bureau of the Census, 1992d). The historical conditions of African Americans started with a long period of slavery, followed by nearly 100 years of legally enforced segregation and discrimination. Despite the passage of voting rights and civil rights laws, attitudinal and institutional discrimination against black Americans persists and reduces their

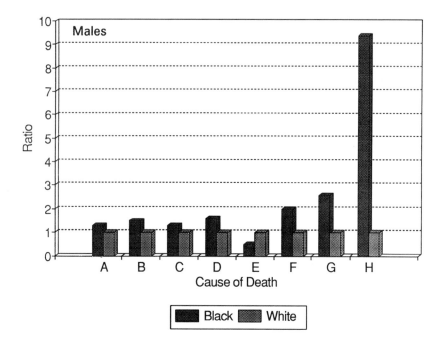

FIGURE 7.2a. The ratio of the black death rate to the white death rate, for eight major causes of death, for males: A, circulatory disease; B, cancers; C, respiratory deseases; D, accidents; E, suicides; F, infectious diseases; G, diabetes; H, homicides. From Rogers (1992, p. 289). Copyright 1992 by the Population Association of America. Adapted by permission.

opportunities in education, employment, economic opportunity, housing, medical care, and most other facets of American life.

The resulting lower status of African Americans leads to higher death rates and a shorter life expectancy for this population. At the beginning of the century the difference in life expectancy between blacks and whites was nearly 15 years (Rogers, 1992). The difference in life expectancy between the races has diminished, but in 1990 it remained at 7 years[2] (National Center for Health Statistics, 1993, Table 4, p. 17). The shorter life expectancy of black Americans is a result of their higher death rates from almost all causes of death. Among the eight major causes of death in the United States, blacks have higher death rates on all but suicide (Rogers, 1992).

Figure 7.2 shows the eight major causes of death and the ratio of the black death rate to the white death rate for each cause. The ratios are shown separately for males (a) and females (b).

[2]The black–white difference in life expectancy was less than 6 years in the early part of the 1980s (5.7 years in 1982), but in the late 1980s the gap increased again.

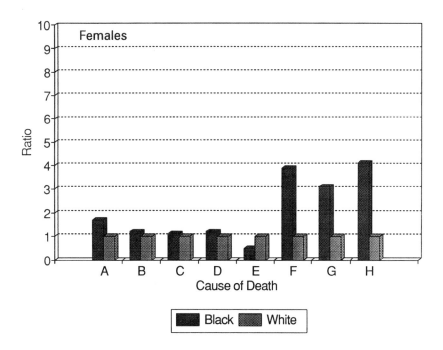

FIGURE 7.2b. The ratio of the black death rate to the white death rate, for eight major causes of death, for females: A, circulatory disease; B, cancers; C, respiratory diseases; D, accidents; E, suicides; F, infectious diseases; G, diabetes; H, homicides. From Rogers (1992, p. 289). Copyright 1992 by the Population Association of America. Adapted by permission.

When all causes of death are combined, the death rate for black Americans is 50% higher than it is for whites—the ratio of black to white death rates is 1.5 to 1 (Rogers, 1992). The two major causes of death in the United States, accounting for nearly two thirds of all deaths, are circulatory diseases and cancer. Blacks have a 30% higher death rate from circulatory diseases and a 50% higher death rate from cancer. These two causes of death account for most of the differences in death between the black and white populations (Rogers, 1992).

The most extreme difference in death rates is found with regard to homicide, for which the black male rate is 9.4 times higher than the white male rate. The homicide death rate for black females is 4.1 times higher than that for white females. Black females, when compared to white females, also have much higher death rates from infectious diseases (3.9 times higher) and diabetes (3.1 times higher).

The causes of black–white differences in mortality are varied, and the issue can be approached at different levels. The higher death rates for circulatory diseases, for example, may be attributable to biological differences, such as a greater tendency toward hypertension among African Americans. But the fact that many black Americans have fewer economic resources and poorer access to good medical care are also contributors to black–white mortality differences. The position of black Americans in the social structure of the United States is clearly implicated in their consistently higher death rates.

One researcher who has studied the differences in black and white mortality has summarized the combined influence of economic status and race succinctly: "The poor suffer higher mortality than the affluent, and among the poor, blacks suffer higher mortality than whites" (Rogers, 1992, p. 299). Later in this chapter we will explore the relationship between social class and mortality more fully.

Life-Style

"Life-style" refers to the choices people make in the conduct of their daily lives. This includes how they eat, drink, and exercise, whether they engage in risky behaviors, and whether they use harmful substances such as tobacco or drugs. Life-style is not a precise scientific concept, but even as a loose concept it has great significance in the contemporary study of mortality. We now recognize that mortality is significantly influenced by the way people choose to live. Life-style is especially important because we are more likely today than in the past to die from degenerative diseases than from infectious diseases.

The significance of life-style is vividly illustrated by a series of studies that have been conducted over the last 30 years on two U.S. religious groups: Mormons and Seventh-Day Adventists. These two religious groups were selected for study because both condemn the use of tobacco and alcohol, and also forbid their members from drinking other potentially harmful substances, such as coffee, tea, and other caffeinated beverages (Enstrom, 1989; Kahn et al., 1984; Lindsted, Tonstad, & Kuzma, 1991). Seventh-Day Adventists are also urged, but not required, to avoid eating meat and highly spiced food (Snowdon, 1988). An equally important characteristic of these two religious groups is that they are tight-knit organizations and their adherents are likely to abide by the official prohibitions. Data from a study of 40,000 Seventh-Day Adventists showed that 99% were nonsmokers and 90% were nondrinkers (Phillips et al., 1980).

In one 8-year study of cancer death rates, Seventh-Day Adventists in California were compared with the general California population. For those cancers that are strongly related to smoking (cancer of the lungs, mouth, pharynx, and esophagus), the Seventh-Day Adventists had a risk of death that was only 20–30% of the general population of similar ages and genders (Phillips et al., 1980). Even in the case of cancers that are less likely to be smoking related, the risk of death among Seventh-Day Adventists was only 65–75% of that for the general population.

Exercise, another life-style characteristic, has also been studied among a sample of 9,484 Seventh-Day Adventist males (Lindsted et al., 1991). While Seventh-Day Adventists generally have similar (and apparently healthy) diets, and low levels of smoking and drinking, they do vary in their level of physical exercise. In a 26-year follow-up study, researchers found that those males who said in 1960 (the time of the first questionnaire) that they got *moderate* or *heavy* exercise had lower death rates over the ensuing years than those men who said they got *no* or *slight* exercise (Lindsted et al., 1991). The effect of exercise was most pronounced in decreasing deaths due to cardiovascular disease (Lindsted et al., 1991).

Studies of Mormons have similarly shown that life-style reduces mortality (Enstrom, 1978, 1989). A comparison between Mormon males who were active in their church and U.S. males in general showed that Mormon men aged 25–64 had death rates only 35% as high as the death rates for U.S. males of the same age (Enstrom, 1978).

In 1979, Enstrom administered a life-style questionnaire to 5,231 California Mormon high priests (religiously active males) and 4,613 of their wives. The same questionnaire was given to a sample of 3,119 white adults in Alameda County, California, which was used as a comparison group. The Mormon sample had no active smokers, and only one or two who used alcohol, coffee, or tea. Over an 8-year follow-up period, data were collected on the deaths of people in these samples. The mortality levels of the Mormon and Alameda samples were then compared to mortality in the white U.S. population. The measure of mortality for the general population was set at 100, and, as expected, the mortality of Mormons was substantially lower. For example, with the cancer mortality level of the general population set at 100, the Mormon male high priests had a cancer mortality level of 47. The Mormon males had a mortality level of 15 for smoking-related cancers and 71 for non-smoking-related cancers. For all causes of death combined, the Mormon males had a mortality level of 47, compared to 100 for the general male population (Enstrom, 1989).

Mormon females also had lower mortality rates than the general population, but they did not gain as much advantage from life-style as the Mormon males. In the Alameda sample, those people who had life-styles similar to Mormons were separated out and they too had much lower mortality rates than the general U.S. population, providing additional evidence on the importance of life-style (Enstrom, 1989).

The greater life expectancy and lower death rates of Mormons and Seventh-Day Adventists are almost certainly due in large part to their life-styles. Of course, it is also possible that members of these two religious groups enjoy a higher-than-average status in society, and thus the first social factor— position in the social structure—may also be giving them a better chance of living. As we move on to consider social factors related to mortality we will often find that it is a combination of life-styles and the advantages or disadvantages of position in society that influence mortality.

SOCIOECONOMIC STATUS AND MORTALITY

Almost all human societies have some inequalities in wealth, power, and prestige. In everyday terms, these inequalities create what are described as social classes. One of the characteristics of social classes is that the people who stand higher in the class structure have better and more "life chances" than those who are in lower statuses. *Life chances* are all the valued and desired things of the society, such as personal comfort, travel, education, and good health. Perhaps most importantly, the people in the higher classes get more of life itself because they live longer than the people in the lower classes.

There is an overwhelming body of evidence to support the claim that people in the higher social classes have lower mortality rates than the people in lower social classes. Antonovsky (1967) carried out a comprehensive review of the historical evidence supporting the relationship between social class and mortality and uncovered only minor exceptions to the general rule. Antonovsky examined and reviewed dozens of separate studies ranging from analyses of British royalty to studies of mortality differences by contemporary occupations. Despite the different research methods used in these various studies and the many different populations studied, Antonovsky was led to "the inescapable conclusion . . . that social class influences one's chances of staying alive (Antonovsky, 1967, p. 66).

Antonovsky's examination of the many studies of social class and mortality did, however, reveal that some differences occurred in the relationship over time. He found that when mortality rates were either very high or very low, the differences among social classes were not as great. It appears that when the overall death rate in a population is very high, all the people in the population are more nearly equal in the face of death. When, for example, infectious diseases were the major causes of death and when the causes of these diseases were little understood, the people in the higher social classes had fewer advantages over the lower classes. When the plague struck Europe, the upper classes might have been somewhat more resistant to the disease, because they had better diets and superior housing conditions, but because no one in that time understood the causes of the plague they too became its victims. The upper classes gained their greatest advantage over the lower classes when improvements in housing, water supply, refuse disposal, nutrition, sanitation, and hygiene were adopted. The higher classes, with their greater economic resources, could more easily afford these improvements in their living environments (Antonovsky, 1967, p. 39).

When the death rates of a population move to relatively low levels, mortality level differences between the social classes diminish somewhat. It seems that under low death rate conditions, nearly everyone is getting access to the improvements in housing, diet, and medical care that provides greater equality in life chances. But even when a society experiences relatively low death rates, some significant differences persist from the highest to the lowest social classes. We will see how much these differences are by examining the social class differences in mortality in the United States.

Social Class and Mortality in the United States

The United States is a country with a relatively low death rate. The crude death rate has been steadily low, at less than 10 per 1,000, for more than 40 years. In 1950 the crude death rate was 9.6; in 1993 it was 8.8 (National Center for Health Statistics, 1994). According to Antonovsky's (1967) analysis, the differences in mortality rates in the United States, from the lowest social classes to the highest, should be relatively small. The evidence is, however, that mortality rate differences between the social classes in the United States are clearly present today, just as they were in 1950.

One early series of U.S. studies was conducted in Chicago in 1950. In this study, the people who lived in the highest income areas lived about 7 years longer, on average, than the people living in the lowest income areas (Kitagawa & Hauser, 1964). A second major study of the relationship between mortality and social class was conducted in 1960, using the educational attainment characteristics of a sample of white Americans who died in that year (Kitagawa & Hauser, 1964, 1968, 1973). For females between the ages of 25 and 64 in this study, the least educated group had a mortality rate that was 16% higher than the most educated group (women with at least 1 year of college). Among the white males between ages 25 and 64, the mortality of the least educated group was 48% higher than that of the most educated group. At age 65 and over, the mortality of the least educated women was 59% higher than that for the most educated women. For males 65 and over, there were only small differences in mortality between the least educated and the most educated (Kitagawa & Hauser, 1964, 1968, 1973).

In the 1970s, Rosen and Taubman (1979) carried out a similar study in which a large random sample of the U.S. adult population was used as the primary data source (a 1973 Current Population Survey—see Chapter 4). By using Social Security numbers, the researchers were able to determine the social and economic characteristics of 3,000 of the people in the survey who had died between 1973 and 1977. The conclusions of this study were much the same as those in Katagawa and Hauser's (1968, 1973) study. The researchers, describing the patterns for white males, reported that "death rates tend to decrease with education and the least educated have a 40 to 60% higher death rate than the most educated" (Rosen & Taubman, 1979, p. 62).

More recent evidence for a continuing relationship between social class and mortality in the United States is found in the work of Navarro (1990). Using data the U.S. government collected on mortality rates for heart and cerebrovascular disease, Navarro found that these major causes of death are convincingly related to the major occupational categories. The occupational categories are, of course, a reflection of social classes in the United States.

Figure 7.3 shows that the mortality rate for heart disease is more than twice as high for the lowest occupational class (operators, fabricators, and laborers) than for the highest occupational class (managerial and professional occupations).

Despite the widely held impression that people in high-status occupations (managers and professionals) have stressful occupations and therefore might have higher rates of heart disease, people in lower status occupations have, by far, the highest death rates for heart disease. The high-status professional occupations may be stressful, but they do not lead to higher rates of heart disease deaths.

Studies of male doctors in both England and the United States have shown that their overall death rates are lower than men in general in their populations. In the English study, male doctors had a death rate that was 19% lower than that for English men generally. However, when compared to other high-status professionals and managers, male doctors did have a 5% higher death rate (Fox & Adelstein, 1978).

The study of U.S. doctors showed that male doctors had an overall mortality rate that was about only 75% as high as the death rate for males in general. Female doctors in the United States fared less well, but their mortality rate was still only 84% of the death rate of all U.S. females (Goodman, 1975).

The relationship between social class and mortality in the United States has been shown in all the studies that have been done over the last 40 years. This pattern has prevailed despite the fact that this has been a period of persistently low mortality.

It is noteworthy that the relationship between mortality and social class has been studied relatively infrequently in the United States (Navarro, 1990, 1991). The vital registration system of the United States does not systematically collect data on the relationship between mortality and the various indicators of social class, such as income, education, and occupation. The

FIGURE 7.3. Mortality rate for heart disease, by occupation, 1986. From Navarro (1990, p. 1238). Copyright 1990 by The Lancet, Ltd. Adapted by permission.

United States is unusual in this regard, being "the only western developed nation whose government does not collect mortality statistics by class" (Navarro, 1990, p. 1238).

Why Do Social Class Differences in Mortality Persist?

There are a number of reasons why the relationship persists between social class and mortality. By definition, lower social classes have fewer economic resources than higher social classes. Having more economic resources can lead to more healthful living conditions and can provide better and more extensive health care. Health care expenditures are what economists call an *elastic cost*, which means that the amount spent on it can be reduced or expanded, depending on how much money people have available (Rosen & Taubman, 1979). For the lower social classes, there may be very little money left for medical care after such necessities as food, clothing, and shelter have been purchased. The wealthy, and even the moderately well-to-do, can always find the money necessary for medical care if their health, and especially their lives, are at stake.

While occupation is often used as in indicator of social class, there are many occupations that influence mortality directly. Two features of occupations may be singled out as direct causes of mortality: stress and physical danger. There is little doubt that lower-status jobs have more physical danger associated with them. Job stress, as we have seen, is often presumed to be highly related to high-status jobs, but research is showing that certain lower-status jobs are just as stressful. We will consider the effects of job stress first, and then return to danger in the workplace.

Occupational Stress

We have already seen that high-status occupations—managers and professionals—do not have higher heart disease death rates. We have also noted that medical doctors have lower overall death rates than the general population.

It is important to note that job stress is found in many lower-status occupations, just as it exists in higher-status occupations and professions. Recent studies in the United States and Sweden have shown that psychological strains associated with occupations are produced by two separate dimensions of work: high workload demands and a low degree of decision making (Karasek et al., 1981, 1988). Jobs that have these two characteristics include work on assembly lines, work with cutting machines, and freight handling. Other jobs that combine heavy workloads with little opportunity for independent action are being a cook or waiter/waitress. Executive and professional jobs may have heavy workloads, but these jobs also have a high degree of latitude in decision making (Karasek et al., 1988, p. 911).

In a study done in the United States, over 4,800 employed males[3] were classified according to the characteristics of their jobs. Those men who had jobs with high work demands *and* a low level of decision making—about 20% of the working population—had the highest prevalence of heart attacks (myocardial infarctions). This result supports similar findings in the earlier study of Swedish male workers (Karasek et al., 1981, 1988).

The stress associated with having a job with a low level of independence and a high workload is also supported by a well-known, long-term study of coronary heart disease and occupations. This study, called the Framingham Heart Study, found one group of women workers to have an especially high risk of coronary heart disease. Women who had clerical occupations, husbands with blue-collar occupations, and children living at home were far more likely to have coronary heart disease than other working women and housewives (Haynes & Feinleib, 1980). These clerical workers often worked as secretaries for managers and executives, where their workloads were heavy and they had few opportunities for independent decision making. The researchers concluded that "the occupations of [these] women, coupled with family responsibilities, may be involved in the development of coronary heart disease" (Haynes & Feinleib, 1980, p. 140).

Danger in the Workplace

While stress may be found in occupations at all status levels, physical danger is more clearly concentrated in lower-status jobs. Blue-collar and manual occupations, such as construction, mining, lumbering, and oil drilling, are widely recognized as dangerous occupations. According to the Bureau of Labor Statistics, the industries with the highest incidence of injuries and illness (but not necessarily deaths) are shipbuilding and ship repairing, meatpacking, iron and steel production, and home building (Hackey, 1991). There is no detailed breakdown of which jobs within these industries are the most dangerous, but there can be little doubt that the manual workers in these industries are the most likely to die from the dangers of their work.

The exact number of work-related deaths each year in the United States is not precisely known. Or, to put the matter somewhat differently, there are widely varying estimates of the number of people who die each year as a result of work-related accidents. In recent years the estimated numbers have ranged from under 3,000 to over 10,000 (Hoskin, 1991; Personick & Jackson, 1992; Toscano & Windau, 1991). Three different federal agencies make estimates of work-related deaths. The Bureau of Labor Statistics conducts a yearly survey of industries that employ at least 11 employees. The Bureau estimated 3,600 deaths in 1989 and 2,900 deaths in 1990. The National

---

[3]The subjects came from two separate data sets, but the analyses, which were done separately, produced the same results.

Institute for Occupational Safety made an estimate of 6,400 deaths in 1985, the last year for which it reported data. The National Safety Council estimated that there were 10,100 work-related deaths in 1989 and 10,700 deaths in 1990.

The figure of around 3,000 work-related deaths per year by the Bureau of Labor Statistics is clearly an underestimate since industries or businesses that employ 10 or fewer workers are not included. The higher estimates of the other two agencies are probably closer to realistic estimates of the total number of work-related deaths each year in the United States.

In addition to workplace accidents that cause death, there are also workplace conditions that are very likely to produce a variety of diseases leading to early deaths. Coal miners and cotton mill workers are known to have a high incidence of respiratory diseases (black lung and brown lung), which lead to higher mortality rates and shorter life expectancy among these groups of workers. The long-term effects of working with asbestos materials, lead, solvents, and other chemicals are almost certainly deleterious to health and probably lead to earlier deaths in some number of cases (Fox & Adelstein, 1978; Smith, 1990; Squires, 1990). The exact number of deaths caused by workplace-related diseases is even more uncertain and speculative than work-related accidental deaths. Again, the National Safe Workplace Institute has made estimates ranging from 47,000 to 95,000 deaths per year (Squires, 1990). These numbers are based on the assumption that 5–10% of all cancer deaths are due to work exposure, as are 2–4% of all lung disease deaths, 1–3% of all kidney disease deaths, and so on (Squires, 1990). Obviously, these numbers are only educated guesses, but, if true, more people die from job-related diseases in the United States than die from motor vehicle accidents ("Occupational Health Neglected," 1991).

Gender Inequality and Female Mortality

A subordinate position in a society can produce higher mortality rates, and that is true whether the subordinate status is based on race, social class, or gender. When females are viewed as subordinate to males (or they are devalued relative to males) their death rates are likely to be higher (than expected) and their life expectancy reduced. Earlier in this chapter we saw that females seem to have a biological advantage over males; thus, in most societies, females have lower death rates and greater life expectancy. Their generally lower death rates are at least partially attributable to their biological advantage; in addition, females may also have lower death rates in many societies because they engage in less dangerous activities—especially during adolescence and young adulthood.

Yet while females generally enjoy a mortality advantage over males, there is mounting evidence that in a number of societies females have higher than expected death rates because of their lower social status and their lesser economic value, compared to males. In a few nations life expectancy for females is actually less than that for males. Data for 1993 show that Bangladesh, Bhutan, Maldives, and Nepal are countries with a greater life expectancy for males than females (Population Reference Bureau, 1993). A 1984 World Bank study found that male life expectancy exceeded female life expectancy in Bhutan, India, and Pakistan.

These countries, with their atypical lower life expectancies for females, provide only a precursory indication of the mortality disadvantage of females in some parts of the world. On the basis of much additional evidence, one writer has made the dramatic claim that there may be as many as 100 million females missing from the world's population (Sen, 1990). They are "missing" in the sense that they should be alive, and part of the population, if their deaths had not been prematurely caused. A more cautious estimate sets the number of excess female deaths at 60 million (Coale, 1991a). In either case the numbers are so immense that the issue deserves more careful consideration.

60,000,000 or More Missing Females

Two kinds of additional evidence support the claim that tens of millions of females are missing because of higher-than-expected female death rates: (1) case studies of particular populations in which there is evidence that females, especially as infants and children, are allowed to die through neglect, inadequate nutrition, or poor medical care; and (2) demographic analyses of sex ratios, which provide numerical evidence about the number of females missing from particular populations.

The case studies of particular populations are the most dramatic and memorable, because they provide illustrations of how females die when they are devalued and unwanted by their families. In many traditional patriarchal societies male children are wanted by parents because sons through their labor on the family farm or in the family business will eventually contribute to the family's economic well-being. Females, by comparison, will typically grow up and marry into other families, where they will contribute their labor to a different household. Further, in many societies a family must provide a valuable dowry for its daughters when they marry. Thus daughters in such societies contribute little to their parents' economic well-being and, in addition, can be costly. This economic reality may, in extreme cases, lead to female infanticide.

In 19th-century India, even though infanticide had been outlawed by the British colonial authorities in 1870, many Indian families believed they had a family duty to allow excess daughters to die. The following dialogue between a British official and an Indian landowner was reported, with the landowner saying:

It is the general belief among us, Sir, that those who preserve their daughters never prosper . . .

Then you think that it is a duty imposed upon you from above, to destroy your infant daughters . . . ?

We think it must be so, Sir. . . . (Miller, 1981)

Even though infanticide is illegal now, just as it was then, there is evidence that female infanticide persists today. One study found that in families where mothers expressed a preference for having no more children, there was a higher rate of female infant deaths than in other families (Simmons, Balk, & Faiz, 1982). In the Punjab state of India, the childhood death rate of females is twice as high as that of males (Das Gupta, 1987). Also in Punjab, females under 5 years who had older sisters had a death rate *twice* as high as that of males under 5. This suggests that parents who already have female children do not want any more, and allow some of those they have to die.

A recent study of 350 districts in India reveals that females aged 0 to 5 have higher death rates in those districts where their sociocultural importance and economic value are low. The highest mortality of female children occurs in those districts where females, when they marry, go to live in their husbands' communities *and* when their labor force participation is low. When female infants and children are seen by their parents as having little value, either socially or economically, the female death rate is higher (Kishor, 1993).

China also has apparent high rates of female infant mortality, but China is a more complicated case. First, Chinese parents often follow the tradition that views a male child as a necessary provider and protector for their old age. When this view is combined with the Chinese one-child population policy (see Chapter 9), there may be pressure to make their one child a male. Press reports coming out of China since the 1980s have indicated that female infanticide was occurring as parents sought to have their only child a male. According to the Chinese policy, if a first child dies, parents can appeal to have a second (Haupt, 1983).

One indication that infanticide is occurring in China can be deduced from "a stern flow of official condemnations of the practice in the Chinese mass media" (Tien et al., 1992, p. 15). If the Chinese authorities did not think it was a problem, there would probably be no warnings about it.

Further evidence for infanticide comes from Chinese fertility data since the beginning of China's one-child policy in the 1980s. Demographers analyzing Chinese fertility data in the mid- and later 1980s noticed an exceptionally high sex ratio among newborns. Normally the sex ratio at birth would be about 105 or 106 males per 100 females, but in many districts of China there were often more than 110 boys born for every 100 girls (Hull, 1990; Johansson & Nygren, 1991). This high sex ratio at birth could have been produced, at least partially, by the infanticide of some first-born females. By allowing a female child to die, or by deliberately killing it, and not reporting the birth to authorities, the parents could try again for a son. Of course, the high sex ratio at birth does not prove that infanticide occurred, but after looking at other evidence, some researchers concluded that infanticide might be causing about 4 deaths for every 1,000 live-born girls (Johansson & Nygren, 1991).

In studies conducted in rural Bangladesh, female mortality was lower than male mortality during the early months of life. This is consistent with the general female biological advantage, but this early period is also a time when infants are typically breast-feeding. During the later part of the first year of life, when Bangladesh children are given supplementary food, the female death rate moved above the male rate. This mortality pattern suggests that additional food was being given to males, but was withheld from females (D'Souza & Chen, 1980).

These studies and reports coming from India, China, and Bangladesh (as well as scattered reports from other countries) have given substance to the assertion that there may be at least 60,000,000 females missing from the world's population. Further evidence for this conclusion is provided by examining the population sex ratios in countries that have relatively low female life expectancies.

Recently, demographer Ansley Coale (1991a) has examined the sex ratios of the populations of a number of countries where the subordinate status of females may have produced their higher death rates. Coale started by noting that in societies where the genders are treated relatively equally, at least where females are given more or less the same health care and nutrition as males, the mortality rates for males are always higher, at every age. This difference in mortality leads to populations in which females outnumber males (low sex ratios). For example, in Europe, the United States, and Japan there are currently between 95 and 97 males for every 100 females. By contrast, the sex ratios in many Asian and a few African countries show the opposite, with males outnumbering females. The sex ratios for some selected countries are: China, 107; India, 108; Pakistan, 110; Bangladesh, 106; Sri Lanka, 104; and Egypt, 105. The entire region of West Asia (roughly the Middle East) has a sex ratio of 106 (Coale, 1991a, p. 519).

The high sex ratios in the countries listed above indicate that females are being lost to their populations through excess deaths. These excess deaths do not just occur in infancy and childhood; according to Coale, "Mortality rates are higher for females than for males in most age intervals from age one to age 50 or 60" (1991a, p. 519). By contrast, in European countries, since the middle of the 19th century, male mortality rates have always been higher than female rates at every age from birth to the highest age attained (Coale, 1991a, p. 518).

To get an estimate of the number of females missing due to excess mortality in these populations with high sex ratios, Coale compared their sex ratios to the sex ratios of European countries when they had similar overall death rates. India, for example, had an actual sex ratio of 108 in 1991, whereas its sex ratio would have been about 102 if it had conformed to the model of European societies. In order for India to have a sex ratio of 102 (relatively more females than there actually are), there would have to be 22.8 million more females in the population. Using this method of estimation, there are 22.8 million missing females in India, 29.1 million missing in China, 3.1 million missing in Pakistan, 1.6 million missing in Bangladesh, 0.2 million missing in Nepal, 1.7 million missing in West Asia, and 0.6 million missing in Egypt. The total number of missing females in these countries alone is over 59 million, which leads to the conclusion that approximately 60 million females are missing from the world's population because of excess mortality (Coale, 1991a). The excess mortality of these females is produced by their subordinate status in the societies into which they were born.

LIFE-STYLE AND MORTALITY

Our earlier introduction of life-style as a social factor influencing mortality described life-style as the choices people make in the conduct of their daily lives. These choices can be ones that increase the length of life, as in the case of Mormons and Seventh-Day Adventists who abstain from the use of tobacco and alcohol, or the choices can be those that reduce the length of life, such as eating high-cholesterol foods, being overweight, or using dangerous drugs. In the section that follows we will examine some of the major life-style choices that influence mortality, especially mortality in the United States. In order, we will discuss the use of tobacco, the behaviors that lead to AIDS and violent death (with a special emphasis on homicide).

Tobacco Use and Mortality

There is no doubt that tobacco use leads to higher mortality and that tobacco use is almost 100% a life-style choice. (The only major exception to this statement is the degree to which "second-hand smoke" increases mortality—in most cases breathing the smoke of others is not a personal choice.) The history of tobacco use, especially after its early discovery in the Americas and its importation into Europe, reveals how tobacco became a life-style choice (Ravenholt, 1990).

When Columbus arrived in the West Indies, he and his crew probably became the first Europeans to observe the use of tobacco. The European explorers and colonists who followed Columbus to the Western Hemisphere themselves cultivated and used tobacco, but it was nearly a century before it was exported and sold commercially to the wealthy and privileged people of

Europe. At about the same time, in the late 1500s, the Portuguese and Spanish sea traders were carrying tobacco to Africa, the countries of Asia, and the islands of the Pacific. Tobacco had started its "Global Death March" (Ravenholt, 1990, p. 213).

The detrimental effects of tobacco on health went largely undetected for as much as four centuries after its discovery. In fact, when tobacco was first introduced into Europe, it was often dispensed by pharmacists as a medical prescription. Many authorities attested to its curative powers. In 1614, when a plague struck London, "doctors declared that steady smokers were less subject to infection than others and recommended tobacco as a disinfectant" (Ravenholt, 1990, p. 216).

Many Europeans in the 17th century sniffed tobacco powder through the nose, which they called "snuffing." It soon became apparent that this practice caused sores and often cancers in the nasal passages. In the United States in the 19th century, when tobacco chewing and cigar smoking were the popular ways to use tobacco, it also became clear that tissues directly exposed to tobacco could easily become cancerous, but smoking itself was not considered exceptionally harmful to health.

The 20th century saw a spectacular rise in cigarette smoking in the United States. In 1900, 4 billion cigarettes were produced in the United States; in 1988, 695 billion were produced. The number of cigarettes consumed per person hit its peak in 1963 when an average of 4,345 cigarettes were consumed for every adult, age 18 and over, in the United States population. By 1988, the average number of cigarettes consumed for every U.S. adult had dropped to 3,096 (Ravenholt, 1990, pp. 222, 223).

Since the rise in cigarette smoking in the 20th century, there has been a corresponding rise in deaths due to smoking (with appropriate lag times). Coronary heart disease, for example, started an upward climb in about 1920, which reflected the rise in cigarette smoking at the beginning of the century. Lung cancer followed the same upward pattern after a 30-year lag time; emphysema moved upward after a lag time of 40 years.

There is some irony in the fact that in the last two decades, as concerns about smoking have increased, there has been an increase again in the use of chewing tobacco and snuff (Rich, 1992). Recent statistics show that there are about 30,000 new cases of oral cancer each year, many of which come from the use of "smokeless tobacco." About half of the new cases of oral cancer lead to death within 5 years. The Surgeon General of the United States predicts that we are likely to see "a full-blown oral cancer epidemic two or three decades from now" (Rich, 1992, p. A16).

The impact of cigarette smoking on death rates is revealed in the way smoking-related deaths among women have followed their smoking patterns. Women in the United States did not start smoking in large numbers until after World War II, especially in the 1950s. In 1965, before the effects of smoking could manifest themselves, only 4% of the deaths among middle-aged women (those aged 35 to 69 years) could be attributed to smoking. Today, 37% of the deaths among middle-aged women can be attributed to smoking (Brown, 1992a).

While it is not surprising that smoking can lead to cancers in the respiratory system, research has now shown that smoking is also associated with cancers in other tissues, including the pancreas, liver, spleen, kidneys, urinary bladder, cervix, prostate, and bone marrow (Ravenholt, 1990, p. 231). Approximately 40% of all cancer deaths in the United States are related to tobacco use.

Cigarette smoking has been identified by the U.S. Public Health Service as the single most avoidable cause of death in the United States (National Center for Health Statistics, 1991, p. 30). Approximately 400,000 lives are lost each year from smoking, and that does not include the adverse effects of cigarette smoking on pregnancy and infant health, or the victims of passive smoke inhalation. By comparison with the two life-style causes of death that we are about to consider, AIDS and homicide, smoking produces eight times more yearly deaths than these two causes *combined*. In 1990, compared to the 400,000 smoking deaths, there were 30,600 AIDS deaths, and 27,400 homicides (U.S. Bureau of the Census, 1993, Table 126, p. 91, Table 131, p. 96).

AIDS and Mortality

The appearance of the AIDS disease in the early 1980s has been called an epidemic, and it is true that in many respects it conforms to the characteristics of other epidemics that have afflicted humans. But in some respects, it does not. AIDS, like many previous epidemics, appeared suddenly and spread rapidly. Epidemics also characteristically affect many people in a population, community, or region, and this too has been true of AIDS in some communities and some areas of the world. Epidemics of the past have often been produced by the spread of highly contagious diseases, however, and in this respect AIDS is different, since HIV is not easily transmitted from one person to another. AIDS also differs from many earlier epidemics in that it is not spread in some unknown or mysterious way (e.g., by bacteria before people were aware of microorganisms).

Soon after the outbreak of AIDS in 1981, medical researchers suspected a virus as the disease's cause and within a few years HIV had been isolated. The modes of transmission were also quickly identified. AIDS, as almost every American knows, is transmitted when the blood or semen from an infected person enters the bloodstream of another person (Hardy, 1992). This occurs in a limited number of ways, including, in order of importance, sexual intercourse, the use of contaminated hypodermic needles (often the case among intravenous drug users), blood transfusions with contaminated blood, and the direct transmission of the virus from a woman to her fetus. The first two of these, sexual intercourse, including anal intercourse, and the use of contaminated needles, account for the vast majority of all AIDS cases, both in the United States and in other countries. Since these causes of AIDS both relate to the more-or-less voluntary actions and behaviors of people, they reflect life-style choices.

There are some cases of AIDS that obviously do not reflect life-style choices. Infants who get AIDS from their HIV positive mothers did not make a life-style choice—there have been over 4,600 such cases in the United States (Centers for Disease Control, 1994). Also, the people who get AIDS when they receive HIV-contaminated blood transfusions cannot be said to have made a life-style choice. By the end of 1993 there had been nearly 10,000 such cases. There have also been a few cases of doctors and health care workers who have gotten infected from the blood of HIV-positive people, and a minuscule number of patients who have been infected by doctors and dentists. The remainder of the 360,000+ AIDS cases in the United States are the result of sexual transmission and intravenous (IV) drug use (Centers for Disease Control, 1995).

The Centers for Disease Control now estimates that about 1 million people in the U.S. population are HIV positive. (C.D.C. National AIDS Clearinghouse, telephone communication, 1995). This estimate is simply the midpoint within a range of estimates, with a low of 800,000 to a high of 1.2 million. These are no more than estimates since only 26 states now have confidential reporting systems for HIV-positive cases, and none of the high-incidence states (especially New York, California, and Florida) are included.

In the world population the numbers who are HIV-positive cases can only be guessed at, but current estimates range from 10 to 14 million (Heilig, 1991, p. 75). The World Health Organization predicts there will be 40 million cases worldwide by the year 2000. Some experts consider this a minumum estimate and expect the number to be much larger (Altman, 1992). The world's total number of HIV-positive people is, like the numbers for the United States, only a best-guess estimate, because there are no firm data.

Surveillance of the AIDS disease in the United States has now been underway for nearly a decade, and we do have reasonably good numbers on the people who have manifested the disease symptoms.[4] Medical facilities and practitioners are required to report new AIDS cases to the Centers for Disease Control periodically. The summaries of these reports are published in the *HIV/AIDS Surveillance Report* (Centers for Disease Control, 1994). Statistical data on AIDS cases in the United States, as of December 1993, are shown in Table 7.2.

The total number of adolescent and adult AIDS cases since 1981 had

---

[4]Beginning on January 1, 1993, the definition of what constitutes AIDS was broadened (primarily to include females who often have different symptoms than males). This new, expanded definition, will, of course, increase the number of AIDS cases in the United States (Centers for Disease Control, 1992).

Table 7.2. AIDS Cases among Adults and Adolescents (13 and Over) by Exposure Category, Cumulative Number of Cases from 1981 to December 31, 1993, and New Cases in 1993

	New cases, 1993		Cumulative cases from 1981 to December 31, 1993	
Exposure category	Number	Percent	Number	Percent
Men who have sex with men	49,963	(47)	193,652	(54)
Intravenous (IV) drug use	29,399	(28)	87,259	(25)
Men who have sex with men and inject drugs	6,098	(6)	23,360	(7)
Hemophilia/coagulation disorder	1,096	(1)	3,133	(1)
Heterosexual contact	9,570	(9)	23,166	(7)
Blood transfusion	1,215	(1)	6,181	(2)
Other/risk not reported or identified	8,649	(8)	19,185	(5)
Total	105,990	(100)	355,936	(100)

Note. From Centers for Disease Control (1994, p. 8).

reached 356,000 by the end of 1993.[5] Nearly 30% of these cases were added in the year 1993, partly as a result of the new, broader definition of AIDS (see footnote 4). Of these 106,000 new cases in 1993, 47% were the result of men having sex with men. This percentage has been going down steadily, suggesting that gay males are changing their behavior and reducing this method of AIDS transmission. By contrast, IV drug users are increasing their share of the new cases; in 1993, 28% of all new cases were due to IV drug use. (Among females, nearly half [47%] of the new cases were due to IV drug use.) An additional 6% of the new cases in 1993 came from men who had had sex with men *and* injected drugs. These three categories constitute 81% of the new cases.

The number and the percentage of new AIDS cases coming from heterosexual contact has been increasing in recent years. In 1988, new AIDS cases coming from heterosexual contact made up only 4% of the total, while in 1993, heterosexual contact accounted for 9%. Over 40% of the heterosexual cases came from having sex with an IV drug user.

Violent Death and Mortality in the United States

About one out of every 15 deaths in the United States is attributable to accidents (both motor vehicle and other accidents), homicides, and suicides. These causes of death are often grouped together as *violent deaths.*

[5]AIDS cases among children under age 13 reached a total of 5,228 by the end of December 1993. Nearly 90% of these were infants infected by their mothers (nearly two-thirds of whom were IV drug users or women who had sex with IV drug users).

Males have higher violent death rates than females, and black males have higher violent death rates than white males. Black males have a violent death rate that is approximately 70% higher than that for white males, 250% higher than that for black females, and 320% higher than that for white females. Much of this excess in violent deaths among black males is due to their exceptionally high rate of homicide deaths (see below) (U.S. Bureau of the Census, 1992e, Table 122, p. 89).

It may be surprising to learn that the highest violent death rates are found among the elderly. The violent death rate is very high for all ages over 65, but increases dramatically above age 75 and is highest above 85. Motor vehicle and other kinds of accidents contribute most to the high violent-death rate among the elderly, though suicide, which is especially high for elderly males, is also a contributor. A study of injury deaths in New York State found elderly death rates due to accidental falls are five to ten times higher than the death rate from falls for other age groups (Relethford, 1991). This study also found the motor vehicle accident death rates of the 75-and-over group to be very high, much like the 15- to 24-year-old group (Relethford, 1991).

Young persons (those aged 15–24 years) have generally higher violent death rates than middle-aged people. The major exception to this pattern is found among young black females who have lower violent death rates than almost all other age groups. This is largely due to lower motor vehicle accident and suicide death rates among young black females.

Motor Vehicle Deaths

In the United States, automobile accidents take many more lives than either suicide or homicide. In 1990, there were 48,000 motor vehicle deaths, 31,000 suicides, and 26,000 homicides[6] (U.S. Bureau of the Census, 1992e, Table 114, p. 82). Males have a motor vehicle death rate (deaths per 100,000 population) that is more than twice that of females (U.S. Bureau of the Census, 1992e, Table 122, p. 89). This difference may reflect the fact that males drive more miles, and also that males often drive more dangerous vehicles—for example, motorcycles and large trucks—and they probably drive more recklessly (Gee & Veevers, 1983).

In the United States, motor vehicle deaths gradually increased in the 20th century as the power and speeds of automobiles increased. Since 1970, however, motor vehicle deaths, as measured by several different rates, have declined (U.S. Bureau of the Census, 1992e, Table 1008, p. 609). A revealing measure for assessing the safety of motor vehicle travel is the number of deaths per 100 million miles traveled. In 1965 that rate stood at 5.3 deaths per 100 million miles. By 1989, the rate had dropped to 2.2, a 60% decline

[6]Homicide statistics include deaths produced by *legal intervention*, which is defined as deaths inflicted by the police or other law enforcement agents in the line of duty.

in 25 years (U.S. Bureau of the Census, 1992e, Table 1008, p. 609). The rate of motor vehicle deaths per 100,000 population has also decreased from its high point in 1970. In that year there were 25.8 motor vehicle deaths for every 100,000 people; by 1989 that rate had dropped to 18.4—a decline of 29% (U.S. Bureau of the Census, 1982, Table 1061, p. 615; 1992e, Table 1008, p. 609).

In international comparisons, the U.S. motor vehicle death rate is relatively high, but not as high as the rates in Portugal, New Zealand, Hungary, and Belgium. Most European countries have substantially lower motor vehicle death rates than the United States, with Norway, Sweden, the United Kingdom, and West Germany (before unification) having the lowest (U.S. Bureau of the Census, 1992e, Table 1368, p. 829). It is probable, however, that many European countries have smaller proportions of their populations who own and drive cars, and that those who do own cars drive fewer miles. Both factors would lower the number of motor deaths per 100,000 population.

Homicide

The United States has few rivals when it comes to homicides: "The homicide rate in the United States is consistently higher than the rates in other industrialized nations of the world" (Fingerhut & Kleinman, 1990, p. 3293). The homicide rate for the U.S. population is 8.7 per 100,000, which is more than two-and-a-half times higher than the country with the next highest rate (Finland, with 3.3) and four to eight times higher than the rates in most other industrialized countries (Fingerhut & Kleinman, 1990).

In an international comparison of the murder rates among young males (aged 15–24), the United States has a murder rate (21.9) that is more than four times as high as the next highest country (Scotland).[7] Figure 7.4 shows the murder rates among 15- to 24-year-old-males for 21 countries (the data come from vital statistics reported in 1987 or 1988). As this figure makes clear, the U.S. murder rate for young adult males is at least 10 times higher than the murder rate for most other industrialized countries. In some international comparisons, there are extraordinary differences: the U.S. murder rate is nearly 44 times higher than Japan's rate, and nearly 22 times higher than West Germany's (before unification).

Within the U.S. population of young adult males, the homicide rate for young black males (those aged 15–24) is 85.6 per 100,000, while for white males in this age group the rate is 11.2. The white male rate is still twice as high as Scotland's, the next highest country, but the black male rate is 17 times as high (Fingerhut & Kleinman, 1990).

---

[7]In international comparisons, some countries include *legal intervention* deaths and others do not. The United States does, but if the legal intervention deaths were excluded the U.S. rate for 15- to 24-year-olds it would still be 21.5 (Fingerhut & Kleinman, 1990).

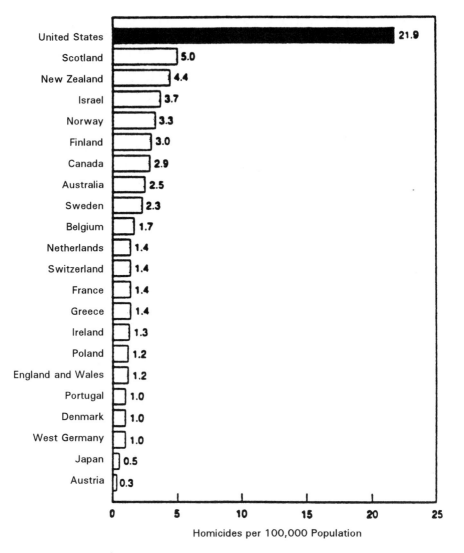

FIGURE 7.4. Murder rates among 15- to 24-year-old males in 21 selected countries. From Fingerhut and Kleinman (1990, p. 3293). Copyright 1990 by the American Medical Association. Reprinted by permission.

About three-fourths of the young adult male homicides in the United States are committed with firearms. In the 14 other countries for which data were available on firearm homicides, less than one-fourth of all homicides were committed by firearms (Fingerhut & Kleinman, 1990). The U.S. policy on firearms, compared to other industrialized countries, allows many more people to have guns, both legally and illegally. Firearms are the cause of death in 15% of all deaths of Americans between the ages of 1 and 34 (Fingerhut et al., 1991). Among teenagers (aged 15–19 years) and young adults (aged 20–24 years), firearms are used in one-fifth of all deaths.

Suicide

Taking one's own life is highly variable from one society to another. In international comparisons of the major countries of Europe, plus Australia, Canada, Japan, and New Zealand, the United States has a moderately low suicide rate (U.S. Bureau of the Census, 1993a, Table 1381, p. 848). Several European countries have suicide rates that are two and even three times the U.S. rate of 12.2 per 100,000 population: Hungary, 38.2; Finland, 27.5; Denmark, 22.4; Austria, 21.7; Belgium, 21.0; and Switzerland, 20.1. European countries with the lowest suicide rates include England and Wales, 7.5; Italy, 7.0; Spain, 7.1; and Portugal, 8.5.

Suicide is related to age, and thus the age composition of a population can influence its overall suicide rate. There is, however, a widespread perception that suicide rates are highest among teenagers, but the opposite is the case. *Teenagers and young adults (15–24 years) generally have the lowest, or near the lowest, suicide rates of all age groups.* This is true in the United States, as well as in the major European countries, plus Australia, Canada, and Japan (U.S. Bureau of the Census, 1993a, Table 1382, p. 849). Almost always in these countries, the highest suicide rates occur among the oldest age groups. Beginning at age 65, and especially after age 75, suicide rates are often especially high.

But there is another way to look at suicide deaths among teenagers and young adults. While their suicide rate is lower than that for other age groups, the proportion of teenage deaths coming from suicides is relatively high (about 13%). Suicide is, in fact, the third leading cause of death among teenagers and young adults. How can one explain this apparent contradiction? The simple answer is that teenagers and young adults are relatively safe from the most common causes of adult deaths, especially cancers and diseases of the heart. They die most often, as we will see below, from their own actions, including suicide, or the actions of others (e.g., homicide).

Suicide is also highly related to gender, with males much more likely to commit suicide than females. In the United States, the male suicide rate is four times as high as the female suicide rate. Suicide is also related to race in the United States, with the white rate approximately double that of the black rate (U.S. Bureau of the Census, 1993a, Table 137, p. 99). The differences in black and white suicide rates have been narrowing over the last 40 years; in 1950 the white suicide rate was about three times as high as the black rate.

One interesting line of research on suicides has found that suicides increase after the well-publicized suicides of celebrities (Phillips, 1974; Stack, 1987). In the most prominent example, after Marilyn Monroe's suicide the U.S. suicide rate increased temporarily by 12% (Phillips, 1974). In a study of the suicides of different types of prominent people (business elites, politicians, artists and authors, and entertainers), the researcher concluded that only entertainment/celebrity suicides increased the U.S. suicide rate. Even motor vehicle accidents appear to increase after highly publicized violent events, including suicides (Bollen & Phillips, 1981). It may be, as this study suggests, that some portion of motor vehicle deaths are in actuality suicides (Phillips, 1980).

Violent Deaths among Teenagers and Young Adults

Throughout this discussion of violent death it has probably become apparent that teenagers and young adults are often among the most likely age groups to die from these causes. This can be seen most easily by examining the 10 leading causes of death in the age group 15–24. These are shown in Figure 7.5. If the percentages for accidents, homicide, and suicide are added, the total is 77.4%. More than three-fourths of all deaths among teenagers and young adults come from these three causes.

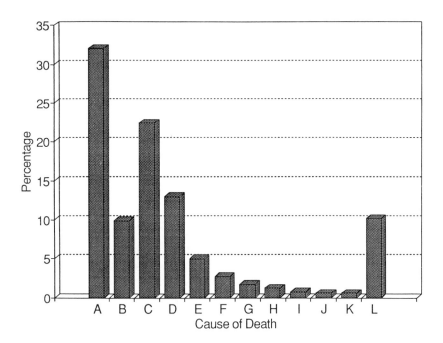

FIGURE 7.5. Ten leading causes of death among U.S. teenagers and young adults (ages 15–24), 1991: A, motor vehicle accidents; B, other accidents; C, homicide; D, suicide; E, cancer; F, heart disease; G, HIV/AIDS; H, congenital anomalies; I, pneumonia and influenza; J, cerebrovascular; K, pulmonary disease; L, all other causes. Data from National Center for Health Statistics (1993, Table 6).

SUMMARY

The limits of life are reflected in life span, the maximum number of years human beings are capable of living, and life expectancy, the average number of years the newborns of a population can be expected to live under prevailing death rates. Life expectancy is produced by life tables. The limits of life expectancy are a subject of debate, with *traditionalists* believing that life expectancy cannot go up much from it present levels, *visionaries* seeing preventative and innovative therapies that will slow down the degenerative process, and *empiricists* expecting that there will be continuing declines in mortality, which may extend life expectancy to age 100 or more.

Biological influences on mortality include one's ancestry (or family line), gender, and possibly even right- or left-handedness.

During the last half century, mortality in the United States can be divided into three distinct stages. From 1940 to 1954, declines in infectious disease death rates were most important in reducing mortality; from 1954 to 1968 the death rate leveled off; since 1968 the death rate has declined again, this time because of reductions in death from degenerative diseases.

Infant mortality often reflects the economic and social well-being of societies, though there are some exceptions to the general rule. The United States continues to have a higher infant mortality rate than many other countries, even though its per capita income is higher than most.

The influence of social factors on mortality rates may be divided into two types: position in the social structure and life-style. Position in the social structure is most often a position in the socioeconomic structure of a society, but if any group, such as a racial minority, is disadvantaged, it will generally have a higher death rate. African Americans have higher death rates than white Americans for all causes except suicide.

"Life-style" refers to the choices people make in the conduct of their daily lives. Studies of two religious groups, Seventh-Day Adventists and Mormons, both of which have restrictions on the use of alcohol, tobacco, and other things, have shown that they enjoy much lower mortality and longer life than the rest of the U.S. population.

There is an overwhelming body of research evidence showing that people in higher social classes have lower mortality rates than people in lower social classes. In the United States, despite relatively low death rates for the entire population for more than 40 years, there is still a clear class difference in mortality. Social class differences in mortality may reflect different levels of economic resources or different occupations, with different exposure to unhealthful conditions and accidents.

When females experience gender inequality, as they do in extreme form in many societies, their mortality may be influenced. Estimates indicate that 60,000,000 or more females are missing from the world's population, primarily because female infants and children are neglected and receive poorer treatment than males. Even in adulthood females often have higher death rates than males in these societies. The sex ratios of a number of national populations reveal the imbalance of males and females produced by higher female death rates.

Several major causes of death in American society reflect life-styles. Foremost among these are tobacco use, which is estimated to produce 400,000 deaths each year. AIDS is a cause of death produced by the spread of a virus. Most HIV infections come from known types of exposure, especially unprotected sex and the use of contaminated needles for drug injections. Violent deaths (accidents, homicides, and suicides) also reflect life-style choices. The murder rate of the United States is higher than the rate in any other country and is generally four to eight times higher than the rates in other industrial countries. Suicide in the United States is moderately low compared to other industrialized countries. Teenagers have a lower suicide rate than any other age group in the population, but more than three-fourths of all teenage deaths result from accidents, homicides, and suicides.

MEASURING MORTALITY

1. The crude death rate (CDR)

$$CDR = \frac{No.\ of\ deaths}{Total\ pop.} \times K$$

The CDR is a measure of mortality based on the number of deaths divided by the total population in a given year. A CDR of 15 means that there were 15 deaths per 1,000 people.

2. The infant mortality rate (IMR)

$$IMR = \frac{No.\ of\ deaths\ to\ infants\ <\ 1\ year\ old}{Total\ no.\ of\ live\ births} \times K$$

The IMR is a measure of infant mortality in a given year. It is derived by dividing the number of infants who died within their first year of life by the total number of live births in that year. A IMR of 45 means that 45 infants died per 1,000 live births.

3. The neonatal mortality rate (NNMR)

$$NNMR = \frac{No.\ of\ deaths\ to\ infants\ <\ 28\ days}{Total\ no.\ of\ live\ births} \times K$$

The NNMR is a more refined measure of the infant mortality rate. Because the probability of dying is greatest during the first 28 days, the infant motality rate is broken down to measure this critical period. A NNMR of 17 means that 17 infants died within the first 28 days per 1,000 live births. The infant mortality rate covering the period from 28 days to the end of the first year is referred to as the postneonatal mortality rate.

4. The age-specific death rate (ASDR)

$$ASDR = \frac{No.\ of\ deaths\ to\ persons\ aged\ x\ to\ x+5}{No.\ of\ persons\ aged\ x\ to\ x+5} \times K$$

The ASDR is a measure of mortality used to compare the death rates of specific age groups. It is derived by dividing the number of deaths among a specific age group (for example, ages 15–19) by the number of persons in that age group. It is computed for each age group. An ASDR of 13 for persons aged 15–19 means that there were 13 deaths per 1,000 people aged 15–19.

8

FERTILITY

Whether or not to have children, how many children to have, and when to have children are a few of the most personal decisions an individual will ever make. These same decisions when extended to an entire population establish the pace for population growth and determine the size and composition of the population. Childbearing, or fertility, is of unquestioned importance for countries, communities, groups, families, and individuals. The governments in some countries want more children and therefore encourage parents to have babies with various incentives; the governments of other countries make strenuous efforts to discourage large families and try to reduce fertility. Some families go to great lengths, and considerable expense, in order to have children; others carefully control their behavior to avoid having children. In this chapter, we shall examine this preeminent population process and the very different public and personal concerns about childbearing.

FERTILITY AS SOCIAL BEHAVIOR

Much of the material in this chapter will illustrate the proposition that childbearing is fundamentally a social behavior. When we say fertility is a social behavior, we mean that this behavior is shaped by the social contexts in which people live. These contexts may be as large scale as the economic or political worlds in which people live or as small scale as the specific circumstances of individual couples who are making decisions about whether to have a baby. Even if one were to accept the unprovable proposition that there is a human "instinct" or genetic "drive" to produce and bear children, such a proposition would explain almost nothing about fertility. It would not tell us why some populations or groups have high fertility and others have low fertility. It would not tell us why some people have large families, others have small families, and some people choose to have no children at all. These are the questions that are more effectively addressed when we begin with the proposition that childbearing is a socially and perhaps economically motivated behavior. As we will see below, the social nature of fertility has not always been recognized by those who have attempted to explain changes and variations in childbearing. Early scientific explanations tended to see changes and variations in fertility coming from biological changes. To understand this view, we need to know the difference between the terms *fertility* and *fecundity*.

The term used to describe the biological capacity to reproduce is *fecundity*. While fecundity can refer to the biological capacity of both males and females to reproduce, the focus is generally on females. Barring temporary periods of sterility or related medical problems, women, from menarche to menopause, are fecund, that is, they are biologically capable of bearing children. Demographers, as we have noted earlier, use the term *fertility* to refer to the actual childbearing of an individual or a population.

MEASURES OF FERTILITY

Many different measures of fertility are used by demographers. When speaking of the fertility of a population, rates are usually employed. These range from the *crude birthrate*, which is the number of births in a given year per 1,000 people in the population, to highly technical and sophisticated measures such as the net reproduction rate. These various birth and other reproductive rates are summarized in the box on page 201.

In addition to rates, other measures of fertility include the number of children ever born to women of different ages and completed family size.

Some demographers, when doing certain kinds of research, use what have come to be called surrogate measures of fertility. A *surrogate measure* is a substitute measure, in the sense that people can be asked about their fertility intentions. It is possible to get an indication of what fertility levels *will be* by asking individuals or couples how many children they expect to have or desire. Their responses are then used as surrogate measures of fertility. These measures of fertility are useful because researchers can assess the personal characteristics of individuals and couples that are related to having different numbers of children, even before they actually have them.

BIOLOGICAL EXPLANATIONS OF FERTILITY

Historically, a number of population theorists have claimed that declining fecundity leads to declining fertility. These theories have all been discredited, because there has been no scientific evidence to support them. But they do illustrate how early demographers attempted to explain changing levels of fertility.

These early biological theories often focused upon some societal characteristic and argued that it affected fecundity, which in turn resulted in reduced fertility. Illustrative of this type of theory is Sadler's (1829) principle that as a population grows more dense, fecundity declines. Conversely, Sadler argued, if a population is sparse, the fecundity of the population increases. Sadler was writing in the early part of the 19th century and was specifically arguing against the Malthusian theory of population growth. His theory ran counter to Malthusian theory because, according to Sadler, populations have a built-in *biological* mechanism that will keep them from growing too large. But there is no scientific evidence for such a built-in control on fecundity.

Another early biological theory was advanced by the 19th-century English scholar Herbert Spencer. Spencer (1873) argued that as societies develop and become more complex, there is a natural decline in the fecundity of the population. The reduced capacity for reproduction, according to Spencer, is directly attributable to the amount of energy expended upon "mental labour carried to excess." Spencer was particularly sensitive to the women of his time, about whom he wrote scoldingly, "The deficiency of reproduction power among them may be reasonably attributed to the overtaxing of their brains—an overtaxing which produces a serious reaction on the physique" (1873, p. 426). According to Spencer, a working woman, especially one who did mental work, was diminishing her capacity to reproduce children. It is true today that there is a tendency for women who are employed, particularly in managerial or professional occupations, to bear fewer children than women who are not employed, but this is a result of social and behavioral factors and personal choice rather than their lack of fecundity.

Another biological theory, this one advanced by Viscount de Lapauge, attributed the decline in fertility in 19th-century France to "racial intermixture." Viscount de Lapauge argued that intermarriage between two "generically different human types" would lead to the sterility or infecundity of the children. He proposed that "fecundity is proportional to the purity of the race, and to the state of stability of local cross-breeding" (quoted in Spengler, 1979, p. 139). Although his theory was not widely accepted, it did help to promote the idea that biological factors accounted for fertility declines.

Some biological theories have emphasized a dietary factor as the cause of reduced fecundity. The first was a theory advanced by Doubleday (1853), who argued that whenever a species or genus is threatened by adverse conditions, nature will compensate for this danger by increasing its fecundity. The reasoning behind this theory was that a shortage of proper food would be such a threat to a population that the fecundity of its females would increase during periods of food scarcity. Conversely, according to Doubleday, during times of abundant food supply, the capacity for reproduction would decline. Again, there is no evidence to support Doubleday's ideas.

In the 1950s, Josue de Castro, a Brazilian nutritionist, advanced a theory in which he argued that the amount of protein in the diet is inversely related to reproductive capacity. Thus, when the proportion of protein in the daily diet is high (as it generally is in modern, industrial countries), fecundity will be low. Alternatively, when the amount of protein in the diet is low, fecundity will be high. The evidence de Castro provided to support his theory was certainly not decisive. Taking countries around the world as cases, he correlated the protein levels in the diet with birthrates. He showed that countries with a high amount of protein in the daily diet had the lowest birthrates and vice versa. Such evidence, of course, only proves that these two factors are related: It does not prove any causal connection. De Castro did describe some experimental evidence produced by studies of rats, but these results were also not conclusive enough to allow inferences to be made about human populations (de Castro, 1952).

Currently, the study of fecundity is generally left to physiologists and biologists. For instance, Frisch (1980) has explored the relationship between amount of body fat and childbearing. She argued that women who have too little or too much fat are likely to have problems with regular ovulation and menstrual cycles, and consequently have lower fertility than other women. Frisch focused especially on the effect of extreme thinness. She noted that ballet dancers, who exert much physical effort and have low levels of body fat, often experience disruptions in their reproductive cycles.

Some demographers who are concerned with fecundity study the social determinants and social consequences for couples who are involuntarily childless or who cannot have the number of children they desire. Other researchers focus on developing technical and statistical models to show the ways in which fecundity may suppress or increase fertility (Larsen & Vaupel, 1993). For example, such models are being applied to studying the way that disease may ultimately affect fertility.

There is clear evidence that disease affects fecundity and thus the fertility of various population groups (Boyd, 1992; Tolnay, 1989; McFalls 1979a, 1979b; McFalls & McFalls, 1984). For example, malaria epidemics in the 19th century led to population declines among two Northwest coast Indian groups, the Chinookans on the Columbia River and the Kalapuyans of the Willamette Valley (Boyd, 1992). First, malaria directly affected the death rates of childbearing women and indirectly affected fertility. Second, during the epidemics of 1838 and 1851, stillbirths and miscarriages were numerous. Third, women who were infected with malaria often developed anemia, which reduced fertility.

In another example, Tolnay (1989) and McFalls and McFalls (1984) argue that the decline of fertility and high rates of childlessness among blacks in the United States between 1875 and 1940 was partially produced by infecundity. Even as late as 1950, 25% of black married women were childless. Much of this childlessness must be attributed to infecundity since there is little evidence that contraceptive methods were widely used by blacks during this period (McFalls, 1979a). The infecundity of African Americans was brought about largely as a result of genital tuberculosis, malnutrition, and venereal diseases. Genital tuberculosis, which is known to cause sterility, was widespread among African Americans, especially those living in urban areas. As the rate of tuberculosis increased among blacks, their fertility decreased. Then as conditions improved for blacks and rates of tuberculosis began to decline, fertility started to increase. In this case, disease affected fecundity and, in turn, fertility. However, Tolnay clearly states, as do McFalls and McFalls, that disease factors are only a partial explanation of fertility levels among black Americans, and that social and cultural factors play a more important part in explaining the patterns of black fertility. For example, as blacks moved to urban areas in greater numbers after World War II, they received better medical care, through private and public health facilities, than had been available to them in the rural South, and this better care would have increased black fertility (McFalls, 1979a).

Other causes of infecundity include genetic factors, stress, nutritional deficiencies, psychopathology, and environmental conditions. Now that the causes of infecundity are better understood, new reproductive technologies are being developed to help couples have children. In turn, issues of surrogate motherhood, *in vitro* and *in vivo* fertilization, artificial embryonation, and artificial insemination are likely to change our ideas about the relationship between biology and fertility (Isaac & Holt, 1987). For example, in the case of surrogate motherhood, the issue of biological parenthood versus social parenthood has the potential to alter our concepts of family relationships and responsibility for children.

SOCIAL EXPLANATIONS OF FERTILITY

An elementary way of demonstrating the effects of social and cultural systems on fertility is to examine *differential fertility*. This term is used by demographers to refer to the differences in fertility between various categories of people in the population. The categories may be social, cultural, or economic.

In the United States, the important subgroup differentials include socioeconomic status (education, occupation, and income), race and ethnicity, and religion. These are important categories because each can have a significant influence on childbearing. In the case of race or ethnicity, it might be that being a member of a minority group or experiencing different patterns of assimilation will influence childbearing decisions. Socioeconomic status may influence childbearing when people view children in terms of what it costs to raise a child or what a woman will lose in terms of earnings if she takes a break from work for childrearing. Religion may influence childbearing when there are norms that support childbearing or reject the use of contraception and/or abortion.

By identifying categories of people in a population that have different levels of fertility, it may be possible to make some inferences about social factors that influence fertility. It is of interest to know which groups and categories in American society have been and are at present demonstrating high levels and low levels of fertility. One would also like to know, whenever possible, what expectations for the future are. Have the existing differentials between racial, religious, and socioeconomic groups changed, grown larger, or diminished? These are some of the questions that will be considered in the following discussion.

Overview of Contemporary Trends in Fertility in the United States

There were over 4 million babies born in the United States in 1990. The last time so many babies were born in 1 year was over 30 years ago, during the baby boom. As Figure 8.1 demonstrates, while the number of births is high, the current fertility rate is at the lowest level it has ever been.

Beginning about 1920, the fertility rate in the United States began to decline. This decline is shown graphically in Figure 8.1. Throughout the Great Depression in the 1930s, the fertility rate remained low, at about 80 births per 1,000 women aged 15–44. As you may recall from our discussion on

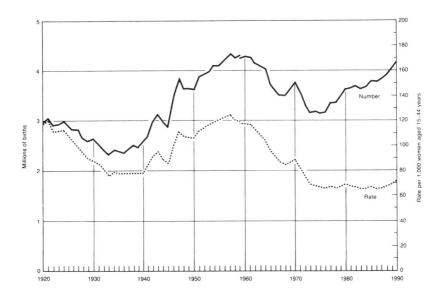

FIGURE 8.1. Live births and fertility rates, United States, 1920–1990. Note: Beginning with 1959, trend lines are based on registered live births; trend lines for 1920–1959 are based on live births adjusted for underregistration. From U.S. Department of Health and Human Services (1993, p. 2).

population composition (Chapter 5), the United States experienced a baby boom following World War II. Both the number of births and the fertility rates increased each year from 1947 to 1957. What is especially important about this time is that the fertility rate increased to 120 births per 1,000 women of childbearing age. On average, women were having three or four children during their childbearing years. Since 1957, the fertility rate has steadily declined. The impact of this decline is evidenced by the fact that average number of children for women is very low—1.9 births per woman. The increasing number of births is explained by the fact that there are more women of childbearing age (because so many were born during the baby boom).

The general pattern of childbearing is important to note. But as we will see, the childbearing patterns of women are affected by social and economic factors. We will discuss these fertility differentials next.

Socioeconomic Status

The relationship between socioeconomic status and fertility has been noted frequently in both the writings of demographers and in the discussions of laypeople. "As socioeconomic status increases, the level of fertility decreases," is simply the scientist's way of expressing the old cliché that ends with the words, "and the poor get children." The scientific and the popular statements are partially valid, but they must be qualified. One kind of qualification is historical, for the evidence is that the relationship between high fertility and lower social class in Western society has been stronger during some periods of history than others. There are certain social, economic, and technological conditions under which the relationship holds, while under others the relationship is weakened, modified, or even reversed.

Traditional Societies

There is much interest in the relationship between socioeconomic status and fertility in traditional societies, especially since today many countries are attempting to lower their population growth rates by encouraging people to have fewer children. (We will discuss this topic in greater detail in Chapter 9). In general, as birthrates decline, the higher socioeconomic groups have fewer children than the lower socioeconomic groups.

There is some evidence that this usual negative relationship between socioeconomic status and fertility is not found in all traditional peasant societies. In traditional societies a large family may be seen as a benefit. Under these economic conditions, the wealthier, landowning couples may have more children than couples who are less well off. Studies from India, Bangladesh, Iran, Nepal, the Philippines, and Thailand show that there is a tendency for higher-status landowners to have higher fertility than other farm families. In these societies, owning land increases the demand for children to work it, both in the immediate future and when the parents become too old to work themselves (World Bank, 1984).

Another case in which fertility is positively associated with higher social class is found in Indonesia (Hull, Hull, & Singarimbun, 1977). Women in the highest classes in Indonesia tend to be better nourished (thus more fecund), tend to have more stable marriages, and tend not to work outside the home. By comparison, women in the lower classes tend to be poorly nourished (less fecund), prone to unstable or disrupted marriages, and work outside the home. Ironically, these women are often employed taking care of the children of the wealthy (Hull et al., 1977).

When education is used as an indicator of social status, in most traditional societies the customary inverse relationship between status and fertility usually remains strong. The higher the educational level, the lower the fertility level. However, in some societies people with no education have lower fertility than people who have had enough schooling to become literate. For example, results from the Demographic and Health Survey in Nigeria reveal that the total fertility rate among women with no education is 6.5, while for women with some primary education it is 7.2. When women have completed primary education, their total fertility drops to 5.6. A similar pattern is noted in Cameroon (Population Council, 1992, 1993). In societies with very low levels of schooling, education acts to increase fertility because education leads to better levels of health and nutrition (Cochrane, 1979).

While a positive association between economic status and fertility does appear in some traditional peasant societies, the opposite is found more often. As income reaches a certain threshold and as education increases beyond the lowest levels of schooling, fertility usually declines. In many societies birth control is practiced by the members of the wealthier classes so they can maintain their favored positions. If, for example, land is in limited supply and a well-off family wants to make sure that their (male) children are provided for, they have an incentive to limit the number of offspring they produce. The Brahmins, a rich, landowning caste in India, have exercised great control over the number of offspring they create (Mamdani, 1972; Douglas, 1966). A large number of children would force a Brahmin family to divide its estate into smaller farms to provide for its sons; were this process to be repeated over just a few generations, a large estate would be transformed into dozens of small plots, and the descendents of what had been a wealthy, upper-class Brahmin family would be impoverished.

Western Societies

The relationship between socioeconomic status and fertility in Western societies during the last century has generally been negative. But there are some exceptions. Coleman and Salt (1992), Wrong (1958), and Haines (1979) have reviewed the experiences of several Western European countries plus

Australia, Canada, and the United States. The relationship between social class and fertility can be divided into three historically significant time periods: from the 1850s to 1910, 1910 to 1940, and 1940 to the present. Rapid declines in fertility generally started to take place in many of those countries sometime after 1870. It was during this period of rapid decline that the relationship between socioeconomic status and fertility became pronounced. For example, in England and Wales between 1852 and 1886, completed marital fertility declined at an annual rate of 1.21% among professionals but at the slower rate of 0.47% for agricultural laborers. According to Haines (1979), class differentials widened during the late 19th century. Coal miners, agricultural workers, and other lower socioeconomic groups did not lower their fertility as quickly as the higher socioeconomic groups.

The inverse relationship between status and fertility, particularly during the period between 1870 and 1910, can be attributed in great measure to the use of birth-control methods in the different strata of society. The upper-middle class started using the available methods of family limitation first, and only later and gradually did the lower classes follow their lead. These class differences in fertility control have been documented best in a study by Banks (1954), who found that the middle and upper-middle classes adopted birth-control techniques during the depression of the 1870s. Later, around the turn of the century, when the use of birth control had become more widespread, the middle and upper classes still had smaller families because they had become more effective users of contraception.

Another explanation for the differences in fertility by social classes during this early period before 1910 is the effect of age at marriage. Since unskilled and blue-collar workers reached their peak earning level at an early age, they married earlier and began childbearing sooner. By contrast, white-collar workers' incomes tended to increase steadily over their working careers. Recognizing this career potential, they delayed both marriage and childbearing (Haines, 1979). The end result was usually larger families among the lower classes because they had been exposed to the risk of childbearing longer.

After 1910, a much less uniform pattern emerged in the relationship between class and fertility. There was, first of all, a general contraction of fertility differences among socioeconomic groups. Furthermore, while the upper classes led the way in declining fertility before 1910, the middle classes began to do so after that date. The declines in fertility in this period tended to be most pronounced among urban wage workers, especially those with white-collar jobs (Wrong, 1958).

During the 30 years between 1910 and 1940, there was a general tendency for the lower classes to have more children, but by the end of that period there were some exceptions to this trend. Some higher status families were moving toward higher fertility. This was the case in Oslo, Norway, in 1930, and for a short period of time in Sweden during the 1930s. The smallest families tended to be found in the middle of the income range (Wrong, 1958).

In some cases involving special populations a positive association between fertility and socioeconomic status has been observed. The most frequently cited case is that found in the Indianapolis study in 1941. The sample group for this study was intentionally restricted to native-born white, urban Protestants. In the final sample, slightly more than 1,000 families were interviewed, and they provided the data for one of the most elaborate fertility studies ever conducted (Whelpton & Kiser, 1950). A small set of couples in this sample had been completely successful in planning their fertility; they had successfully planned the number and spacing of their children. Among these particular couples, those with the highest socioeconomic status had the highest fertility. The higher the socioeconomic status, the more children. It must be reemphasized, however, that the positive relationship was found only among those couples who planned *both* the number and the spacing of their children. The relationship did not hold for those couples who were successful just in planning the number of their children.

While this type of positive relationship between social status and fertility has been observed in some societies, or with particular populations, it has certainly not replaced the more familiar negative relationship. Even though the differences found between classes today tend to be much smaller than those found in previous historical periods, the most recent data show how, overall, the lower socioeconomic categories continue to have the highest fertility. The data presented in Table 8.1 show the number of children ever born to wives aged 35 to 44 (a group believed to have completed their childbearing). The educational attainment of the wives is used as one measure of socioeconomic status: As educational level of wives increases, their fertility decreases. The major occupation categories of husbands and family incomes were also used as measures of socioeconomic status. Again, the negative relationship generally holds. Both high-income families and families in which the husband holds a high-status job tend to have fewer children than other families. The significant exception occurs in the lowest income bracket (under $10,000), in which women typically have fewer children than women in families earning $10,000 to $20,000 (U.S. Bureau of the Census, 1991d).

The total effect of what we know about the relationship between socioeconomic position and fertility may be summed up in a few brief statements. Historically, the inverse relationship was most pronounced in Western societies when the birthrate started to fall rapidly in the last part of the 19th and the first part of the 20th centuries. Higher status couples were the first to adapt the idea of family limitation and to use birth-control techniques. As birth-control methods became available and were accepted throughout a society, new patterns of fertility emerged. Fertility has sometimes been found

TABLE 8.1. Children Ever Born to Wives Aged 35–44 in Married Couple Families, by Selected Characteristics

Subject	Children ever born, per 1,000 wives
Years of school completed by wife	
Not a high school graduate	2,783
High school: 4 years	2,166
College	
1–3 years	2,077
4 years	1,965
5 or more years	1,739
Total	2,139
Major occupation of husband, civilian occupation	
Managerial or professional	1,990
Sales or technical	2,014
Service	2,142
Farming, forestry, or fishing	2,588
Craft, repair, or production	2,206
Laborer or operator	2,297
Total	2,119
Family income	
Under $10,000	2,513
$10,000–$14,999	2,648
$15,000–$19,999	2,628
$20,000–$24,999	2,255
$25,000–$29,999	2,320
$30,000–$34,999	2,287
$35,000–$49,999	2,130
$50,000 or over	2,092

Note. From U.S. Bureau of the Census (1991d, Table 3, p. 1).

to be lowest among the middle classes and highest among the lower and upper classes. There are a few cases where the association between social class and fertility has been positive. However, contemporary evidence indicates that the inverse relationship still exists, even though it is less pronounced than it was around the turn of the 20th century.

Women's Employment and Fertility

The relationship between women's employment status and fertility has received much attention in recent years. As more and more women enter the labor force, and as they remain in the labor force throughout their child-bearing years, the number of children women have has declined and the age when women begin childbearing has increased.

Stated as a general relationship between women's employment and fertility, women who are employed tend to have fewer children than women who are not in the labor force. Table 8.2 shows that this relationship is true for women regardless of age, race, or ethnicity. In every comparison, women in the labor force have fewer children than those not in the labor force.

The direction of the causal relationship is difficult to establish. Do women who work have fewer children because they work or do they work because they have fewer children? (Waite & Stolzenberg, 1976; Stolzenberg & Waite, 1977; Smith-Lovin & Tickamyer, 1978). One answer to these questions is that it depends on whether one is interested in the long-term or the short-term effects of employment (Cramer, 1980). Cramer believes that in the short run, fertility affects employment (women often drop out of the labor force during the later stages of pregnancy and while their children are infants). In the long run, employment affects fertility (women will curtail fertility in the pursuit of careers or simply to continue in their jobs). It is probably the case that each affects the other in a reciprocal kind of relationship.

The only circumstance in which the employment of women is not related to lower fertility occurs when women are employed in occupations that are compatible with childrearing (Dixon, 1975; Stycos & Weller, 1967). Farming families have the greatest number of children. Women who work in agricultural areas can apparently combine work and childrearing. Women (and men) who are employed in agriculture are likely to be working at or near the home, where childrearing and work can occur at the same place and time. Not only do we find that employment in agriculture and fertility are compatible but also that these women are likely to have very high fertility. Perhaps it is still the case that farm children may share in the family workload.

In spite of the incompatibility between women working and having children, there has been a continuing increase in the numbers of women with preschool age children who work. Since the 1980s, over half of all women of working age have been in the labor force. Today this includes 53% of all

TABLE 8.2. Children Ever Born per Woman, by Employment Status, Age, and Ethnicity

Age and employment status	White	Black	Hispanic	Total
15–24				
In labor force	0.2	0.5	0.4	0.3
Not in labor force	0.4	0.7	0.6	0.5
25–34				
In labor force	1.1	1.5	1.5	1.1
Not in labor force	1.9	2.3	2.1	1.9
35–44				
In labor force	1.8	2.1	2.4	1.8
Not in labor force	2.3	2.7	2.9	2.3

Note. From U.S. Bureau of the Census (1990b, Table 2, pp. 19–26).

women aged 18–44 who have a child under 1 year of age (U.S. Bureau of the Census, 1991d). The increase in labor force participation by women with preschool-age children and especially those with infants does not imply that childrearing and employment have become more compatible. The need for child care services in the United States is greater than ever before. Studies show that women would work more hours if child care were available at a reasonable cost; women also say they would have additional children if child care was more readily available (Mason & Kuhlthau, 1992).

Women's labor force participation also affects the timing of childbearing. Women who work tend to begin their childbearing at later ages than women who are not in the labor force (O'Connell, 1991; Rindfuss, Morgan, & Swicegood, 1988). The postponement of childbearing into the late 20s and early 30s is related to the increasing educational attainment and labor force activity of American women. This issue and others related to the timing and the transition to parenthood will be discussed in greater detail at the end of this chapter.

Race

An examination of the fertility rates of blacks and whites in the United States since the middle 1800s shows a clear and consistent pattern between race and fertility. African American fertility has always been higher than white fertility. Estimates of black fertility and family size during the 19th century are found in the work of Tolnay (1981). The total fertility rate was, for blacks, 7.26 in 1882, 6.32 in 1892, 5.37 in 1902, and 4.84 in 1907. During the same time, the total fertility rate for the white population was 4.25 in 1882, 4.01 in 1892, 3.38 in 1902, and 3.52 in 1907 (Coale & Zelnik, 1963).

Throughout the 20th century, the total fertility rate among blacks has continued to be higher than that among whites, but similarities in the trends are quite striking. During the Great Depression of the 1930s, when fertility was declining, the total fertility rate for black women was 2.73, while that for white women was 2.51. Then, during the baby boom, when fertility was increasing, the total fertility rate for blacks reached a high of 4.54; for whites the rate was 3.53 (U.S. Bureau of the Census, 1980). Total fertility rates have continued to decline since the peak years of the baby boom, but as in the past the rates are higher for black women than for white women.

Even though the trend lines are generally similar, several issues about African American fertility have engaged the interests of demographers. First, there is the question of what accounts for the persistent difference in the rates of childbearing between blacks and whites. Second, there is the question of whether the rates will someday converge. Whether or not there might be a convergence of black and white fertility depends on changing behaviors of both groups.

Early analyses on the differences between blacks and whites focused on black fertility broken down by social class. One study in the 1960s revealed that black women with at least a college education, particularly if they lived in urban areas, had *lower* fertility than white women with a comparable education (Kiser & Frank, 1967). These results suggested that if blacks continued to improve their general socioeconomic position in society, particularly if economic, educational, and residential discrimination could be reduced in urban areas, their overall fertility level would move toward the level of that of whites.

More recent analyses have found that the persistence of differential fertility is more complex than socioeconomic status. There are several factors producing high levels of fertility among blacks, but other factors have the potential to push their fertility level down. One factor pushing toward high fertility is that black women tend to begin childbearing at earlier ages (O'Hare et al., 1991). The tendency for black women to have their children at earlier ages than white women is, in part, responsible for producing higher completed fertility among black women. The difference in fertility is most apparent among teenagers. As Figure 8.2 shows, the fertility rate among black teenagers is considerably higher than the rate among white teenagers. Moreover, unmarried black teenagers have a birth rate of 74 per 1,000, while the rate for white teenagers is 17 per 1,000 (National Center for Health Statistics, 1990).

As we noted in our earlier discussion, socioeconomic status is an impor-

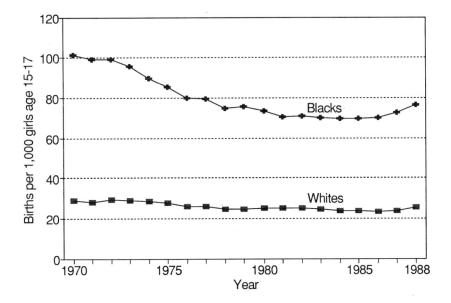

FIGURE 8.2. Fertility rate (per 1,000) by race for girls aged 15–17, 1970–1988. from National Center for Health Statistics (1990).

tant differential in fertility. Thus, is it important to note that the average socioeconomic status of African Americans is lower than that of whites. The fact that the overall socioeconomic status for blacks remains below that for whites is another partial explanation for the higher fertility rate among blacks. If there is no change in these areas, then black fertility will remain higher than white fertility.

On the other side, pushing toward lower fertility, black women are likely to be in the labor force, which is associated with low fertility. And black women with some college education have completed fertility rates that are similar to those of white women with the same educational level. In fact, as Table 8.3 demonstrates, black women who have a graduate or professional degree have fewer children than white women and similar rates of childlessness. Evidence suggests that as more black women complete their college education, overall fertility rates could converge.

But as we will see below, there are some factors that may begin to offset these differences. The childbearing behavior of whites may be changing in such a way that the racial disparity will become less evident.

Religious Differentials

In the United States Catholics, Jews, and Protestants are the broad religious groupings that have typically been studied for signs of differential fertility by religion. At this level, the pattern of differential fertility has been fairly clear and consistent. Catholics have had higher fertility rates than Protestants and Jews. Further, people who do not have any religion generally have had the lowest fertility. While these religious differences have prevailed in the past, these long-standing patterns have been changing. Most importantly, recent research suggests that Catholics may no longer have the highest fertility.

To begin with, the task of studying differential fertility by religion is made more difficult by the fact that there are no data on religion coming from the U.S. Census Bureau. Nor have there ever been. Therefore we must rely on survey data that have often been collected for other purposes. Even with available survey data, it is difficult to establish that it is religion per se that influences fertility. The problem of interpretation has always been difficult because religion is related to so many other variables, which are themselves related to fertility.

Protestants

With regard to Protestants, denominational differences in fertility have been observed. However, since the denominations have such great variability in their social and economic makeup, it is difficult to assess the effects of denominationalism alone. In fact, two classic studies of fertility, the Indianapolis study of Social and Psychological Factors Affecting Fertility (Whelpton & Kiser, 1950) and the Princeton study of Family Growth in Metropolitan America

TABLE 8.3. Children Ever Born per 1,000 Women Aged 35–44 and Percent Childless, by Race and Education, 1992

Education	White women		Black women	
	CEB	% Childless	CEB	% Childless
Total	1,890	18.2	2,226	12.6
Less than high school	2,621	9.5	2,760	7.8
High school graduate	1,963	14.5	2,174	11.5
College, 1 year +	1,694	22.4	2,061	15.3
Some college, no degree	1,842	17.9	2,394	9.6
Associate degree	1,797	18.6	2,071	12.9
BA degree	1,657	24.4	1,814	20.2
Graduate degree	1,352	31.5	1,266	30.5

Note. From U.S. Bureau of the Census (1993c, Table 2, p. 3).

(Westoff, Potter, Sagi, & Mishler, 1961) drew the similar conclusion that denomination was of little importance after other related variables, such as social class, had been controlled.

Although denomination affiliation does not appear to affect fertility, religious participation does. Mosher, Williams, and Johnson (1992), using data from the 1988 National Survey of Family Growth, show that the total fertility rate among Protestant women who attend church on a weekly basis is 2.24, compared to 1.75 for Protestant women who attend church less often. Protestant fundamentalists tend to have more children than other Protestants, a fact that may be related to an earlier age at marriage among fundamentalists.

Jews

Since the early 1900s, the birthrate of Jews in the United States has been consistently lower than the birthrate of the total white population. At the same time, Jewish fertility has responded to the same period effects as the general population, and therefore fluctuations in Jewish fertility have been similar. For example, the total fertility rate in 1935 was 1.25 among Jews and 2.13 for the total white population. After this low point in fertility, the rates for all segments of the population began to rise as the baby boom gained momentum. By 1955, the total fertility rate was 2.80 for Jews and 3.47 for all whites. As the baby boom passed in the 1960s, the fertility of both groups fell, but again the total fertility rate among Jews was lower than the rate for all whites. In 1970, this rate was 1.47 for Jews and 2.24 for all whites (Della Pergola, 1980); by 1988 the estimated rate was 1.54 for Jews, consistently lower than other religious groups (Mosher et al., 1992).

A major factor influencing fertility among Jews is their educational level. When the educational level of a group is above average (as it is among Jews in the United States), its fertility is lower. Education probably affects fertility indirectly since higher education usually leads to later marriage and more effective use of contraception (Della Pergola, 1980).

Religious involvement also accounts for some of the fertility differences among Jews, and in turn leads to an overall low fertility level. Frequent synagogue attenders from the Orthodox and Conservative denominations have higher fertility than other Jews. However, they comprise only one-fourth of the Jewish population. The lowest fertility occurs among Jews who infrequently attend synagogue and have high occupational status. Forty percent of all American Jewish couples fit this description (Lazerwitz, 1980).

Catholics

Two important issues relate to Catholic fertility: first is the degree of variation among Catholics, and second is the degree of variation between Catholics and other religious groups.

Dealing first with the variations *among* Catholics, factors related to religiosity (or the importance of religion) are important in explaining fertility differences among Catholics (Williams & Zimmer, 1990; Janssen & Hauser, 1981; Westoff & Ryder, 1977). Catholic women who consider religion to be very important or who frequently receive communion have higher fertility than Catholic women who consider religion to be fairly important or receive communion less than once a month. This is similar to the pattern observed among Protestants and Jews. But there is one significant difference. Compared to Protestants, Catholics who attend church weekly have *fewer* children than Protestants who attend weekly (Mosher et al., 1992). Why this may be so needs further elaboration.

Contraception among Religious Groups in the United States

If Catholic fertility in the United States is no different today than Protestant fertility—indeed, if Catholic fertility is lower—we may reasonably ask whether contraceptive use differs by religion. Contraception does differ by religion, but not always in the way one might expect and certainly not to the degree one might suppose. We get our information on contraception among American women from the 1988 National Survey of Family Growth, which is a national random sample of 8,450 American women between the ages of 15 and 44 (Goldscheider & Mosher, 1991).

In this survey American women were asked about their current contraceptive status and their religious affiliation. Table 8.4 shows contraceptive use among four major religious categories in the United States: Catholics, Protestants, Jews, and Others.[1]

---

[1]The category "Other" in this table refers to those women with "no religious affiliation." It does not include members of less frequently represented religions in the United States (Muslims, Hindus, Buddhists, etc.).

TABLE 8.4. Current Contraceptive Status and Method Being Used by White, Non-Hispanic Women, Both Married and Unmarried, Aged 15–44, in the United States

	Protestant	Catholic	Jewish	Other
Percent practicing contraception[a]	63%	63%	61%	63%
All contraceptors	100%	100%	100%	100%
Sterilization				
Female	30%	18%	12%	20%
Male	15%	14%	14%	14%
Pill	28%	34%	14%	32%
IUD	2%	1%	1%	2%
Diaphragm	5%	7%	26%	12%
Condom	13%	18%	23%	17%
Rhythm	2%	3%	0%	1%
Other	5%	6%	9%	3%

Note. From Goldscheider and Mosher (1991, pp. 102–115).

[a]Those women who are not practicing contraception include women who are not having intercourse, or are sterile, pregnant, or seeking pregnancy.

The overall percentage of white, non-Hispanic women practicing contraception is virtually the same for the four religious categories. Catholic women practice contaception at the same level as Protestant and "Other" women (63%) and slightly more than Jewish women (61%).

There are some differences in the types of contraception used (as shown in the bottom part of the table), but these differences do not seem to reflect religious prohibitions. Catholic women are only very slightly more likely to use the rhythm method (the only officially approved method for Catholics) than the other three religious categories. Catholic women use sterilization less than Protestant women, but more than Jewish women.

Going beyond the information in Table 8.4, Hispanic women, both Protestant and Catholic, are less likely to use contraceptives than non-Hispanic women. Among Hispanic Protestants, 56% used contraceptives, while among Hispanic Catholics, 49% did so. It appears that being Hispanic, especially in combination with being Catholic, does lower the level of contraceptive use. A similar pattern is found among black American Protestants and Catholics. Fifty-seven percent of blacks who are Protestant (and who are not Hispanic) practice contraception; 53% of Catholic (non-Hispanic) blacks practice contraception (Goldscheider & Mosher, 1991). Hispanics, both Protestant and Catholic, but especially the latter, are slightly less likely than other religious/ethnic categories to practice contraception. But it appears being Hispanic is more influential than religion with regard to use or nonuse of contraception.

Even within Protestant denominations there are some differences in contraceptive use. Fundamentalist Protestants have relatively low levels of contraception (52% of the women between ages 15 and 44 use some contraceptive method). Mormon women have an especially low level of contraceptive use: Only 49% of Mormon women use contraception (Goldscheider & Mosher, 1991). This accords with the high fertility found among fundamentalist Protestants and Mormon women.

In general, American women with different religious affiliations use contraception differently, but the differences are not simply a matter of Catholic women not using the methods prohibited by the Catholic church. Catholic women who are not Hispanic use contraceptive methods just as much as the women of other religions (or of no religion).

To sum up the results of these various studies, it appears that the relationship between religion and fertility in the United States is compounded by issues of socioeconomic status, ethnicity, and religiosity. Religous differences persist, but many of the differences are quite small when other factors are taken into account. In general, women with no religious affiliation have the lowest fertility, while fundamentalist Protestant, Hispanic Catholic, and Mormon women have the highest fertility (Mosher et al., 1992).

Immigration and Fertility

As the overall level of fertility has declined in the United States, differences in completed family size between various groups have declined as well. Future differentials in fertility may well rest on the childbearing experience of immigrants to the country.

In general, immigrants who move to the United States are young adults. The number of children immigrant couples have depends on several factors. Among the most important are the cultural norms regarding family size in their countries of origin. The rate of assimilation, the adaption to U.S. cultural norms, and the general socioeconomic characteristics of the immigrants will also influence fertility. These factors taken together suggest that as length of residence increases, the fertility of immigrant groups should begin to approximate the fertility of native U.S. residents.

Recent research comparing Mexican immigrants with native-born women of Mexican heritage and non-Hispanic white women lends support to the idea that immigrants adjust their fertility over time. In fact, Stephen and Bean (1992) found that among young immigrant women, fertility might fall below the average of native-born women. They attribute the lower fertility among young immigrant women to the disruption in life-style that follows immigration. But a word of caution is necessary. Young immigrant women have more years remaining in which to have children. Thus, findings of lower current fertility may be temporary and the result of delayed childbearing. After the disruption of immigration, women may have higher completed fertility. Indeed, Bachu (1991) has noted that the age-specific fertility rates among women aged 25 to 29 are greater for foreign-born women than for native-born women. Moreover, the age-specific fertility rates remain higher among foreign-born groups throughout the childbearing years.

In another study, Kahn (1994) found that immigrant fertility was higher than native-born fertility. The reason for the higher fertility is associated with compositional factors (age, education, etc.) and with declining fertility among the native population. Kahn's research also shows that fertility expectations, that is, how many children people say they expect to have over time, show that eventually immigrants will adjust their fertility levels to the U.S. norms.

In sum, it seems that when immigrants come from countries where the fertility level is high, for example, from Latin America and Mexico, the fertility of the immigrants remains high and then declines. If immigrants come from countries where the fertility level is low, then there is little variation between immigrant women and native-born women. Thus, it is the selectivity of the immigrants by compositional factors that affects fertility.

THE TRANSITION TO PARENTHOOD

The existing evidence on fertility differentials describes the patterns in childbearing at the aggregate, or group, level. Missing from the discussion is a framework to explain why the patterns exist. This section describes a framework for understanding the process of family formation and the transition to parenthood.

An elementary sociological proposition about fertility serves as the basic principle of this framework: Every social system, if it is to survive, must have an institutional structure and a value system that ensures the continuing production of new members. Without institutional and cultural forms supporting reproduction, a society would not replace its members lost by death, the population would decline, and eventually the society would disappear. An extension of this proposition is that the institutional structure and the cultural values that societies have developed to ensure reproduction have been developed under conditions of high mortality. The short average length of life that prevailed in earlier historical periods indicates that many people died before they reached reproductive maturity. Under these conditions, only reproductive norms that produced very high fertility could provide the replacements necessary for the society to maintain or increase its population size.

It is also necessary to note that people must not only be motivated to bear children, but, because of the helplessness of humans through the first years of life, they must also be motivated to care for and rear children. People must receive satisfactions from the parental role. The social system usually provides such rewards.

Fertility as a Result of Decision Making

It is a basic assumption of the framework introduced here that fertility behavior in modern, urban–industrial society is the result of a decision-making process by individuals and couples. Through most of history, fertility may have been largely determined by fate or chance. It is unlikely that people made decisions not to have children; more commonly, they would have them unquestioningly. However, increasingly, particularly in modern societies, this submission to fate is no longer the most prominent feature of childbearing behavior. Individual couples are making decisions about the children they have.

While fertility can be viewed as the outcome of decisions made by individual couples, we must also recognize that the actual decision will not always be made on a conscious level, and it certainly will not always be a single, unitary act at a given moment in time. Nor can we assume that decisions regarding childbearing are always rational decisions. Some decisions may be made rationally in the sense that people weigh alternatives and make decisions that best meet their needs or objectives. Thus, for example, women who choose to abort an unwanted or unplanned pregnancy and couples who plan both the timing and spacing of their children are making rational decisions. However, it is just as likely that people drift into parenthood. In this instance, couples may not care whether a pregnancy occurs or they might just assume that having children is "the thing to do." Knowing that a decision can have important consequences for their lives, individuals may avoid making decisions or make them by default. For example, although a couple may have contraceptive techniques available to them, they may not use them regularly. This behavior is not the same as the conscious decision to conceive a child, but the end results are often the same, and the decision not to take action is still a decision (Back, 1967).

The second feature of the actual decision-making process is that rarely is there a single occasion when a decision to have a child is made. More commonly, the ultimate outcome is the result of a process or a series of minor decisions. The decisions of some voluntary childless couples are a good illustration of this point. What begins as a temporary decision to postpone having a child often becomes a series of postponements until the couple arrives at a point where they realize that they have in effect already made the decision not to have children (Rindfuss, Morgan, & Swicegood, 1988; Veevers, 1980).

With these considerations in mind, it is possible to suggest the following broad and ideal outline of how family formation occurs in modern society. Rindfuss et al. (1988, pp. 6–13) refer to this pattern as "on-time parenthood," as opposed to teenage or delayed parenthood. When a couple marries, the two people may discuss very carefully and quite explicitly the number of children they want and when they want them. If they do and if they have a workable contraceptive technique available to them (in both a physical and a moral sense), then they may well be able to plan the number and spacing of their children. This idealized version of fertility decision making is not likely to be the most common type, however. The more likely process can be conceptualized in the manner shown schematically in Figure 8.3.

This figure represents, from left to right, the number of children a couple may have. Assume the couple enters marriage in a childless state and before a child has been conceived. The movement of a couple from childlessness to having a child will be produced by various forces that might be conceived of as vectors. The total effect of these vectors on a married couple will determine if and how soon they have their first child. In the same way, moving from one child to a second will be determined by the relative force of the vectors. When the vectors moving in the direction of having an additional child are exceeded by the vectors going in the opposite direction, then childbearing for the couple, in general, will cease.

The vectors shown in Figure 8.3 are basically of three types. They include the general "cultural press" for having children, and two sets of "situational-specific factors." One set of situational-specific factors exerts pressure for having a child, while the second set pushes against having an additional child.

The Cultural Press toward Parenthood

Every society has some values and disvalues for family formation. These values shape the normative framework for childbearing. The set of values and norms influencing childbearing decisions comprise the *cultural press* toward parenthood. This single label covers all the institutional and cultural supports for childbearing. It is best understood as a set of values that specify the cultural context of when and how many children it is proper to have.

The cultural press operates within the socialization process. The socialization process, if it is sufficiently effective, causes most people to accept the virtues and values of having children and when to have them without question. For example, traditional gender roles include the expectation that women will become mothers and men will become fathers. Women and men who choose to remain childless are likely to have at least entertained the idea of having children (Rindfuss et al., 1988).

Support for the cultural press of family formation can be documented by noting the predominant patterns of childbearing. Most people in the United States get married, and most married couples have children. Despite the fact that childbearing norms in the United States are changing, most childbearing takes place within marriage: 73% of all births in 1989 were to married couples (Ahlburg & DeVita, 1992).

The timing of the first birth is also affected by the subtle influence of

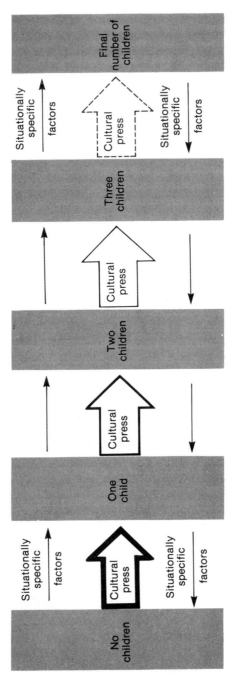

FIGURE 8.3. The process of family formation.

the cultural press. There have been major societal changes that have affected the timing of the first birth. Ryder (1990) suggests that there has been a normative transformation, which allows women to weigh alternatives to motherhood. The most important societal change that has prompted this transformation is the increase in employment opportunities for women. This has led to a delay in marriage. As the average age at marriage has increased, so too has the percentage of women who delay having children. Today, the median age at marriage for men is 26 and for women it is 24. The peak years for childbearing occur to women between the ages of 20 and 29, but the number of first births to women over age 30 has been increasing (Ahlburg & DeVita, 1992).

There are some variations in the normative structure of family formation. The number of children who are born to single women and the number of premarital conceptions have increased dramatically during the last several years. High rates of teenage childbearing and of unwed parenthood suggest that the cultural press regarding marriage is not as strong as it once was. However, the cultural press toward parenthood remains quite strong.

Finally, the cultural press influences family size. The average family size in the United States is two children. Once couples enter into the parenthood role, they are likely to have two children. An interesting way to see the effect of the cultural press on family size is to observe the seating available in family restaurants. During the 1950s when the average family had three or more children, tables in family restaurants were set up to easily accommodate five or more people. Today, a quick look at the table arrangements show that most tables are set up for four people. A family with several children will certainly pick up cultural clues that their family size is larger than the norm. The cultural press exerts subtle influence on when to stop having children as well as when to have children.

The subtle influence of the cultural press is most often taken for granted, and the evidence is likely to be more qualitative than quantitative. That the cultural press is important is noted by Rindfuss, Morgan, and Swicegood (1988) in their study on the transition to parenthood:

> Our empirical research makes no attempt to document this normative pressure. We measure neither expectations regarding the need to become a parent, nor the sanctions brought to bear when one does not become a parent as soon as expected. Rather, the normative pressure to become a parent is taken as a strong and central assumption. (p. 20)

For most people, the process of family formation is governed more by the general cultural press than by any careful, considered, rational decision making. It is only after one or two children have been born that more specific considerations come into play, and decision making in a more familiar sense begins to operate.

Situational-Specific Factors

The discussion thus far has focused on the traditional cultural values labeled "cultural press." Little has been said about the situational-specific factors that are represented as vectors in Figure 8.3. Some of these factors move a couple toward having more children, while others act to stop the couple from having more children than they already have. The term *situational-specific* refers to the conscious and explicit reasons that individuals may have for wanting or not wanting a child at any particular stage of life. These explicit and conscious motives are directly related to the specific goals or objectives of the individual or couple. While these situational-specific factors may be infinite in their variety—dependent as they are on the particular circumstances of each individual—it is possible to suggest a few that are likely to be widespread.

First, some of the positive situational-specific factors (those moving a couple toward having a child) might include (1) the desire to have a child (or a male child) to please parents or other kin (possibly a particularly prominent factor pushing toward having the first child), (2) the explicit desire of a couple to have their children while they are young, (3) the desire to have a sibling mate for an already born child (most pronounced when the couple has one child and wishes to avoid the "only-child syndrome"), and (4) the desire to have a child as a way of trying to strengthen a shaky marriage.

Examples of the negative situational-specific factors include (1) the desire to complete the husband's or wife's education before having a child (or another child); (2) the desire of a couple to have personal freedom to work, travel, and the like before having a child; (3) the desire to achieve some specific economic goal before having a child; and (4) the desire to give a small number of children particular advantages that would not be possible with more children.

This list of both positive and negative factors could be expanded indefinitely and in ever-increasing detail, but future empirical research will have to establish which factors or sets of factors are most important in determining family size. The research results available now are limited, since few studies have been conducted longitudinally (following the same couple over time) to see just how the factors influencing fertility change in response to changing circumstances.

There are two empirical studies that illustrate the way that these situational-specific factors influence the decision-making process. Teachman and Schollaert (1989), using information from the 1973, 1976, and 1982 National Survey of Family Growth, studied how gender of children affects the timing of additional children. Is there a preference for sons, for daughters, or both? They found that the balance of gender composition affects the timing or interval between births. Women are more likely to have a third child sooner if they currently have either two sons or two daughters. Women who have both a son and a daughter wait longer to have an additional child. Thus the decision to have a third child and the timing of having the third child are dependent on the gender balance of children already born.

The second study that illustrates the importance of situational-specific factors was conducted by Leslie Whittington (1992). She used data from the Panel Study on Income Dynamics to determine the effect of tax exemptions on fertility. The federal tax structure in the United States allows people with children to deduct a set amount of money from their taxable income. This deduction reflects the federal government's willingness to partially subsidize the cost of raising a child. Whittington found that when the value of the tax exemption increases, the probability of some couples having another child increases.

The general theme reflected here is that there are two major determinants in family formation: The cultural press is the most important force until some children are born, and then situational-specific factors (both positive and negative) enter the decision-making process. Thus, the decisions individuals and couples make about the timing and number of children they want are shaped by their culture. Cultural norms and subcultural variations by race, ethnicity, religion, and socioeconomic status, in turn, influence the way situational-specific factors are evaluated. When the negative factors outweigh the positive ones, most couples will stop having children.

The ability for people to make deliberate choices about childbearing has been enhanced by the availability of contraceptives. The wide range of choices, including sterilization, the pill, various barrier methods, and abortion, provides people with the means to plan the timing and spacing of their children. Research in this area shows that almost all married couples go through many years of fecundity without producing children. Stopping fertility during the fecund years is, of course, accomplished by practicing birth control, but most precisely, it is the improved use of contraception that keeps couples from having more children than they want. As couples have the number of children they want, they turn to the most effective contraceptive means available. In fact, voluntary sterilization in the United States is the most popular method of contraception used by older couples (Mosher & Pratt, 1990a).

On the other hand, new reproductive technologies (*in vitro* fertilization, artificial insemination, and surrogacy) allow some couples the opportunity to have children that they otherwise would not have been able to have. Although the percent of women who are not able to conceive a child has not increased, the use of infertility services *has* increased (Mosher & Pratt, 1990b). The increased use of infertility services attests to the strength of the cultural press pushing people into having children and to the situational-specific factors that aid couples to make these important decisions.

This suggests again that the early part of family formation is often the result of a general cultural press that places great value on the production of children. It is only after the cultural value has been satisfied by the birth of at least some children that more specific considerations begin to play a greater role. When the negative value of having more children exceeds the positive value, most couples will stop having children.

SUMMARY

Childbearing is fundamentally a social behavior, shaped by institutional structures and cultural values. A distinction is made between the terms *fertility* and *fecundity*. Fertility refers to actual childbearing behavior; fecundity refers to the biological capacity to reproduce.

Demographers have traditionally demonstrated the social nature of fertility by studying societal variations in fertility. Fertility differences by socioeconomic status, women's employment status, race or ethnicity, religious ideology, and immigrant status have provided additional information. These studies document historical trends and provide insights into the future of fertility behavior.

To understand fertility behavior today, there is a need for a frame of reference that focuses on decision making and the process that shapes the transition to parenthood. During the early stages of family formation, the principle influence on the decision (or nondecision) to have children is the general cultural press for having children. In later stages, more situational-specific factors emerge. Eventually, the situational-specific factors determine the ultimate family size.

MEASURING FERTILITY

1. The crude birthrate (CBR)

$$CBR = \frac{\text{No. of births}}{\text{Total pop.}} \times K$$

The CBR is a crude measure of fertility based on the number of births divided by the total population in a given year. A CBR of 22 means that there were 22 births per 1,000 people in a given year.

2. The general fertility rate (GFR)

$$GFR = \frac{\text{No. of births}}{\text{No. of women aged 15–49}} \times K$$

The GFR is a more refined measure of fertility than the CBR. It is based on the number of births divided by the number of women in their childbearing years for a given year. A GFR of 45 means that there were 45 births per 1,000 women aged 15–49 in a given year.

3. The age-specific fertility rate (ASFR)

$$ASFR = \frac{\text{No. of births to women aged } x \text{ to } x + 5}{\text{No. of women aged } x \text{ to } x + 5} \times K$$

The ASFR is a measure of fertility used to compare the childbearing behavior of specific age groups. It is derived by dividing the number of babies born to women of a specific age (e.g., 20–24) by the number of women of that age. It is computed for each age group. For example, an ASFR of 100 for women aged 20–24 means there were 100 babies born per 1,000 women aged 20–24 in a given year.

4. The total fertility rate (TFR)

$$TFR = \Sigma \, (ASFR \times 5)/1,000$$

The TFR is an estimate of the average number of children that would be born to a woman at the end of her childbearing years if she conforms to the current age-specific fertility rates. The computational formula above calls for the age-specific fertility rates of all women in their childbearing years (each 5-year interval between 15 and 49) to be summed, then multiplied by 5 and divided by 1,000. The number 5 used in the computational formula represents the age interval often used for age-specific fertility rates. If single-year ASFRs were used, the number would be 1. A TFR of 2.2 means that each woman would average 2.2 children during her lifetime.

5. The net reproduction rate (NRR)

The NRR is used to indicate generational replacement. It measures the number of daughters that would be born to a woman (or group of women) at the end of her lifetime if she conformed to the current age-specific fertility rates and mortality rates. The calculation of this rate is complex and requires information on fertility and mortality. However, it is easy to interpret. A NRR of 1 means that a population will replace itself but will not grow. A NRR of less than 1 indicates that the population is not replacing itself, and if the rate continues, the population will decline. A NRR that is greater than 1 means that the population is not only replacing itself but is also growing.

9

WORLD POPULATION GROWTH

In 1994 there were over 1 billion people in China. The total fertility rate was 2.0 and the infant mortality rate was 31 per 1,000. It will take China 61 years to double its population (Population Reference Bureau, 1994). It is difficult to imagine what 2 billion people will mean to a country such as China in the year 2055. More difficult to imagine is what China might have looked like if it had not taken direct action to slow its rate of population growth. In 1962, the population of China was 653,302,000. At that time, the birthrate was 41 per 1,000 and the mortality rate was 14 per 1,000. The infant mortality rate was 89 per 1,000 live births (Banister, 1987). If those rates had remained constant, the population would have reached 1.1 billion by the mid-1970s (Tien, 1992), and with a rate of natural increase of over 2%, the population of China would be approaching almost 2 billion people *today*.

We will broaden our demographic scope in this chapter, as we turn our attention to world population growth. First, we will discuss population growth in the past among developed countries. We will see that population growth in the past was associated with economic development. But the conditions that led to a positive relationship between population growth and economic development in the past are not the same today. Many countries in the developing world have mixed views about the relationship between economic growth and population growth. And the responses to population growth from the worldwide community are as varied as the world's many countries themselves.

THE DYNAMICS OF POPULATION GROWTH

If we set aside migration for a moment, only differences between birthrates and death rates can produce changes in population. Populations can only grow in size when there are more births than deaths. If the crude birthrates and death rates are at the same level, regardless of whether they are high, medium, or low, a population will not grow in size. But if the birthrate goes up while the death rate remains constant or if the death rate goes down while the birthrate remains constant, the population will grow. When demographers study population growth, they are primarily interested in the difference in the rates of birth and deaths. The greater the difference between these two rates, the greater the population growth.

As we noted in Chapter 1, during the long course of history population growth was quite low. Through the hundreds of thousands, even millions, of years of human existence, birthrates and death rates were apparently similar. Probably both were also at high levels. Whenever these vital rates are high, the potential for growth is greater than when the rates are at low levels. This is true (as we will discuss in more detail below) because a decline in the death rate is likely to occur first and faster than a decline in the birthrate. When this happens, the gap between the rates may become quite large.

Demographic Transition Theory

One of the first attempts to study population growth in a comparative framework was carried out in 1929, by American demographer Warren Thompson (1929). He published a paper in which he provided what proved to be the basis for a theory of population growth that concentrated on transitions in vital rates. This early paper was not an attempt to set forth a completely formulated theory but was written to summarize and categorize what was then known about vital rates in countries around the world. Thompson observed that the countries of western Europe and North America (labeled Group A countries) had death rates and birthrates that were declining, but the death rates had reached a fairly low level and seemed to be slowing their rate of decline. The birthrates were moving downward more rapidly and toward the level of the death rates.

Another group of countries, found mostly in eastern and southern Europe, had death rates that were falling quite rapidly, but their birthrates remained high or were declining only slightly. These countries (labeled Group B countries), since they had a considerable gap between their birthrates and death rates, were growing rapidly.

Finally, a third group of countries (Group C), constituting much of the world's population and covering most of Africa, Asia, and Latin America, had different demographic conditions. As far as could be determined from the limited data available at the time, both the birthrates and death rates were very high, and thus the rate of population growth in these areas was quite low.

These three groups of countries, with their distinctive vital rates, later became associated with three different stages in a demographic transition. The Group C countries, with high birthrates and high death rates, were in a premodernization stage, much like Europe might have been in 1600. The Group B countries, with falling death rates and relatively high birthrates, were viewed as comparable to the early stages of modernization that occurred in northern and western Europe in the 17th, 18th, and 19th centuries.

The Group A countries were, in fact, the northern and western European countries (plus Canada and the United States) that were well along the way toward modernization. Their death rates, as we have noted, were at low levels by the 1920s, and their birthrates were moving down toward the same low levels.

From these and other similar observations during the two decades after Thompson wrote his seminal article, a more abstract conceptualization of transition theory emerged. Theorists assumed that the Group A countries were a model whose population patterns could be used to predict future patterns in the Group B and Group C countries. These theorists believed that, with regard to the European countries, the transition from relatively high death rates to relatively low death rates had occurred first, and that declines in birthrates followed sometime thereafter. These declines in vital rates appeared, at least in a general way, to occur concurrent with economic development. Thus the transition came to refer not only to changes in vital rates but also to changes in economies from primarily agrarian to more urban, commercial, and eventually industrial systems. The economic changes were believed to account for the changes in the demographic rates. The transitions in death rates and birthrates over time are shown graphically in Figure 9.1.

The Group A, B, and C countries were viewed as representative of stages in a continuous process. This continuous process can be visualized more clearly when the transition is broken down into three stages:

Stage 1: A premodern stage of economic development. The birthrates and death rates are both high. Population size is relatively stable.

Stage 2: The beginnings of modernization and economic development. The death rate is dropping, but the birthrate remains high. Population size is growing.

Stage 3: Modernization and economic development are well advanced. The birthrate is now low, approaching, or reaching, the low level of the death rate. The population is stable or growing very slowly.

We would emphasize that the depiction of the demographic transition in Figure 9.1 is an idealized model. The demographic and economic processes described by transition theory are not inevitable, nor do the details of the process work out in exactly the same way in every population. Exceptions of various types have been found in a number of countries. For example, in both France and Hungary, there is evidence that birthrates started falling before death rates (Demeny, 1968; Goubert, 1968). Indeed, as these and other exceptions have been studied more extensively, questions have been raised

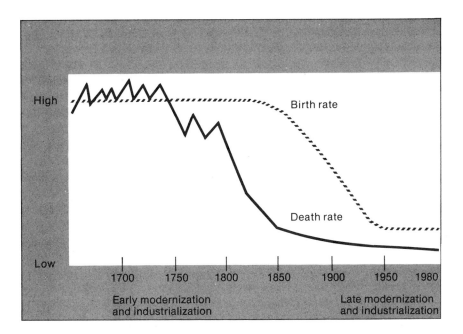

FIGURE 9.1. The demographic transition.

about the utility of the demographic transition theory (Coale & Watkins, 1986).

The criticisms are threefold. First, as noted, there are areas or regions in Europe where birthrates started falling before death rates. Second, in some less-developed countries today, the declines in death rates and birthrates have occurred prior to any significant economic development. Third, in other countries where the basic model seems to fit, for example, Japan, the time frame in which the vital rates fell from high to low levels has been compressed, so that there is no consistent indication of how much time must pass before the transition is complete.

In spite of these criticisms, demographic transition theory has earned its place in the history of what we know about population growth. As a model, it has guided current research. In the next sections, we will examine the way mortality declined in the past and the current state of mortality decline.

THE TRANSITION TO LOWER MORTALITY

To understand how population growth is produced by declines in mortality, we will first examine how and why death rates declined over the last two centuries. The focus of our attention will be on Europe since the 18th century and the United States in the 19th and 20th centuries. After we examine the historical record of mortality decline, we will examine evidence from contemporary developing countries that shows much more rapid mortality declines.

Two Centuries of Decline in the Death Rate

The declines in mortality in the United States and western Europe occurred gradually over a 200-year period. Figures 9.2 and 9.3 reveal how steadily mortality in England and Wales and the United States declined during the last century. The steady decline in mortality was concurrent with modernization and economic development. But economic development alone does not account for the decline in mortality. Behavioral and social changes were also important.

Two complementary themes will emerge from this examination of the British and American mortality experiences: (1) the substantial declines in the death rate that occurred in the 18th and 19th centuries in Great Britain and in the 20th century in the United States are not primarily attributable to medical advances and practices, and (2) the decline in death rates has been more directly attributable to environmental and social factors. Environmental factors include improvements in nutrition due to more abundant food supplies and the introduction of hygienic measures such as the purification of water, more efficient

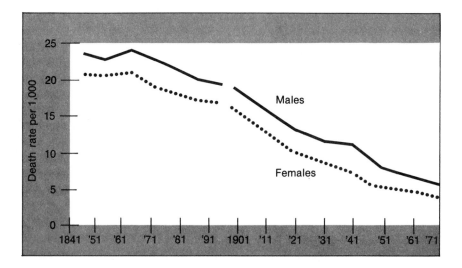

FIGURE 9.2. Death rates (standardized to 1901 population), England and Wales. From McKeown (1979, p. 31). Copyright 1979 by Princeton University Press. Reprinted by permission.

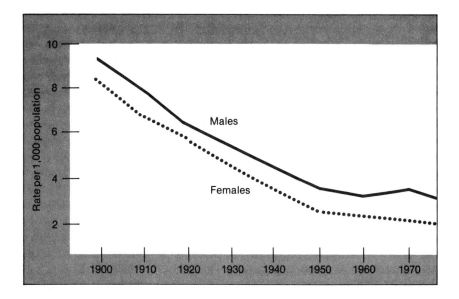

FIGURE 9.3. The trend in mortality for males and females separately (using age-adjusted rates) for the United States, 1900–1973. For these and all other age- and sex-adjusted rates, the standard population is that of 1900. From McKinlay and McKinlay (1977). Copyright 1977 by MIT Press. Reprinted by permission.

sewage disposal, and improved food hygiene (McKeown, 1976). It is also possible that increased personal hygiene and cleanliness contributed to a reduction in mortality (Coleman & Salt 1992; Razzell, 1974). Social and cultural factors, in particular changes affecting the status of children, may also have lowered the death rate (Caldwell, 1990).

The scholarship of Thomas McKeown (1976, 1979) has been particularly influential in establishing that environmental factors have been more important than medical practices, techniques, or therapies in the decline of death rates. McKeown's examination of the causes of death in Great Britain during the 1800s and 1900s led him to conclude that declines in mortality were primarily attributable to the lessening of the effects of infectious diseases. He particularly singled out the declines in deaths due to airborne infections, which accounted for 40% of the decline in mortality rates between 1848 and 1971. Foremost among the airborne infections were tuberculosis, bronchitis, pneumonia, and influenza. Lesser reductions in the mortality rate were attributable to declines in deaths due to scarlet fever, diphtheria, whooping cough, and measles (McKeown, 1976).

In addition to the airborne infectious diseases, there are those that are waterborne or foodborne. Declines in the waterborne and foodborne diseases also contributed to a reduction in the mortality rate, especially in the 19th century. The principal reductions were produced by the declines of cholera, diarrhea, dysentery, nonrespiratory tuberculosis, typhus, and typhoid.

The declines in deaths from infectious diseases did not generally occur because of medical discoveries, treatments, or practices. In many instances the deaths due to a particular disease were already declining long before the most effective medical intervention was discovered or developed (McKeown, 1976, 1979). Tuberculosis provides a striking illustrative case. In 1850, tuberculosis was the single largest cause of death in England and Wales. However, the tuberculosis death rate was declining by 1850 and continued to decline through the rest of the century. It was not until 1882 that the tubercle bacillus was identified and isolated. It is true that there were medical treatments for tuberculosis in the 19th century, but it is now known that they were ineffective therapies. When President Andrew Jackson's wife was diagnosed as having tuberculosis in the 1820s, her doctor prescribed smoking (Remini, 1984). Not until 1947, with the introduction of streptomycin, was an effective chemotherapy available. By that time, the death rate due to tuberculosis was only one sixth of what it had been a century earlier (McKeown, 1976).

McKeown even calls into question the widely accepted idea that inoculation and vaccination led to greatly lowered 18th- and 19th-century death rates. He argues that 18th-century smallpox inoculations may actually have increased the death rate by spreading the infection through the population (McKeown, 1976). However, McKeown's position with regard to smallpox has not gone unchallenged. It may be that the decline of smallpox, even before the 19th century, is one exception to his general principle that medical practices have not generally been responsible for lowering the death rate. Razzell (1974) has assembled some evidence to support the position that inoculation procedures did reduce smallpox deaths in the 18th century. And McKeown himself frequently acknowledges that the vaccination technique, which was developed after inoculation, had the effect of lowering the smallpox death rate in the first half of the 19th century (McKeown, 1979).

McKinlay and McKinlay's (1977) study of mortality declines in the United States in the 20th century also addresses the issue of whether medical practices and therapies contributed to the decline in deaths from infectious diseases. As Figure 9.3 shows, the mortality rate for the United States went down rapidly after 1900. The decline was especially pronounced between 1900 and 1950. A closer examination of this decline reveals that most of the decreases occurred because of reductions in 11 major infectious diseases (typhoid, smallpox, scarlet fever, measles, whooping cough, diphtheria, influenza, tuberculosis, pneumonia, diseases of the digestive system, and poliomyelitis). In 1900, about 40% of all deaths were accounted for by these 11 infectious diseases; today, they account for only a very small percentage of deaths.

McKinlay and McKinlay identified the date of the most prominent medical intervention for each disease and then calculated what part of the death rate for that particular disease came after that date. For example, penicillin became available in 1946 as a treatment for scarlet fever, but only 1.75% of the declines in the scarlet fever death rate came after 1946. Put slightly differently, more than 98% of the decline in the scarlet fever death rate came between 1900 and 1946. Thus penicillin as a medical therapy cannot receive credit for reducing the scarlet fever death rate. Only the death rates of influenza, whooping cough, and poliomyelitis showed substantial declines (over 25%) after the discovery of an effective medical intervention. The overall conclusion of McKinlay and McKinlay is that only 3.5% of the decline in the overall death rate since 1900 can be explained by medical intervention in the major infectious diseases.

If we accept the argument that medical techniques and therapies have had much less effect on lowering the death rate than is usually assumed, what did cause the death rate to drop? There are several alternative answers to this question, depending somewhat on the time period under consideration and on whose scholarly arguments one accepts. McKeown believes that in the 18th century reductions in mortality were primarily attributable to improvements in nutrition due to greater food supplies. With more nutritious diets, the people became more resistant to infectious diseases. The increased food supply was produced in part by the Agricultural Revolution, which involved a more efficient use of land and the scientific breeding of animals.

A second factor that improved the diets of Europeans during this period was the widespread adoption of two foods that had been brought from the American continents: the potato and maize (corn). Both added to the nutritional value of the European diet. Many scholars feel that the potato in particular, which is a very nutritious food, was a very important element in the improvement of health and resistance to infectious diseases in many European countries (Drake, 1969; Salaman, 1949).

Razzell (1974) argues that improvements in personal hygiene and cleanliness were important in reducing disease, and thus death. By contemporary standards, the English people in the 18th century (and probably most other people as well) had very low standards of cleanliness. Even among the well-to-do, it was very uncommon to take a bath. One woman of that era (an American) noted in her diary that she withstood a shower bath "better than expected, not having been wet all over once for 28 years" (quoted in Enos & Sultan, 1977, p. 191).

But beginning in the 19th century, personal cleanliness became much more important in the everyday lives of people. Washstands became capable of holding soap in the 1770s, even though the washbasins were still very small by our standards. However, the bathtub, as a fixed part of the bathroom, did not come into being until the end of the 19th century. One other evidence of greater cleanliness was the increased consumption of soap. In the first 40 years of the 19th century, soap consumption per person doubled (Razzell, 1974, p. 16). The increased use of cotton for clothing and bedding may also suggest greater personal cleanliness, since cotton goods could be washed more easily than either wool or linen clothes and bedding. In sum, there is reason to believe that some early decreases in mortality, but especially the decreases coming in the first half of the 19th century, were due to increased personal hygiene and cleanliness.

A third factor in the decline of mortality was the public health and sanitation measures introduced in many European countries in the latter half of the 19th century. The waterborne and foodborne infections were greatly reduced when water was purified, sewage was treated and disposed of properly, and foods were somewhat more hygienically prepared. The pasteurization of milk, which was introduced late in the 19th century, was a particularly important development in improving food safety. Milk without bacteria undoubtedly lowered the death rate, especially among infants.

There is also evidence that the changing status of children had an effect on the decline in mortality. Infant mortality rates declined in England during the last half of the 18th century. Coleman and Salt (1992) attribute the decline to more favorable attitudes toward children. Books on childcare began to appear and were readily received. Childcare practices that were detrimental to children, like swaddling, which restricts mobility, were no longer recommended. Breastfeeding, which does improve the life chances of infants, was adopted by more and more families.

Mortality Declines in the Developing World

The declines in mortality in the United States and western Europe were occurring gradually over a 200-year period. The economically less-developed countries of the world have had declines of about the same magnitude, but in a much shorter period of time. The last 50 years have seen substantial death rate declines in almost all developing countries and truly dramatic declines in some. Between 1950 and 1990, life expectancy in developing countries increased by 19 years, from 41 years to about 60 years (El-Badry, 1991). There are, of course, great variations from one continent to another and from country to country. For example, the life expectancy at birth in Africa is only 55 years, while throughout Latin America life expectancy is 65 years for men and 71 for women (Population Reference Bureau, 1994).

Africa is the continent with the highest overall death rate today, with many countries there having crude death rates at or over 20 per 1,000 (Population Reference Bureau, 1994). Infant death rates are especially high in Africa, averaging 92 for the continent and sometimes exceeding 140 (more than 140 babies of every 1,000 born die in the first year of life) (Population Reference Bureau, 1994). Even these numbers, high as they are in terms of contemporary international comparisons, represent significant declines from 50 years ago.

Among the developing countries, the Latin American nations have the lowest crude death rates. Brazil, the Latin American country with the largest population (151 million in 1992), has a death rate of 7 per 1,000; Mexico, with a population of 88 million, has a death rate of 6 per 1,000. Infant mortality in these two countries stands at 69 and 47, respectively, close to the average of 54 for Latin America, but high compared to the United States and western Europe. Even so, the infant death rates of Mexico and Brazil are only 50% of what they were in 1950.

The developing countries in Asia are quite varied in their death rates, but the overall rate is 10. The most impressive improvements for the largest population occurred in China, which since 1949 has lowered its death rate faster than any other major country in history (El-Badry, 1991). Taiwan and Sri Lanka have also had very large decreases, with crude death rates of 5 and 6, respectively.

Even though the death rates of some developing countries are still high (as in Africa), the countries of the developing world have clearly started to reap the benefits of the scientific and technological advances that were made in Western society over a long, two-century period. Death control measures have been adopted, whenever possible, in the developing world and have produced rapid population growth that is, as yet, unstemmed.

Caldwell (1990) has studied the mortality transition in developing countries and has attempted to explain the differences between the mortality transition in developing countries and that of developed countries in the past, and the unique causes for the current mortality decline. His underlying assumption is that behavioral changes have contributed to the more recent mortality declines.

Caldwell (1990) conducted fieldwork in West Africa, south India, and Sri Lanka, and has offered four reasons for the mortality transition in developing societies. In general, he asserts that the rapid decline in mortality, now found throughout the developing world, results from structural changes that profoundly affect the family. For example, the gradual displacement of agrarian economies with market economies transforms the family structure. Second, increases in educational attainment, especially among women, further contribute to the decline in infant mortality. Third, the importation of Western technology in health care excelerates the rate of mortality decline. And fourth, Caldwell suggests, that many people undergo a "Westernization" or shift in attitudes that also accounts for the decline in mortality. This shift in attitudes is one in which individuals realize that they have some control over mortality. Taken together, these changes lead to behavioral changes that cause people to lead healthier lives.

The essence of Caldwell's argument is that mortality declines in less-developed countries are the result of behavioral changes—changes that are occurring rapidly in the less-developed countries. Women's increasing education has an important part to play in these behavioral changes. Caldwell cautiously points out that the medical and economic changes are *not* as important as the social and behavioral changes that individuals make in bringing about the mortality transition. But again, it is the combination of the two that is necessary to explain the unprecedented rapid decline in mortality.

One general conclusion about mortality declines is that when conditions for improving one's chances of living become available, they are quickly accepted by the people. This is an important element in the workings of the mortality transition. It seems that all existing societies place a high positive value on life, which means that it is thought to be good to keep life going once it is in existence. To put the matter somewhat differently, the people of most societies can easily accept death control measures and actions because they are consistent with the prevailing value that life is good. But it also means that producing new life is considered a good thing. This latter part of the value placed on life will have an important impact on fertility, as we will see below.

There is another important reason for the early decline of mortality as modernization begins. It is quite possible for the death rate to go down in a population without any particular conscious action on the part of individuals or families. For example, when the city officials in 19th-century London put in place a sewer system, it had the effect of lowering the death rate of the city (McKeown & Record, 1962). Yet many of the people of London were totally unaware that the safer disposal of sewage could reduce deaths. In recent times, the chlorination of water supplies has been an important step in eliminating typhoid fever as a cause of death. The people using the chlorinated water supply may be unaware of its presence or its effects. Most importantly, they need make no personal decisions in their own lives in order to have a more healthful and possibly a longer life. On a worldwide basis, medical scientists have successfully eliminated smallpox from the earth. The only remaining smallpox germs are safely held in a few medical facilities. These general improvements translate into lowered death rates without individuals making any conscious changes in their lives. As we will see, the case of fertility is quite different.

THE FERTILITY TRANSITION

Much has been written about the fertility transition in developed countries, especially in countries like the United States and England. As a general rule, the decline from high birthrates to low birthrates occurred in tandem with economic development but *after* the mortality rate fell. Under the assumptions of the demographic transition model, the drop in fertility was seen as a natural and rational process in response to changing societal, and especially economic, conditions. As we have already noted, new research has shown that fertility sometimes declines before economic development, and sometimes even before mortality declines. The exceptions to the theory and the large amount of consistent evidence need to be examined in more detail.

According to transition theory, modernization brings about a decline in fertility, which typically comes after the decline in mortality. One reason why fertility declines come later than mortality declines is that the characteristics of fertility are almost the opposite of the characteristics of mortality. Whereas lowering mortality is completely consistent with the value placed on life, lowering fertility is not. The value placed on life makes producing life a positive act. Having children is, in society after society, an act that is valued, praised, and greeted with joy. If fertility in a society is to be lowered, it is necessary to some degree to act in opposition to that value. The values for producing children are embodied in customs, habits, laws, and doctrine. Not to have any children, to postpone having children, or to have few children is to run counter to the currents that exist in many societies and cultures. Even when it might be advantageous for an individual couple to refrain from having children, these prevailing values, customs, and norms are strong pressures pushing the couple to have children anyway.

The second feature of childbearing that makes changes in fertility come more slowly than mortality is that any changes in a population's fertility rates must come about as a result of personal or couple decisions and acts. While the death rate may decline if the government decides to chlorinate the water supply, the birthrate cannot be so easily affected by any government action. While there have been fanciful discussions of putting contraceptive substances in the water supply of societies with high fertility, there is no evidence that it has ever been done. Childbearing is the product of very personal behavior, so it is much less likely that the fertility of a population can be directly lowered by the actions of a government or central authority in the same way that the death rate can be lowered.

As historical demographers reconstruct the fertility transition from the past, and as demographers study the ongoing fertility transition in less-developed countries, two things become clear. First, fertility declines because of a combination of many factors, not because of any single factor. Second, the relationship between fertility and economic development is much more complicated than demographic transition theory suggests. In the following section, we will describe the fertility transition of the economically developed countries. Then we will turn our attention to the ongoing fertility transition in the less-developed countries.

The Fertility Transition in Developed Countries

The European experience shows clearly that birthrates have declined over the last 200 years. The analysis of Knodel and van de Walle (1979) leads them to the following general conclusion:

> Over the long run, and at the highest level of generality, broad developmental changes that transformed Europe from a predominantly rural–agrarian to a predominantly urban–industrial society accompanied the transition from high to low levels of fertility (p. 217).

While this conclusion is supportive of demographic transition theory, Knodel and van de Walle hasten to emphasize that this very general support must be tempered by a more detailed examination of European fertility declines. They point out that studies in 17 countries and many provinces reveal that no specific level of economic development was associated with declines in fertility. For example, Knodel (1987) has documented that the fall in fertility throughout Germany occurred in many different regions, even those that remained predominately rural. Similarly, Ireland experienced a decline in fertility at the end of the 19th century even though it remained a largely agricultural economy (Ó'Gráda, 1991).

What factors would explain the decline in fertility that generally accompanies economic development? We can identify four major explanations. These include contraceptive technology, declines in infant mortality, changing societal values and norms, and changes in the role expectations for women.

Contraceptive Technology

One of the simplest and perhaps most widely believed explanations, especially among the lay public, is that declines in fertility were produced by the availability and use of modern contraceptive techniques. But that does not seem to have been the case in the European declines in fertility.

In Europe, the declines in fertility were remarkably concentrated in a 40-year period between 1880 and 1920 (van de Walle & Knodel, 1980). Contraceptive technology at that time was primitive by contemporary standards. The method available to most people was withdrawal (coitus interruptus). Abortion was also likely to have been widely used as a means of birth control (Knodel & van de Walle, 1978, p. 234). In the 1820s, in England, the vaginal sponge had been promoted as a contraceptive in an unsigned pamphlet written by Francis Place and in a book on contraception written by Richard Carlite titled *Every Woman's Book* (Coleman & Salt, 1992). However, it is doubtful that contraceptive use was widespread when fertility declined in Europe at the end of the 19th century.

In the view of van de Walle and Knodel (1980), what spread across Europe during that period was not information on specific contraceptive techniques but rather the idea that family limitation was possible and acceptable. They argue strongly for the diffusion of *ideas* as an important element in the decline of European fertility.

In the case of England, the first family groups to limit family size were those of doctors, clergy, and lawyers. The idea of family limitation then filtered down through the other social classes (Coleman & Salt, 1992). In areas where cultural, linguistic, and religious differences are small, diffusion occurs more rapidly (Watkins, 1991; Coale & Watkins, 1986). In Belgium, for example, part of the population spoke French and another part spoke Flemish. Since France had experienced an early fertility decline, it is notable that the French-speaking parts of Belgium had declines in fertility much earlier and faster then the Flemish-speaking areas (Knodel & van de Walle, 1979).

Returning to the issue of how important contraceptive technology was in Europe's fertility decline, it must be concluded that much of the decline occurred without the benefit of modern contraceptives. Contraceptives may facilitate fertility decline, but they are not the cause.

Declines in Infant Mortality

The idea that declines in infant (and childhood) mortality will bring about a decline in fertility is a long-standing idea in demography. Knodel and van de Walle (1978) report that in German literature the idea appeared as early as 1861. Yet recent reviews of historical fertility declines show that fertility often drops before infant mortality does. When infant mortality drops first, the declines are not as great as would be expected to cause fertility declines (Donaldson, 1991; Cleland & Wilson, 1987).

At least three mechanisms have been postulated to explain how declines in infant mortality might reduce fertility. First, there is the *physiological* mechanism, which operates through the effects of breastfeeding on ovulation and conception. It has been established that as long as a woman is lactating, and while she breastfeeds a young child, she is less likely to become pregnant again. Though individual women may become pregnant while breastfeeding, most breastfeeding women will not become pregnant. Thus breastfeeding reduces the birthrate. Whenever infant mortality declines in a population, and when breastfeeding is the primary method of providing nutrition, there will be more time between births.

The second mechanism by which declining infant mortality may reduce fertility has been labeled either the *replacement strategy* or the *insurance strategy*. While slightly different, these two strategies operate in nearly the same way. The assumption is that a married couple will intend to produce a number of children sufficient to take care of them in their old age, or for other reasons. Thus, couples want a certain number of children alive when they finish reproduction. When a child dies in infancy or childhood, the couple will go on to have a replacement birth. Obviously, with this reproduction strategy, the more infant and child deaths, the more births a couple will have. The insurance strategy simply extends the point a bit. Because of the uncertainties of life under high mortality conditions, parents may have more children than they really want, or need, as a precaution against the possibility of early deaths. This too would mean that conditions of high infant and child mortality would lead to high fertility. Conversely, as infant mortality drops, the number of replacement or insurance births would be reduced.

A third mechanism might be called the *excessive survivors* effect. This mechanism differs from the others since it would probably operate through two or more generations instead of in the reproductive lives of a single couple. Beaver (1975, p. 8) suggests that individuals may control their fertility as a "simple defensive reaction to the burdens of families larger than existing social and economic structures can readily handle." When, for example, the inheritance system gives equal shares to all survivors, a large number of children will have to divide the family wealth and property into insignificantly small amounts. Or, if every daughter must be provided with a dowry, a large number of surviving daughters could be an extraordinary financial burden to a family. These burdens, produced by excessive survivors of one generation, may cause the next generation to bring down its fertility.

While all of these mechanisms are plausible explanations for why fertility declines may follow declines in infant and child mortality, the empirical support for these hypotheses, though in evidence, are weak. In fact, Knodel and van de Walle (1978) show that declines in infant mortality in some European provinces *did not* precede declines in fertility. They argue that perhaps it was high fertility that caused high infant mortality, not the other way around. They develop and support the argument that before fertility started declining in Europe, there were many unwanted children. This being the case, parents were more likely to let their unwanted babies die. There is social–historical evidence that many parents abused their children, did not feed them, left them in filthy conditions, and suffocated them when they slept in the beds of parents. Infanticide by neglect may have been resorted to by parents with large numbers of unwanted children. If this were the causal relationship, it would be understandable that infant mortality might go down only after fertility declines.

Changing Societal Conditions, Cultural Values, and Norms

Another explanation for the declines in fertility can be found in the cluster of societal and cultural changes that often accompany modernization and economic development. "Societal changes" refers to the new ways of life that people experience. Among the most prominent are living in cities and urban areas rather than in small communities and on farms, living in geographically mobile nuclear family units rather than in more stationary family systems, and attaining formal education as a basis for entering the labor force rather than learning a way of making a living from one's parents. The education of children, it should be noted, in many countries has been made compulsory, and at the same time, laws have been passed prohibiting children from entering the labor force. Also, with modernization, women are more likely to be in the paid labor force. These kinds of societal changes are a few of the most prominent that accompany modernization and economic development. Their significance for the fertility transition lies in the fact that they all may reduce fertility.

"Cultural change" refers to changes in the values of a society. Values are basically ideas about what is good, what is right, and what is proper. As societies in Europe became more developed and modern, one value change played an important part in the decline in fertility. This was the change from *traditionalism* to *rationalism*. Traditionalism and rationalism are alternative bases that people have for taking action. When traditionalism is the primary basis for action, people do essentially what the members of their group have done in the past. The customary ways of the group or the society take precedence over all other considerations. When rationality becomes the basis for human action, however, individuals behave in ways that further their own self-interests. With a rational approach, people select the goals or objectives they wish to achieve, consider which means or methods will best achieve these goals, and work to achieve them. The emphasis is on efficiency and effectiveness, and these are determined by rational calculation.

Relating traditionalism directly to childbearing, it might be said that people will have the number of children that custom dictates in their social group. In extreme high-fertility cases, this might be unlimited fertility or what is usually called *natural fertility*. This concept simply means that married couples make absolutely no effort to control or limit the number of children they have. The numbers are determined entirely by their fecundity, or their reproductive capacities. In pretransition Europe, there is reason to believe that unlimited or natural fertility was the prevailing way of life (Knodel & van de Walle, 1979).

As people move toward a more rational approach to childbearing, they have an increasing tendency to assess how many children they need or want. Coale (1973) has said that the spread of a tendency to consciously think about how many children one wants is a critical precondition for any decline in fertility. In his words, "Potential parents must consider it an acceptable mode of thought and form of behavior to balance advantages and disadvantages before deciding to have another child" (Coale, quoted in Teitelbaum, 1975, p. 422).

Norms and Role Expectations with Respect to Women

The role of women is a cultural element that takes on enormous significance in the decline of fertility (Knodel & van de Walle, 1979). Women must invariably bear the burden of unlimited fertility, and therefore they are likely to be much more receptive to family limitation than men. But if women are isolated from broader communication networks and subordinated to males, they are more likely to have larger numbers of children. When women obtain higher levels of education, and especially when they enter the labor force, they are exposed to broader communication networks. Further, these two factors are also likely to increase their status relative to men, and this, in turn, will make them more influential in childbearing decisions. When women have more voice in childbearing decisions, fertility is likely to decline.

A substantial amount of empirical evidence comes from historical demographic studies of European nations and other developed countries, as well as from contemporary developing countries, to suggest that societal and cultural changes are influential factors in the decline of fertility. These societal and cultural factors also provide us with some insights about the kinds of changes that might be made in countries where there is an interest in decreasing fertility. If these factors have been important elements in the fertility declines in Europe, they are likely to contribute to declines in countries that still have high fertility rates.

The Fertility Transition in Less-Developed Countries

Birthrates continue to be high in many parts of the world. In Africa, the crude birthrate is 42 per 1,000 and the total fertility rate is 5.9 children. In much of Asia and Latin America, the crude birthrate is above 30 per 1,000 (Population Reference Bureau, 1994). These rates are high compared to those in the developed countries, but the rates throughout Asia and Latin American have recently declined and the rates are beginning to decline in Africa. Recent attempts to formulate a model to explain the ongoing fertility transition in less-developed countries have drawn upon the historical tradition, but must be modified to reflect contemporary conditions (Caldwell, 1982; Cleland & Wilson, 1987; Hess, 1988; Donaldson, 1991; Basu, 1992). Social and cultural change (with or without economic development), and direct action by governments to limit fertility, are among the most important factors that now influence fertility.

Recent evidence coming from the World Fertility Surveys suggests economic development need not necessarily precede fertility decline. After reviewing much of the literature on the fertility transition, Cleland and Wilson (1987) conclude:

> We do not deny that in the long term economic modernization, with its need for skilled, capital-intensive labour, progressively strips children of their productive functions and tends to raise the direct costs of their upbringing. However, it is quite a different matter to assign to this change a key causal role in inducing demographic transition. We can see little

evidence to support the view that structural changes affecting family economics are responsible for initiating fertility decline. The evidence is more consistent with ideational forces. (p. 27)

What, then, are the ideational forces that exert such strong pressure on families to change their family size? Cleland and Wilson suggest that it is cultural change and education, and most importantly the willingness to accept new ideas, that change behavior and lead to declines in family size. Caldwell suggests that these ideational forces are associated with parents' aspirations for their children. Caldwell's theory relies on the family's economic aspiration, and the flow of wealth from parents to children, not the overall level of a society's economic development, to explain the fertility transition. Using a similar scheme, Donaldson (1991) suggests that it is changing group behavior in the use of time that accounts for declining fertility.

Donaldson's analysis of the fertility transition in 10 countries provides an empirical test of these various factors. First, using the traditional economic factors that have been most often used to explain fertility declines, she shows that fertility is related to increases in per capita income (a measure of economic development). Fertility declines are also related to declines in mortality. But the fact that fertility often remains high or even increases along with changes in these two variables means that they are not sufficient as explanations. Fertility decline, according to Donaldson, is the result of changing cultural norms affecting family life-styles and the use of time. Time is used here as a cultural variable. When the structure of a society changes, the life-styles of families will be affected. The cultural norms about the time that needs to be spent with children, balanced against the time needed to produce desired goods and services, interact to constrain fertility.

Donaldson and the other theorists just discussed do not discount the importance of economic change. They merely point out that the causality is complex and that the fertility transition in less-developed countries today is more fully explained by changes in culture and social behavior than by the overly simplistic assumption that macroeconomic development alone is its cause.

Additional evidence for the independent importance of cultural change as a predominant predictor of fertility change can be found by surveying the stage of the fertility transition in various parts of the world. Many countries throughout the Middle East have experienced economic development. The oil-producing countries have high per capita income. But cultural resistance to Western ideology, and the confirmation of the traditional role of women and children in the family, has kept the birthrate in the Middle East high (Sadik, 1991). For example, in the United Arab Emirates the per capita GNP is $22,220, while the total fertility rate remains at 4.1 children (Population Reference Bureau, 1994). In Africa, where most governments actively promote programs to reduce fertility, the fertility transition is beginning to occur, but economic development is lagging far behind. As Caldwell, Orubuloye, and Caldwell (1992) suggest, the African fertility transition will be different from all others because cultural traditions supporting high fertility are deeply embedded. Only in countries where the population is motivated to seek lower fertility will we see fertility declines.

WORLDWIDE POPULATION GROWTH

To understand the significance of current rapid population growth in many parts of the world, the reader must be aware of the following trends:

1. Mortality rates have gone down in the less-developed countries of the world.
2. While part of the mortality decline can be attributed to some beginnings of modernization, most is attributable to the importation of Western technology, often through the mechanism of Western aid and assistance.
3. The resulting mortality decline has occurred over a very short period of time, compared to the time it took for the European transition.
4. The birthrate in many of the less-developed countries remains high.
5. The birthrate remains high because the cultural changes that were apparently responsible for lowering fertility in Europe have not yet occurred in the less-developed countries.

The worldwide decline of mortality has been so great that the world death rate is now estimated to be less than 9 per 1,000. The important point is that the less-developed countries of the 20th century have been able to lower their death rates with great rapidity compared to the time it took for European countries to lower theirs. This is shown vividly in Figure 9.4 which shows in two separate graphs the declines in death rates and birthrates for Sweden and Sri Lanka. The same time scale is used in both parts and shows how the decline in mortality in Sri Lanka occurred over a much shorter time period than that in Sweden. The result is a much larger gap between the birthrate and the death rate in Sri Lanka. It is the distance between these two vital rates that tells the story of how fast the population is growing. In 1940, in Sri Lanka (then called Ceylon), the death rate was over 20 per 1,000. By 1950, the death rate dropped to 11.5 per 1,000 people, while the birthrate remained at nearly 40 per 1,000. A difference of this size leads to a growth rate of nearly 3% a year, which doubles the population in just 23 years.

Sri Lanka is only one example of the kind of growth pattern that has emerged in the less-developed countries since the middle of the 20th century.

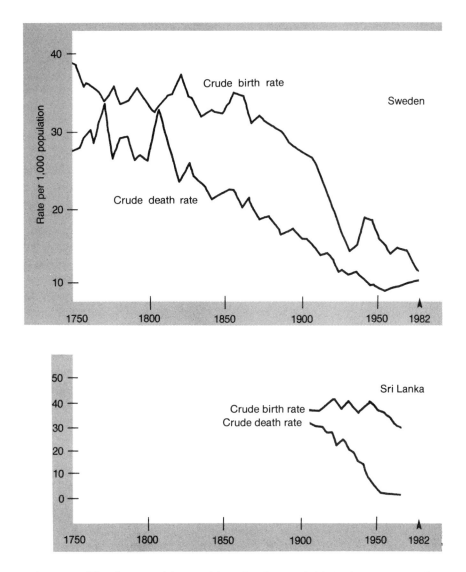

FIGURE 9.4. The demographic transition, Sweden, and Sri Lanka. From Mosher (1980, pp. 395–412). Copyright 1980 by the Population Association of America. Adapted by permission. Also from Omran (1977). Copyright 1977 by the Population Reference Bureau. Adapted by permission.

For example, for all of Latin America—though there are country-to-country variations—the death rate has dropped to 7 per 1,000 in the 1990s. Yet the birthrate remains at a relatively high 28 per 1,000. The resulting growth rate is 2.1%, which doubles the population in 34 years. If the Latin American population of about 453 million in 1992 continues to grow at its present rate, it will climb over 600 million by the year 2010.

Some of the less-developed countries of the world have not experienced quite such sharp declines in their death rates. As a case in point, the countries of Africa below the Sahara desert still have death rates of nearly 20 per 1,000 in the 1990s. But the African countries illustrate, in the extreme, the other critical demographic factor affecting growth, namely, the birthrate. A substantial number of African countries continue to have birthrates of 40 per 1,000 or greater.

The result of these factors is that great portions of the world's population, especially in Asia, Africa, and Latin America, are growing very rapidly. What should be our response to the rapid rate of population growth in many countries of the world? If the objective is to lower the rate of population growth in many countries of the world, how can this be accomplished? Indeed, can it be accomplished? These questions will be considered in the remainder of this chapter.

RESPONSES TO POPULATION GROWTH

The idea that it is desirable to slow down, or stop, the population growth rate of the world has gone virtually unchallenged. Leaders from around the world and from numerous organizations, including the World Bank and the United Nations, regularly speak of the need to slow the rate of population growth. Leaders at the Third African Population Conference held in Dakar, Senegal, in 1992, adopted the goal of "bringing down the regional growth rate from 3.0 to 2.5 percent by the year 2000 and 2 percent by the year 2010" (Third African Population Conference, 1993). Ehrlich and Ehrlich even suggest for the world as a whole, we should "embark on a slow population shrinkage" (1990, p. 180). In the United States and most other developed countries, the view is nearly universal that the rate of world population growth, especially in the less-developed countries, should be slowed down by decreasing fertility rates. While there are a few leaders of particular countries who favor continued growth for their populations, for example, Libya, these views reflect special nationalistic interests. However, there are a few contrary voices. From the United States, economist Julian Simon (1981, 1990) has challenged the view that there is a need to curtail growth. Simon views people as the "ultimate resource," a vast pool of imagination, skill, and industry that we should not try to limit.

People: "The Ultimate Resource"

Simon has vigorously challenged the widely held view that there are, or will soon be, too many people on earth. He argues that the economic evidence coming from the past suggests that population growth leads to the improvement of living conditions, not to their deterioration. He also sees population growth as leading to more plentiful supplies of natural resources, including so-called finite resources, not to their depletion.

As evidence for his positions, he cites economic data for various finite natural resources such as copper, iron, lead, and aluminum. For each of these, the long-run trend has been for their cost per unit to go down over time, indicating not scarcity but a more plentiful supply. He attributes the decline in mineral prices to improvements in acquiring and processing minerals.

Simon believes that there are two features of population growth that make the acquisition and processing of natural resources and raw materials (including foodstuffs) more efficient, and therefore supplies more plentiful. First, the very pressures produced by the needs of a population will stimulate activity to meet the pressures. For example, agricultural methods (hybridization, fertilization, irrigation, etc.) will be improved to meet food needs. Second, the more people there are, the more minds will be available to apply themselves to the relevant questions.

Simon's argument has many other dimensions and features, but his essential position can be summed up in three sentences:

1. Population growth in the past has been associated with more, not fewer, resources and raw materials.
2. Since that has been true in the past, the best prediction is that it will continue to be true in the future.
3. From a policy standpoint, if one accepts the first two points, it follows that efforts to reduce fertility should be minimized or stopped entirely.

In general, demographers and many others concerned about population have been critical of Simon (Barlow, 1994; Timmer, 1982; Sirageldin & Kantner, 1982; Preston, 1982; Willson, 1981). The most significant criticism concerns Simon's tendency to see population growth as beneficial in the long run, while paying little attention to the short run. Simon acknowledges that in the next 50 years there may be some problems created by population growth, but he argues that eventually adjustment mechanisms will take care of these problems. His critics point out that the lives of many millions of people now living might be better if we added fewer people to both their families and their countries. The 50-year "short run," which seems to be of little interest to Simon, is, in the words of two reviewers, "A span longer than the life expectancy of most people born on earth" (Sirageldin & Kantner, 1982, p. 172). These reviewers fear that if Simon's view becomes widely accepted by policymakers, the lives of many of the world's people may be harder and more miserable than they might otherwise be.

Simon's work is interesting and provocative, but it is a minority viewpoint. Most observers believe that something should be done to slow population growth in many countries to help avert starvation, disease, and poverty. There are three perspectives about what should be done and what will be effective. We turn to these views next.

"Economic Development Is the Best Contraceptive"

This slogan conveys the essence of one popular approach to the problems of rapid population growth. The obvious meaning of the slogan is that economic development will carry with it those things that have been found in the past to reduce fertility. These are, of course, the social structural and cultural factors involved in the process of modernization that seemed to make fertility drop in the last stage of the demographic transition. The advocates of the development-is-the-best-contraceptive position seem to accept the basic principles of demographic transition theory. That is, they apparently believe that the economic development of Europe reduced fertility and that the same process will occur again in the high-growth, less-developed countries of today.

This view suggests that the emphasis should be placed on economic development because then the population problem will take care of itself. In due time, there will be enough development that fertility will move downward (as it did so often in Europe). This approach also places some emphasis on the causal linkage between declining mortality and declining fertility. Earlier in this chapter, we reviewed both the plausible arguments and the empirical evidence for the suggestion that when the infant death rate of a society falls, the fertility rate will also fall, though somewhat later. This idea receives empirical support from both historical and contemporary demography, but there is enough uncertainty to raise questions about this view. The major problem with this approach is that the absolute numbers of people and the rate of population growth in the developing countries are much greater today than they were during the transition period of developed countries. In today's developing countries population growth is often so great that it severely hinders economic development.

Another version of the development-is-the-best-contraceptive approach rests on the idea that certain political and economic changes can hasten or stimulate development, with the result that desired declines in fertility will follow. Frequently, the leaders of new nations, especially those that had previously been subject to colonial control and exploitation, have been advocates of this position. Upon taking over the leadership of their newly independent nations, leaders were inclined to say that whatever the problems of a country were, they could now be resolved. These leaders had plans and programs for their new countries, and they often felt that as soon as these could be implemented, economic development would follow quickly. As population growth continued, the validity of this position was weakened and most government leaders reversed their positions.

The Family Planning Approach

Most people who think about controlling population growth advocate the family planning approach because it is the most direct and visible method of reducing fertility. The aim of family planning programs is to provide the means for individuals to have only the number of children they want and to plan how they want to space their children. The aim is not simply to convince people to use contraception or to limit the size of their families; family planning programs also provide assistance to those couples who are infertile and want to have children. Indeed, some argue that the primary justification for family planning programs is to improve the health of mothers and children (Sadik, 1991).

For the family planning approach to be effective at reducing the growth rate of populations, either many couples must already want to limit the size of their families or they must be convinced to limit their family size. Family planners believe that if contraceptives are provided, individuals will use them to reduce the number of children they have, and/or to control the spacing of births over time.

Ronald Freedman (1990), in a review of family planning programs, argued that this approach works best when it is linked with social and economic change. This means that family planning is most effective when couples have already decided that they want fewer children because economic, social, and cultural conditions point them toward wanting fewer children. Family planning, then, may be responsible for increasing the speed of the decline, but is not responsible for the decline itself. This point is also supported by reviews of family planning programs written by Mauldin and Ross (1991) and by Pritchett (1994).

Mauldin and Ross compared family planning programs in 100 developing countries. They divided these countries into four groups ranging from those that had very strong family planning programs, to those that had very weak programs. They found that in countries with strong family planning programs, the total fertility rate declined, on average, 33% between 1975 and 1990, compared to only 5% in countries with weak program efforts. The countries showing the greatest declines were in east Asia. But these were also the countries that had also experienced the most social change and economic development. Mauldin and Ross show that family planning programs can have an independent effect on reducing the fertility rate, but once again the evidence suggests that without other societal and economic changes family planning programs will not be as effective. Pritchett (1994), after reviewing data from World Fertility Survey reports and from the Demographic and Health Survey, stated a similar conclusion.

The consensus seems to be that family planning programs only work where the *motivation* to have fewer children is already in place. This has led to another response to population growth, the structural approach.

The Structural Approach

The structural approach begins with the assumption that in a general but pervasive way all societies are set up to make sure that new children will be brought into the population. In other words, childbearing is a socially motivated behavior. The social institutions and the cultural values of societies will in some substantial way encourage people to have children. Advocates of the structural approach recognize this fundamental fact and take the logical next step by saying, if the objective is to lower fertility, then these institutional structures and cultural values must be changed.

Blake (1972), in one of the early statements reflecting this view, said that it would be necessary to reduce or eliminate the "pronatalist coercive" character of most societies. She pointed to the many ways in which society pressured people to have children. Couples in many societies have no real choice. In order to begin to reduce fertility in any society, these pronatalist features have to be minimized.

At the very least, cultural norms must change. People must be willing to accept the idea that making choices about family limitation is possible. People must be able to express a preference for a certain number of children. van de Walle (1992) refers to this as numeracy.[1] If women do not know the number of children they want, or if they assume the answer is "up to God," then childbearing behavior will not change. But what could be changed in a society that could lead to new norms that will promote conscious choice about childbearing? The answer would, of course, vary from society to society, but a few examples that have some generalizability can be offered. We will consider two: the role of women and the economic value of children.

The Role of Women

When women are isolated from participating in activities outside the home, their status (or worth) rests largely on their ability to bear and rear children. Under these societal and cultural conditions, where the primary role for

[1]The concept of numeracy refers to whether or not women will respond to survey questions with a precise number of children wanted or desired. In most survey analysis, a response of "don't know" will be coded as missing data. In fertility research, this response is noted separately. For further discussion, see Riley, Hermalin, and Rosero-Bixby (1993).

women is motherhood, fertility is likely to remain high (Youssef, 1974; Dixon-Mueller, 1993). When opportunities exist for women outside of the home, then the need for women to have children as the basis for their status is reduced.

To decrease fertility, one obvious societal change would be to change the status of women. Making societal changes that will raise the status of women will give women more access to broader communication networks and reward women for doing things other than childbearing. Two societal changes that are likely to improve the status of women are to expand educational and employment opportunities for women. In study after study, the one factor that is strongly associated with declining fertility is the increase in women's educational attainment. This is true in countries where the fertility decline is well under way and more recently, in Indonesia (Gertler & Molyneaux, 1994), and in Africa, where the fertility decline is just beginning (Caldwell et al., 1992).

Equally important is the changing role of men. Axinn (1992) found that in Nepal the experiences of women are important in reducing fertility, but he also found that the activities of men are important. When husbands work in wage-labor rather than in family-based activities, when men live independently from relatives prior to marriage, and when leisure activities expand, couples are more likely to limit family size. The role of husbands may be especially important in Nepal because family planning programs have promoted a particular male contraceptive method: vasectomy.

The Economic Value of Children

In many societies children contribute to the economic well-being of their family. At a very early age, children are able to work in the fields, tend livestock, or perform other income-producing chores. Rural people especially benefit from their children, and historically, rural–agrarian people have had the highest birthrates. When child labor laws and compulsory education laws are passed (societal and structural changes), birthrates decline. As birthrates decline, another structural change is likely to take place. Children from small families are more likely to attend secondary school than children from large families. As Knodel and Wongsith (1991) have shown with regard to Thailand, when family size decreases, the population's overall educational attainment will increase. This is an important structural change related to the role of children.

The second way in which children are beneficial in many societies is that they provide economic support for their parents when they grow old. When children provide economic support for their parents, it is not in the economic interest of the parents to reduce family size. Given the lower life expectancy of males in many countries, women who become widows may be especially dependent on their sons for support.

While this argument may be used to justify having large families, evidence coming from Thailand shows that old-age support is not necessarily affected by fertility decline (Knodel et al., 1992). In Thailand, as a result of the fertility decline, it is likely that aging parents will now have only two children to help them in their old age, not five as was the norm in the past, but parents will still be cared for by their children. Thus, the decrease in family size, per se, does not affect the assistance elderly parents have come to expect.

All of these factors, when taken together, have the potential to lower the growth rate in developing countries. With limited resources and different cultural traditions, each country must develop its own plan to reduce the gap between deaths and births and lower growth rates. There is one country, China, that has attempted to reduce the rate of growth in a dramatic way, combining an extensive family planning program with structural changes. We will review the Chinese experience in the remainder of this chapter.

CHINA: A CASE STUDY OF DEMOGRAPHIC CONTROL

Over the last 25 years, many efforts have been made throughout the world to reduce population growth rates. It has often seemed that the obstacles to be overcome are so many and so deeply entrenched that reducing fertility and controlling population growth is an unreachable goal. The very populations where fertility control is most needed seem to be the most impervious to the efforts of family planners. Some of these populations are very large, running into the hundreds of millions. The people for the most part live in poverty, often at the very edge of survival. They are uneducated, illiterate, and bound by custom, tradition, and superstition. These are populations in which the family is typically the strongest institution and a woman's childbearing capacity is the badge of her social status. Under these conditions it is not surprising that too often efforts to introduce family planning measures fail or are only moderately successful.

But one country may already have provided a demonstration that all the obstacles described above need not make population control impossible. That country is China, whose population is the largest in the world. At the end of World War II, after being ravaged by the Japanese during their occupation, the masses of Chinese people were rural, illiterate, and living in poverty. Theirs was an economy of subsistence. Many Chinese lived or died solely on the basis of the yield of the yearly rice crop. When droughts or floods caused crop failures, famine quickly followed. The death rate in China was very high and its birthrate was equally high. In short, it was a society with an economy and population that was very much like those of other preindustrial societies, both past and present.

What has happened in China since World War II, especially demographically, has been truly remarkable. In 1949, a Communist revolution changed the political order of China. The Communist leadership has had intermittent turmoil, but its policies have had a striking effect on the Chinese population.

We may begin by considering the death rate, because we know that a declining death rate produces rapid population growth. China's official death rate statistics from 1950 through the 1970s are considered by most demographic experts to be of dubious validity. During that time, the Chinese government suppressed population statistics unless they reflected favorably on the country (Banister, 1987). In fact, population specialists were refused access to the statistics. In 1949, the crude death rate was reported as 38 per 1,000. It declined steadily to 22 per 1,000 in 1959, but then rose again to 44 per 1,000 during the Cultural Revolution in the 1960s. Since the 1960s the death rate has fallen rapidly to its low of 7 per 1,000 today. The substantial decreases in the Chinese death rate since 1949 may be attributed to several factors. In the 1960s China began spraying insecticides on rice fields and in households in an attempt to reduce infectious diseases. The government also began a vaccination program for children. But much of the credit for improving the health of the population must go to the "barefoot doctors." Barefoot doctors were local residents who were given short courses in basic medical care. They were trained in "first aid, disease prevention, management of common illnesses, eclectic use of Western and Chinese medical techniques and drugs, and family planning services" (Banister, 1987, p. 61). The barefoot doctors program, supplemented by cooperative medical care programs, provided an inexpensive method of basic health care for the Chinese masses.

The best assessment of China's success in reducing mortality since 1949 has been made by Banister (1987) and Banister and Preston (1981). They estimate that in 1949 the life expectancy at birth in China was less than 30 years. By 1972–1975, life expectancy in China had risen to between 61.7 and 64.4 years. It is estimated that "China has probably gained an average 1.5 years of life expectancy for every calendar year since then" (Banister & Preston, 1981, p. 107). China's progress in lowering the death rate, and thus extending life, has been two or three times as fast as is generally expected for a developing country. Only a few countries (notably Chile, Taiwan, and Sri Lanka), with much smaller populations, have managed to achieve the same impressive reduction in mortality.

Since China apparently reduced its mortality at a record rate, we would expect that its population growth rate would have been exceptionally high. Certainly, that would have been the case if Chinese fertility had remained high. The Chinese estimate of the birthrate in 1949 was 44 per 1,000. The Chinese population grew from 559 million in 1949 to over 1 billion people at the time of the 1982 census (Banister, 1987). An increase of perhaps 440 million people in 33 years is large, but it is not as large as it would have been if the Chinese had not substantially reduced fertility. The Chinese experience shows that even in the face of great obstacles to fertility reduction, it is possible to lower the birthrate. As we will see, it was done in part by taking a structuralist approach to reducing fertility.

China's Attempts to Reduce Fertility

China's efforts to control fertility directly did not come immediately with the revolution in 1949. In fact, in 1949, the Chinese leader Mao Zedong specifically denied that China had a population problem. This denial was consistent with his ideological belief that under socialism a rapid increase in population would not impede economic development. However, several societal changes that accompanied the revolution probably had some negative impact on Chinese childbearing. The new regime made a concerted effort to break down the patriarchal family system. In addition, women were given positions of greater importance, as they were brought into the political system and the labor force. This improvement in the relative status of Chinese women would, according to both theory and documented reality in a number of societies, have the potential for decreasing fertility. One additional societal change the Chinese made early was the promulgation of the marriage law of 1950. Under this law, women were granted equal status with their husbands. This law, which was designed to eradicate the age-old Chinese custom of marrying off (or selling) very young daughters, raised the legal age of marriage to 18 for women and to 20 for men. This change, which was later supplemented by a propaganda campaign promoting the positive value of late marriage, also had considerable potential for lowering fertility (Banister, 1987; Tien, 1992; Ye, 1992).

All of these societal changes were made largely for the purpose of either solidifying the revolution or eradicating the injustices and abuses characteristic of the previous regime. Direct action to decrease fertility did not begin until about 1954, a year after China had conducted its first census. Many believe that this count of the people, which revealed a population of 600 million people, served as the impetus for starting a program to curb fertility. However, Tien (1963) has argued that efforts to control reproduction had been approved by the State Council (the highest administrative unit in China) some 10 months before the official census count was announced. Whatever the reasons may have been, there was general surprise in the rest of the world when China started an intensive nationwide family planning program in 1956. By 1957, the program was in full swing, with the blessings of the top Chinese leaders. It was a classic family planning program, combining the distribution of contraceptive devices, educational programs on reproductive physiology, and a propaganda campaign extolling the virtues of small families (Tien, 1973).

What China created in the 1950s was, in effect, a population policy that combined structural changes and a family planning program. The combined impact of these two approaches were much like those that demographers would have expected. Fertility declined in this period. Figure 9.5 shows that between 1953 and 1960 there was a substantial decrease in the crude birthrate in China. Some of the declines of 1959 and 1960 can probably be attributed to serious food shortages, but the effects of societal changes and an intensive family planning program are clearly in evidence. It was during this time that all legal barriers to contraceptives were repealed. Although reliable contraceptives were still not readily available to most Chinese, the absence of legal restrictions for contraceptive use set the stage for their widespread use in later years (Banister, 1987).

However, beginning in 1961–1962, China's birthrate went up again at a very rapid pace. This increase reflects the abrupt abandonment in 1958 of the family planning program due to political factors and the onset of Mao's "Great Leap Forward." This upsurge is of some possible significance since it suggests that the societal changes alone were not enough to keep the Chinese birthrate down. It was also necessary to have a family planning program.

By 1962, the government started returning gradually to its family planning policy and program (Banister, 1987; Tien, 1963; Chen & Kols, 1982). This second campaign saw the central government playing a greater role, especially through the establishment of a birth planning office. The program was primarily directed at people living in the cities, for whom later marriage and the use of contraceptives were strongly encouraged. In 1966, this program was interrupted by the Cultural Revolution. Much of the administra-

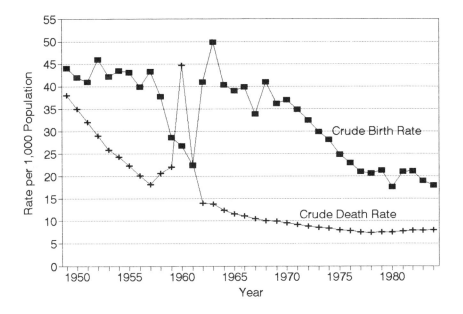

FIGURE 9.5. Crude birth and death rates, China, 1949–1984.

tive and distribution apparatus of China was disrupted by the activities of the Red Guard, and the birth planning system was affected along with other government agencies.

The third stage in China's population control program started in 1971. There is strong evidence that declines in fertility had already started among the better educated people living in the cities (Lavely & Freedman, 1990). But the government program greatly accelerated fertility decline. The official Chinese program promoted three reproductive norms. In Chinese, these are *wan*, *xi*, and *shao*, which in translation are *later*, *longer*, and *fewer*. These refer to later marriage, longer spacing between children, and fewer total children (Tien, 1980, 1983, 1992). Specifically, China's leaders asked people to delay marriage until their middle or late 20s. While the legal age of marriage was still 18 for women and 20 for men, there was much pressure on young people to wait far beyond the legal age. Compliance with the law varied by rural and urban residence. In rural areas, the average age at marriage was lower than in urban areas. In some urban areas, local authorities tried to encourage later marriage by stating that the combined age of the partners should be 50 years of age (Banister, 1987). In 1980, the legal age was increased to 20 for women and 22 for men. Given that the age at marriage had already been higher in China's urban areas, the effect of the marriage law of 1980 actually lowered the age at marriage in the urban sector, but raised it in the more populated rural areas. Overall, the average age at marriage for women in 1981 was 24.8 in the urban areas and 22.3 in the rural areas (Banister, 1987).

The rationale for encouraging later marriage is that as a result of delay the age at which women begin childbearing will also increase. Pressure to have children at 3-year or longer birth intervals and to limit families to two or three children was exerted by local birth planning committees. Couples who wished to have children had to request permission to do so from these committees. When permission was granted, by means of a planned-birth card, the couple could then have a child. The planned-birth card was necessary for the woman to receive obstetrical care and other privileges. In addition to this kind of control, there were incentives, such as paid vacations, for couples who opted to have surgical procedures to avoid future conceptions. In some areas of China, the rewards and punishments for controlling fertility were more pronounced. Food and clothing rations and work points, a form of income in agricultural communes, were granted or withheld when couples met, or failed to meet, family size objectives. Also, contraceptive use was encouraged; the intrauterine device (IUD) and sterilization were the most widely used methods, with abortion as a backup measure (Tien, 1992).

But the most dramatic and widely publicized stage in China's fertility control program started in 1979 when the "one-child" campaign was launched (Banister, 1987; Goodstadt, 1982; Chen & Kols, 1982). Since

China's birthrate had gone up dramatically in the 1960s when the first family planning program was abandoned, a large cohort of potential new parents was moving into its childbearing years in the 1980s. Chinese leaders believed that if this large cohort produced even two children per couple, their plans for China's economic development would be endangered. The Central Committee of the Chinese Communist party announced: "Women who give birth to one child only will be publicly praised; those who give birth to three or more will suffer economic sanctions" (cited in Banister, 1987, p. 184).

The policy was implemented under a variety of regional and local laws. The one-child policy was to be a voluntary agreement. For couples who agreed to have only one child, benefits would be provided; for couples who had a second child, there was a policy of neutrality; for couples who had a third child, penalties could follow. The penalties included withdrawal of the benefits given with the one-child certificate. The following benefits illustrate the types of incentives a one-child couple might receive:

INCOME:
- In urban areas, parents receive a monthly stipend, ranging from 5 to 8 percent of the average worker's wage, that continues until the child reaches age 14.
- In rural villages, parents receive as much as an extra month's work points a year until the child reaches age 14.
- In rural communes, the child receives an adult's grain ration and counts as 1.5 to 2 persons in the allocation of private farming plots.

HOUSING:
- Living space equal to that for a two-child family.
- Preferential treatment when applying for public housing in urban areas.

HEALTH:
- Two weeks of extra paid maternity leave.
- Highest priority in receiving health care for the child.

OLD AGE:
- In urban areas, a supplementary pension over and above that provided by the current labor protection law.
- In rural areas, a guaranteed standard of living equal to or higher than the local average.

CHILDREN'S EDUCATION:
- Highest priority in admission to nurseries, kindergartens, and school programs.
- Exemption from tuition costs and extra expenses from primary school through senior middle school.

CHILDREN'S JOBS:
- Highest priority in receiving desired job assignments when they grow up.
(Chen & Kols, 1982, p. 601)

The success of the one-child program has been problematic from its inception. Initially, the policy to encourage one child drew opposition primarily from rural people and the military. (Later, military personnel were among the most likely to become one-child families.) By 1982, charges of coercion against couples who had a second child were heard, and punishments rather than incentives were being used to enforce the policy. Officials continued to insist that the policy was voluntary. Most recently, charges of forced sterilization or abortions have been reported in newspapers (see Kristof, 1993; WuDunn, 1993).

To what extent the policy has been coercive is difficult to judge. There is mounting evidence that where the preference for males is quite strong, female infanticide has taken place (see Chapter 7) or female children have been hidden from the authorities. The government has responded by allowing some couples to have a second child if the first child is a female, or where having only one child presents a hardship. This is especially likely in the rural areas (Banister, 1987). But one thing can be said about the program: The fertility rate has declined, which was the government's goal.

Recent assessments of the policy of later marriage (*wan*) and of spacing (*xi*) are possible because the Chinese government has released data from the 1982 census and other survey data (Feeney & Feng, 1993; Ye, 1992; Yi et al., 1991; Coale, 1989). For example, Ye (1992) found that the policy of later marriage has increased the mean age at marriage, and as a related consequence the age at first birth is higher. But couples who in the past married and waited for 1 to 2 years before having a child have simply decreased the interval between marriage and childbearing. In another study, Feeney and Feng (1993) found similar results. Their overall assessment is that the marriage and childbearing behavior of Chinese couples have responded positively to the Chinese government's policy. Couples are marrying later, and having fewer children than before the policy was instituted; this is especially true in the cities. However, among couples who have more than one child, the policy of longer birth intervals—a minimum 4-year birth interval—has never been realized.

We can learn several things from the Chinese program. To begin with, China has now had more than a quarter of a century of experience with fertility control programs. Whenever these programs were temporarily abandoned, they were usually resumed in a more potent form. A second factor is that the Chinese political system does have an infrastructure that makes it possible for policy established at the top to be implemented throughout Chinese society. This is more true of some parts of the society than others, for it is clear that policies are often more effectively implemented in urban areas than they are in the countryside. Certain provinces of China have also been notorious for not meeting official government goals. Yet, on the whole, China's political system has made it possible to implement a strong fertility control policy (Cooney & Li, 1994).

But there is one basic and underlying factor that makes it possible for China to achieve what many would have thought a demographic miracle 25 years ago. That factor is the combined set of institutional and societal changes the Chinese have made that simultaneously reduced the motivation for childbearing. Some changes were made to ensure the success of the revolution (breaking up the patriarchal family), some were made for ideological reasons (greater equality for women, increased age at marriage), and others were made specifically to reduce motivations for fertility. In this latter category, the Chinese have recognized that people often have children because they provide economic security in old age. Providing extra income and supplementary pensions for couples who agree to have only one child is a pointed counterbalance to this age-old motivation for having more children.

The lesson to be learned from the Chinese is clear. They started with the largest national population in the world, a population that presented every conceivable obstacle to reducing fertility. The death rate of this population dropped very, very fast, which should have produced an exploding population growth rate. Yet as of the 1990s, the growth rate is in the range of 1.3%, which is low by the general standards of nations around the world and exceptionally low for a country that is still striving to achieve economic development. The official Chinese goal is to slow the rate of growth to meet the target of 1.294 billion people by the year 2000. To attain that goal, the total fertility rate must fall to 2.1 by 1995 and below 2 after that. If the total fertility rate remains the same as in 1990 (2.3), China will have 2 billion people by the year 2050 (Tien, 1992) (see Figure 9.6). The population goals of China may not be achieved, but if the present policies are continued (or, if they are strengthened), they may well be reached. In any case, the Chinese experience shows clearly that fertility declines can be achieved as the result of a strong government policy (which China has only inconsistently had), a concerted family planning program, and structural/societal changes that reduce the social and economic motivations for childbearing.

The implications of the Chinese case for world population growth are several. First, since the Chinese constitute at least one fifth of the world's population, and a much larger proportion of the less-developed countries' population, whatever happens to its population growth rate will greatly influence world figures. Second, the Chinese can serve as an example of what can be achieved demographically under very adverse conditions when a societal approach is combined with a family planning approach. (Incidentally, it should be clear that societal changes of the type made by the Chinese are not contingent upon a socialist revolution. Other political systems could initiate the same kinds of changes. It is true that having a highly centralized political system, where decisions are made at the top and promulgated down through the various levels of society, can be more effective in making changes than a more democratic system. But even democratic systems can take the kinds of actions necessary to reduce fertility [greater equality

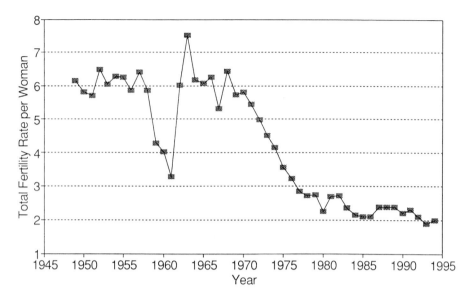

FIGURE 9.6. Total fertility rate, China, 1949–1984.

for women, rewards for having fewer rather than more children, freely available abortion and sterilization, etc.].)

Third, the Chinese case sheds some light on the effects of slowing population growth. The Chinese are working under the assumption that controlling their population will enable them to speed up their economic development. This view is the generally prevailing view among economists, demographers, and political leaders. The Chinese case will allow us to see whether economic development is speeded up when a smaller proportion of capital resources is needed for dependent children.

The Chinese may, sometime, change their course, as they have done before for brief periods. As we will see in the next chapter, what happens in the realm of population policy is often shaped by political events, and political events are shaped by the efforts of individuals and groups to gain or retain power.

SUMMARY

To understand the dynamics of recent population growth, we need to study the fundamental features of the demographic transition. A comparative framework, first proposed by Warren Thompson in 1929, suggests that changes in the vital rates of a society, its fertility and mortality, are linked to changes in economic development. As a society begins to make the transition from a rural–agrarian economy to an urban–industrial economy, mortality declines while fertility remains high. As long as high birthrates and low death rates prevail, there will be a high rate of population growth.

The experiences of countries in Europe, of the United States, and of many of the countries in the developing world show that the relationship between economic development and transitions in mortality and fertility are more complex than proposed by the demographic transition theory. The transition from high death rates to low death rates, and from high birthrates to low birthrates, are related to many social factors and occur under varying stages of economic development.

While population growth did occur in Europe, the growth rate was much less rapid there than that being experienced by many countries today. These latter countries have been able to adopt death control measures and bring about decreases in mortality without decreasing fertility in any similar way.

There have been many responses to population growth. Not all people agree that population growth is a problem. However, others argue that in the short term, excessive population growth is a terrible problem. Many programs to reduce population growth have as their core the introduction and distribution of birth control devices. But for birthrates to decline, attention must be given to the features of the target population's social structure and culture that promote fertility. Improving the status of women and reducing the economic value of children are two of the structural features that may affect a couple's motivation to have a large number of children.

China has developed several population policies in an attempt to reduce its rate of population growth. The most dramatic and widely publicized program is the one-child policy. Other programs have been implemented, all with the intent to slow population growth and promote economic development.

10

POPULATION POLICIES AND POLITICS

Policies and politics are two different things, and yet they are closely related. In brief, *politics* is primarily concerned with gaining and retaining power over the actions of government, including policy making. *Policies* are the actions taken by governments to achieve their goals and objectives. Very often, policies are adopted because they are politically useful or expedient, which means that they help persons or groups in their efforts to gain or retain political power. This chapter will examine population policies and the politics of population, and we will see that the two are very often intertwined.

POPULATION POLICIES

Since policies are the actions taken to achieve the goals and objectives of governments, *population policies* can be defined as governmental actions aimed at influencing the growth, decline, size, composition, or distribution of a population. Some governments want their populations to grow faster and thus may initiate policies to increase the birthrate, decrease the death rate, increase immigration, or some combination of the three. Other governments may wish to slow down population growth, so their policies will be intended to lower the birthrate and/or to reduce immigration. (Increasing the death rate is not generally a policy option, except in those lamentable cases in which a government targets some part of the population for extermination or genocide.)

Because of widespread concerns about overpopulation in recent decades, many countries have been trying to lower their rates of population growth. To do this they often take steps to lower their birthrates. Nonetheless, these same countries may also have policies aimed at lowering their death rates, which would, of course, add to the population.

238

Almost every government has some policies aimed at reducing mortality, usually implemented by means of public health and safety programs. Many countries have comprehensive health care systems that ensure at least minimal medical care for all residents. Often, countries have laws requiring vaccination and inoculation for major infectious diseases. Many countries have laws and regulations that enhance occupational safety and reduce environmental hazards. In the United States, some states and many localities have passed laws and ordinances prohibiting smoking in public places as a way of reducing the health dangers caused by inhaling secondary smoke.

But population policies (or policies that affect populations) are often not as clear cut and focused as these examples might suggest. Governmental actions of all kinds can influence population processes—births, deaths, or migration—and often in unintentional ways. For example, in some countries governments control the allocation of housing for citizens. When housing is in short supply and there is little hope of receiving a larger unit to accommodate a growing family, many people will limit their family size (David, 1982). Especially in countries with planned economies, where governmental policies can expand or contract the supply of residential housing, fertility rates are likely to be influenced by the housing supply.

Population policies may be divided into three types: direct, indirect, and disguised. To understand the full range of population policies we will consider these three types and offer illustrations of each.

Direct Population Policies

Population policies are *direct* when the policy is clearly and explicitly directed at influencing or shaping the size or composition of a population or of specific population processes. The public health measures described above are direct population policies since they are unambiguous in their objectives. Public health and safety policies are aimed at reducing the death rate, which means keeping more people alive and in the population longer.

Another example of a direct population policy is the one-child policy of China described in the previous chapter. This policy of fertility limitation was explicitly acknowledged by Chinese leaders as an effort to limit their country's population growth during the last two decades of the 20th century. In the 1970s the Chinese leaders recognized that half their population was under age 21. With approximately 500,000 young people poised to move into their reproductive years, the Chinese government recognized that if all couples had two or more children the growth of China's population would be staggering. The Chinese government adopted the one-child policy to avoid this anticipated addition to their country's already large population. Its action was a direct and explicit policy aimed at slowing population growth.

The immigration laws of most countries are also direct population policies since they either keep people or certain categories of people from entering a country or, in some cases, make it easy for them to do so. Countries that want more people in the population, perhaps to settle open lands, will allow almost all immigrants to enter the country. In the 19th century the United States had vast areas of land to settle and before 1875 placed no restrictions on people entering the country. The only regulation was that ship captains had to keep a list of alien passengers arriving at U.S. seaports (Espenshade, 1990a). We will discuss subsequent U.S. immigration policies later in this chapter.

Indirect Population Policies

Many governmental actions may have an incidental or, in some cases, unintentional influence on demographic processes and population characteristics. These are *indirect population policies*, which can be defined as government policies that are not directly aimed at population matters, but which nonetheless have incidental, and sometimes unexpected, influences on population factors. Housing policies, as described above, illustrate indirect population policies since the primary aim of such policies is not to influence fertility but to allocate limited housing to the people of the society. The effect on fertility is incidental, and may even be unanticipated.

A striking example of an indirect population policy occurred in the United States in the 1970s with the imposition of a 55 mph speed limit on the nation's highways. This national speed limit was introduced after the oil shortage of 1973 as a way of lowering gasoline consumption. While this law did reduce the use of fuel, it had the additional effect of lowering the accidental death rate on United States highways. From 1973 to 1974, the motor vehicle death rate, as measured by the number of deaths per 100,000 miles traveled, dropped by 15% (U.S. Bureau of the Census, 1979).

As the oil shortage eased in the 1980s, Congress considered legislation that would allow states to increase the speed limit again. During this debate, some argued that increasing the speed limit would raise motor vehicle death rates again, which was a recognition that speed limit policy had implications for the death rate. When this argument was raised, the speed limit policy moved from an indirect population policy to, at least partially, a direct population policy. (Others continued to base their arguments for a continuation of the speed limit on the goal of conserving gasoline.)

The recent increase in the Swedish birthrate, which we discussed in Chapter 3, is another example of indirect population policy. In our earlier discussion we noted that the upturn in Swedish childbearing after 1986 was probably the result of laws giving employed women (and men) paid maternity leave (Hoem, 1990). Maternity leave policy is part of Sweden's general policy of making it easier for women to combine careers with parenthood. The maternity leave policy was also designed to improve infant and child care in Sweden, and to involve fathers as well as mothers in caring for their children. But, as one observer has noted, these policies are "largely indirect pronatalism" (Hoem, 1990, p. 740).

When the maternity leave law of Sweden was modified in 1986, it became economically advantageous for couples to have a second child born within 30 months of their first child. This law probably led to increased fertility in Sweden as couples modified their childbearing plans to take advantage of the economic benefits of the law (Hoem, 1990). Many couples probably had a second child much sooner than they might otherwise have had one.

An historical example of indirect population policies has been identified by Johansson (1991). Johansson believes that European fertility declines in the 19th century were the result of "implicit"[1] government population policies, and uses the historical case of England and Wales to demonstrate his view.

In Great Britain, by the 1870s, the government was promoting the idea of responsible parenthood, and emphasized that parents, "not God or fate, must bear responsibility for their child's welfare" (Johansson, 1991, p. 397). This view of parenting differed from an earlier view that placed much less importance on parental responsibility. As British parents began to accept this new standard of parenthood, they experienced social pressure to curtail childbearing in order to meet their new responsibilities.

More importantly, in the decades after 1870, the British government passed laws prohibiting child labor and making schooling compulsory. These actions were aimed primarily at improving the welfare of children, but they had the indirect effect of lowering fertility in England and Wales. The child labor and compulsory education laws made children less valuable as potential wage earners for their parents. When children are barred from employment and no longer can make economic contributions to the family, married couples in the working classes generally reduce their family size (Caldwell, 1982). Compulsory education laws, especially when combined with laws prohibiting child labor, were important deterrents to large family size in 19th-century Great Britain. While these laws may have been passed with the explicit aim of improving the welfare of children, they had the indirect effect of lowering British fertility (Johansson, 1991).

The importance of mass education as a means to reduce fertility can also be seen in contemporary developing countries (London, 1988, 1992; London & Hadden, 1988). A recent study has found that when the females of developing countries are given more schooling, the country's fertility rate is likely to decline (London, 1992). The opposite side of this relationship will be considered later in the chapter when we discuss how India's lax enforcement of school attendance laws helps to keep Indian fertility relatively high.

---

[1]Johansson's use of the term *implicit* policies is nearly equivalent to our use of the term *indirect* policies.

Disguised Population Policies

The assessment and evaluation of government population policies is further complicated by the fact that governments sometimes deliberately disguise their population policies. A *disguised population policy* is a policy that is explicitly directed toward one objective, when, in fact, it is aimed at something different. The most common instance of disguised population policies takes place in the area of population control or limitation. Governments are often reluctant to tell their people that there are too many of them, or that they should reduce their fertility. Thus, when government leaders introduce population control measures, they may justify doing so by providing some other reason. The most prominent example of disguising objectives is found in connection with abortion policies.

The availability of abortion has proved to be a very effective way of reducing and controlling birthrates, and a number of countries have adopted abortion-on-demand policies for that purpose. Yet the official and thus legal justification for allowing abortion is often couched in terms of maternal health. In Chapter 3 we described how the Japanese needed to reduce their fertility after World War II and used a very liberal abortion policy to achieve that objective. But the official justification for their liberal abortion policy was not to decrease the birthrate, but instead to *improve maternal health*. The Japanese policy, which almost certainly was aimed at reducing fertility and limiting the population, was disguised as a maternal health policy (Oakley, 1978).

When state legislatures in the United States started liberalizing abortion laws in the 1960s, their actions were almost invariably justified on the basis of contributing to the physical and mental health of women. While it is difficult to prove, it is probably true that some legislators were equally interested in controlling fertility and population growth, especially among some segments of the population. During this period there was widespread concern about the high and increasing costs of social welfare payments, including aid to dependent children (Berkov, 1971). Since unwanted and illegitimate children were assumed to be among those most likely to receive governmental aid, some legislators probably favored more liberal abortion policies as a way of reducing welfare costs. But again, this was not usually the justification offered. The liberalization of abortion laws in the various state legislatures was done ostensibly for the purpose of improving maternal health.

INTERNAL AND EXTERNAL POPULATION POLICIES

Population policies can also be divided according to whether they are internal or external. An *internal population policy* is a policy of a nation, or some other political jurisdiction, that primarily affects its own population. When a nation initiates and supports a family planning program among its people, it is executing an internal population policy. An *external population policy* is a policy of a nation that affects the population of another nation or group of people. When, for example, a nation, through its foreign aid and international assistance programs, assists another country with its family planning program, that action is part of its external population policy. Or, taking a recent example, when in 1992 various countries, and the United Nations, provided food and assistance to the starving and ill people of Somalia, they were carrying out an external population policy. The policy was aimed at reducing Somalian deaths due to starvation.

External Population Policies

External population policies are often played out in the arena of international politics, where individual nations or blocs of nations vie for power and control in international affairs. At the international level, competing ideologies or national interests often cause nations to take opposing positions. Nowhere can this be seen more clearly than in the United Nations, where for several decades there were debates about the world's "population problem." These debates effectively delayed actions that might have addressed a rapidly growing world population much sooner and more effectively.

The first phase of the U.N. debates over population lasted from approximately 1953 to 1965. During this period the questions debated were: Is there a population problem? And if there is, what should be done about it?

United Nations Debates of the 1950s and 1960s

While many demographers and other scientists were becoming concerned about rapid world population growth in the early 1950s, most political leaders in the United States and elsewhere were neither interested nor concerned. At the United Nations, this lack of concern was manifested in the very tentative and limited way in which population was approached. A U.N. Population Commission was established in 1946, but its main role was "to undertake studies and make available the best technical knowledge [about population]. It was not intended to be a decision-making body" (Symonds & Carder, 1973, p. 42). In short, the U.N. Population Commission was only a data-gathering agency, and as such it did begin to assemble statistical evidence showing that the populations of many countries were growing at a rapid rate.

Despite this evidence, there was a long-running debate in the United Nations about whether a population growth problem existed, and whether the organization should get involved in activities designed to slow it. As this debate continued into the 1960s, two groups of nations took opposite sides.

On one side, the United States,[2] the United Kingdom, Sweden, and India took the position that the world was facing a serious population growth problem. On the other side, an unusual coalition of Catholic and communist countries argued that there was no population problem. The Catholic and communist countries, so often opposed on many issues, were closely aligned on the issue of population, though for very different reasons.

The traditional position of the Catholic church on contraception is well known because papal pronouncements have persistently renounced the use of any method of birth control other than the rhythm method (McQuillen, 1979; Murphy, 1981). In 1994, Pope John Paul II assembled 114 Catholic cardinals from around the world for a meeting that reaffirmed the church's "unbending position against artificial birth control and abortion" (Von Drehle, 1994, p. A1).

With respect to the population debate of the 1960s, the Catholic church advanced another view about population pressures, one that is generally less well known. Pope John XXIII (1958–1963) acknowledged that many countries were impoverished, and while he did not attribute their poverty to overpopulation, he did express the view that it was the duty of the rich nations of the earth to assist the poor nations. This position was echoed by Pope Paul VI in a papal encyclical issued in 1967, in which he acknowledged that population problems did exist, but argued that their solution might best be found in the economic development of the poorer nations.

The communist or socialist position was remarkably similar to the Catholic position, but on one point it was diametrically opposed. The communists viewed birth control of all types, including abortion and sterilization, as acceptable, and indeed as a basic human right.

The fundamental communist position on population came directly from Marxist writings, which were largely a reaction to the work of Malthus (see Chapter 2). While Malthus believed that human misery and poverty were caused by populations growing faster than resources, Marx attributed these ills to the exploitative nature of capitalism. In his writings Marx emphasized the idea that overpopulation is just an excess of labor supply that is artificially created by the capitalists as a way of controlling and exploiting the masses. In a socialist economy, according to Marx, there could be no overpopulation.

In the U.N. debates of the 1960s the coalition of Catholic and communist countries opposed and impeded every step that would have moved the organization closer to taking action on the rapid population growth that was occurring in many countries. However, it became increasingly

[2]The United States came rather gradually to the view that there was a population problem and that something should be done about it. During the 1950s, under the presidency of Dwight Eisenhower, the U.S. position on population control was very timid. But increasingly through the 1960s successive administrations recognized the seriousness of the population problem (Donaldson, 1990).

difficult for the United Nations to ignore or refuse the requests of a number of member states for assistance in curbing their birthrates. As Symonds and Carder (1973, p. 202) put it in reference to India, "Once the Indian government had adopted a population policy and asked for international aid, it became increasingly difficult for the problem to be ignored by the United Nations agencies."

For reasons that are not apparent, communist opposition to U.N. population control activities started to diminish around 1965. After that time the United Nations, through the World Health Organization, began to provide some economic support for population control and family planning activities. While these efforts have since been helpful in curbing population growth in some countries, the United Nations was guilty of at least a decade of delay in tackling the population problem because of the politically inspired debates of the 1950s and early 1960s.

The next stage of the international debate about world population growth was played out in three world conferences, one in Bucharest, Rumania, in 1974, the next in Mexico City, Mexico, in 1984, and the last in Cairo, Egypt, in 1994 (Finkle & Crane, 1975; Haupt, 1984; Mauldin et al., 1974; Stycos, 1974).[3]

World Population Conference: Bucharest, 1974

By 1970, the population specialists of the United Nations considered it an appropriate time to organize a population conference that would formulate and adopt a world plan of action for controlling population growth. From 1970 to 1974, preliminary regional meetings were held in different parts of the world. On the basis of the discussions at these meetings, a Draft World Population Plan of Action was prepared for the upcoming 1974 conference in Bucharest, Rumania. The draft plan, in its essentials, stated that rapid population growth constituted a serious impediment to economic development in many countries. It went on to argue that individual nations and international agencies (such as the United Nations) should commit themselves to carrying out population control and family planning programs (Finkle & Crane, 1975). While the draft plan did not ignore the connection between economic development and population growth, it did place greater emphasis on population and family planning programs as the way to achieve economic development.

The organizers of the conference wanted the focus to be limited to population control and family planning programs and not spread into broader economic, political, and social issues. But their hopes in this matter were soon dashed. Instead of approving the draft plan as it was written or merely amend-

---

[3]There had been two previous U.N. meetings on world population—Rome, 1954, and Belgrade, 1964—but these were comprised largely of population experts and the focus was on technical demographic matters more than population policy.

ing it, the conference participants engaged in acrimonious debate. The population issue suddenly became, once again, highly politicized.

The two opposing groups at the Bucharest conference can be labeled the "incrementalists" and the "redistributionists." The incrementalists saw population growth as the major inhibitor of economic development. They believed that nations would move toward economic development by taking incremental steps, the first of which would be to limit their population growth. The countries most aligned with the incrementalist position were the United States, the United Kingdom, Canada, and West Germany[4] (Finkle & Crane, 1975).

The redistributionists stressed the idea that population problems were produced by economic underdevelopment and argued that the most effective way to achieve development would be to redistribute economic resources among the nations of the world. The essential point of the redistributionist position was that the rich countries of the world had long exploited the poorer, underdeveloped countries and that it was now time to redistribute some of the world's wealth (Finkle & Crane, 1975).

On the redistribution side of this debate were most of the Third World countries. India, for example, which had previously aligned itself with the population-control position, now took the side of the redistributionists. Argentina, a Catholic country that had long opposed population control efforts, was also on the redistributionist side, as was Algeria, a country with a socialist regime.

In general, the European socialist countries were on the redistribution side of the debate. They felt that colonial/capitalist countries had exploited the less developed countries and that therefore they should make amends by giving these countries the money they needed.

It should not go unnoticed that the redistributionist position was consistent with the Catholic church's position as it had been enunciated by a succession of popes. Catholic countries, especially those in underdeveloped parts of the world, such as in Latin America, were understandably aligned with the redistributionist position.

The compromise plan of action that eventually came out of the Bucharest conference was very far from what the organizers had anticipated prior to the conference. It was a compromise between the two sides of the debate recognizing the importance of both economic development and family planning programs. While it did not stop efforts at population control, this world conference did introduce a new political dimension into the population debate.

World Population Conference: Mexico City, 1984

In the summer of 1984, 10 years after the Bucharest conference, the nations of the world assembled in Mexico City for another population conference.

---

[4]This occurred prior to the reunification of West Germany and East Germany.

The position of the United States at this conference was particularly interesting because it was a near reversal of the position it had held 10 years earlier. The United States had for decades supported family planning programs,
but in 1984 the United States opposed them. The Reagan administration
announced that the United States would no longer support any family planning organizations or programs that promoted abortion.

A second major tenet of the U.S. position at the Mexico City conference was that a greater emphasis should be placed on economic development,
especially development in a "free economic environment" (Haupt, 1984,
p. 8). This view echoed the redistributionist position of a decade earlier, which
also stressed the primacy of economic development over family planning
programs. However, unlike the redistributionist position, which was based
on the idea that the wealthier nations should give more economic aid to the
less wealthy nations, the new U.S. view argued for free-enterprise economic
systems as the best route to economic development. The abrupt turnaround
in the position of the United States in the 1980s is another example of the
way in which politics can influence population policy.

But that was not to be the last sudden change in the U.S. position, for in
1993 when Bill Clinton became president U.S policy reverted back to nearly
what it had been in the pre-1980s period. Three days after taking office, the
new president signed an executive memorandum overturning the "Mexico City
policy." President Clinton stated that the restrictiveness of the policy had "un-
dermined efforts to promote safe and efficacious family planning programs in
foreign nations" ("Clinton Overturns Mexico City Policy," 1993, p. 1).

The Clinton policy withdrew the restrictions on giving foreign aid to
organizations that were engaged in abortion-related activities. However, a
U.S. law passed in 1973 still prohibited U.S. funds from being used "to pay
for the performance of abortions as a method of family planning, or to
motivate or coerce any person to practice abortion" ("Clinton Overturns
Mexico City Policy," 1993, p. 1).

World Population Conference: Cairo, 1994

When the most recent international population conference on world population met in September of 1994, the United States was again strongly in
support of family planning programs. But, President Clinton, and the American delegation led by Vice President Gore, also stressed the importance of
equality for women, especially equality in educational opportunities, and
urged more international support for maternal and child health.

The major controversies at the Cairo conference swirled around opposition to abortion by the Vatican and the concerns of some Islamic fundamentalist countries that the proposed family planning and abortion policies
might weaken Islamic families. But, in the end, compromises were reached
and most observers concluded that the 180 nations of the world had reached

a higher level of accord on population policy than ever before (Kalish, 1994; Rensberger, 1994).

The major new perspective to come from the Cairo conference was an emphasis on the empowerment of women. In developing countries, women have often been viewed as "the problem," because of their high fertility rates. The Cairo conference sees the empowerment of women, through better education, economic opportunities, family planning, and improved health, as the agents for the solution to problems (Kalish, 1994, p. 2).

Internal Population Policies

As we have seen in the preceding section, political considerations—actions taken to gain or retain political power—have often played a part in the external population policies of nations. But political influences are even more in evidence when we turn to the internal population policies of individual countries. To illustrate this point we will describe the population policies of three nations: India, Iran, and Singapore. These countries have not been chosen at random; they were selected because each illustrates the influence of politics on population in different ways. Later in this chapter we will focus on the population policies of the United States, giving special attention to immigration policy.

India

The Politics of Population in India. India has been trying to lower its birth-rate for over 40 years; for four decades it has been official government policy to slow population growth. Yet the census of 1991 revealed that 160 million people had been added to the population in the 10 years since the 1981 census (Chaudhry, 1992). India's population is growing at such a rapid rate that in the next century it will be vying with China to be the country with the largest population in the world.

The continuing high growth rate of India's population has recently been described as "an unmitigated disaster for the country" and "a devastating indictment of the family planning programme" (Banerji, 1992, p. 883). Many observers share the view that the population control program of India has been only minimally successful (Narayana & Kantner, 1992), and some have judged it to be an "abject failure" (Demerath, 1976, p. 80). In the light of these evaluations, we may ask why has the Indian population control program had such limited success? For the answer, we need to look at the way the policy has been supported and carried out by the country's political leaders, and also at the indirect population policies of India that continue to support high fertility.

When India gained its independence from Great Britain in 1947 it had been a crown colony for 70 years and had been under British control for a century and a half. India's first prime minister after independence was Jawaharlal Nehru, who initially announced that his country did not have a population problem. While admitting that there was widespread poverty in India, he attributed this poverty to the exploitation of the Indian people caused by British colonial rule. In this respect, he was responding in the style of many political leaders of "new" countries. When new countries come into being, either through revolution, political coup, or the peaceful withdrawal of a colonial ruler, their new leaders are usually reluctant to advocate limiting the size of the population (Stycos, 1963). It is simply not good politics for a new leader to begin a new government by urging reductions in the number of the nation's people. New leaders are likely to blame the leadership, or the regime, that preceded them for whatever problems the country may have.

By 1952, Prime Minister Nehru had changed his position and had finally initiated a nationwide program of population control (Visaria & Visaria, 1981). From the very beginning of India's program of population control, however, and with few exceptions thereafter, it has been limited, tentative, and cautious (Demerath, 1976). For example, in the first year of India's population control program only $125,000 was spent, and even this meager amount did not use up the total budget allocation (Narayana & Kantner, 1992).

Population control in India has, from its beginnings, been a highly sensitive political issue. Many Indian politicians have been reluctant to support family planning; it is usually an issue that can damage, but not enhance, a political reputation (Cassen, 1978). Or, the population and family planning issues have been unexciting and uninteresting to politicians; in the words of two recent observers, "population control is an indifferent matter for most politicians" (Narayana & Kantner, 1992, p. 73). Thus, while India has had an explicit policy to control population growth, it has not generally pursued that policy vigorously.

Only in the late 1970s did India's government have, for a short period of time, an aggressive population control policy. Beginning in 1976, under the leadership of Indira Gandhi, the government announced a new national population policy (Cassen, 1978). At the national level, the parliament legislated that henceforth each Indian state would receive government grants and other funds and have representation in the national parliament on the basis of its 1971 population. This program was designed to assure the different states that they would not lose their already established degree of political power or funding if their populations should decline more than those of other states. The new policy also stressed the importance of female education, nutrition, and basic health services (Cassen, 1978). In addition, the minimum age of marriage for females was raised to 18 and the minimum age for males to 21. The policy also called for government departments and employees at all levels of government to help in efforts to motivate citizens to adopt responsible reproductive behavior.

But the most controversial parts of India's 1976 population policy had to do with sterilization. First, the financial incentive for having a sterilization operation was increased and put on a graduated scale (e.g., parents of two children who submitted to sterilization received more than parents of four children). Second, and much more controversial, the law gave state legislatures permission to mandate compulsory sterilization. Third, states were urged to develop systems that would reward government employees for adopting the small-family norm, and for punishing them if they did not.

Under this new policy, India moved toward the greater use of male sterilization, which was pushed with great vigor. Indeed, some have said with excessive vigor (Narayana & Kantner, 1992). Since the individual Indian states could establish their own sterilization programs to carry out the government's policy, programs and practices varied greatly across India.

A journalist gave an account of one shocking incident, which may have been typical of others too. In a Muslim village near Delhi, at 3:00 A.M., all men over 15 years of age were roused from their beds and told to report to a central road. The men were then sorted into groups according to their appropriateness for sterilization. About 400 men were taken to various police stations, where they were charged with crimes—possession of illegal arms and suspicion of threats of violence—and then taken to clinics where they were sterilized (Gwatkin, 1979).

These kinds of actions were not routine, but there is evidence that in their zeal to carry out the national policy, some local and state authorities occasionally resorted to extreme measures. Such extreme actions, however, even if sporadic and infrequent, were enough to turn public opinion in India against the government's population control policy.

In 1977, when elections were called in India, the government of Indira Gandhi was under heavy criticism, although it must be noted that the population policy, and the abuses that may have been related to it, were only part of the cause for this criticism. As early as 1975, Gandhi had declared a national emergency and had adopted an extremely repressive regime (Cassen, 1978). It is not, therefore, surprising that Indira Gandhi's political party lost the election and she was unseated as the prime minister.

What happened next is a classic example of how politics can shape population policy. The incoming majority party, the Jenata party, led by Prime Minister Desai, disassociated itself as much as possible from the family planning movement and especially the sterilization program. In the last months of the Gandhi regime, nearly a million sterilizations had been performed. In 1977 and 1978, under the Janata party government, the number fell to about 50,000 per month (Nortman, 1978). Prime Minister Desai took special pains to disassociate his government's policy from the previous program of Prime Minister Gandhi by announcing that "compulsion in the area of family welfare must be ruled out for all times to come" (quoted in Nortman, 1978, p. 728).

The Gandhi government had set an objective of reducing the crude birth-rate of India to 25 by the early 1980s. The Desai policy set a less ambitious goal of 30 by 1983 (Nortman, 1978). In 1994, the crude birthrate of India stood at 29.

The events of 1977–1978 in India clearly show how politics influences population policy. The Desai government apparently considered the family planning program too dangerous to pursue with vigor. But the events of the years that followed illustrate even more clearly how political considerations can take precedence over demographic realities.

In 1980, India again held national elections. This time the Congress party, still under the leadership of Indira Gandhi, achieved an overwhelming victory. Even though she was once again the Prime Minister, Indira Gandhi did not revive her previous family planning and sterilization policy (Narayana & Kantner, 1992). Her son, Sanjay, was quoted in the Indian press as saying that the previous program would be dropped because "people don't like it" (quoted in Kocher, 1980, p. 310). Gradually, in the years since, India has resumed a family planning program, but never with the same intensity, or with the same results, as in the 1976–1977 period (Narayana & Kantner, 1992).

Critics of the current Indian family planning program are making the same charges of weak leadership with limited competence as were being made nearly 20 years ago (Banerji, 1992; Demerath, 1976). The recent ministers of health and family welfare, who are charged with administering the family planning program, have been described as "political lightweights or discards" who are shuffled in and out of the Indian cabinet (Banerji, 1992). If the leaders of the family program are weak and incompetent, the program is not likely to be successful. But there is another factor that contributes to the less-than-impressive results of India's family planning program: Some indirect policies and practices are operating in ways that are keeping fertility high in large parts of the population.

Indirect Population Policy in India. Earlier in this chapter we saw that compulsory mass education combined with laws prohibiting child labor have the effect of reducing fertility (London, 1992). When children can no longer add to their parents' income by working because they are compelled to remain in school and cannot be employed, parents tend to reduce their fertility. In India, Myron Weiner (1991) has found, the opposite situation prevails. Vast numbers of Indian children are still working, and the compulsory education laws of India are widely ignored.

Less than half of India's children (48%) between the ages of 6 and 14 are attending school. Rural children have especially low attendance. They stay at home and tend the cattle, take care of younger children, collect firewood, and work in the fields. Urban children are often employed in cottage industries (e.g., making fireworks), in restaurants, and as household workers (Weiner, 1991, p. 3).

According to Weiner, a combination of forces in India actually conspires to keep enrollment and attendance in the schools very low. Small businesses, the educational establishment, and the parents of school-age children are all involved in a subtle collusion that keeps many children out of school. Parents and employers both find child workers economically beneficial and profitable. Employers get a docile and cheap workforce, while poor parents collect the meager earnings of their children. The schools benefit when large numbers of school-age children are absent from the classroom because then there will be fewer children to teach with the same facilities and resources.

The higher social classes of India and many political leaders of India also contribute to the lax enforcement of compulsory education and child labor laws. Members of these groups generally believe that education is more important for children who will work with their minds when they reach adulthood, while children who grow up to work with their hands have little to gain from formal education. As long as the upper classes see no great advantage to enforcing the compulsory education laws or the child labor laws, poor parents will be allowed to keep their children out of school and working.

These beliefs and behaviors are an indirect stimulus to fertility, since parents generally continue to have large families as long as children are a source of economic value. This is the principal tenet of Caldwell's (1982) *wealth-flow* hypothesis, and what Lesthaeghe (1992, p. 15) has called the *economic utility of children*. Under social and economic conditions where children, even at a young age, provide labor for their parents or contribute the wages of their labor to their families, family size will tend to be relatively large. Thus, the policy of not enforcing compulsory education laws keeps fertility relatively high and acts in opposition to the official government policy of reducing fertility.

Iran's Population Policy and the Islamic Revolution

Iran conducted its first modern census in 1956 and numbered the population at 19 million. Despite suffering a large number of casualties during its drawn-out war with Iraq during the 1980s, Iran today has a population of more than 61 million. Over the last decade, Iran's growth rate has been well over 3% a year, a rate of growth that will, if it continues, double the population in a little over 20 years (Aghajanian, 1991). By the year 2010, Iran could have a population of over 100 million if it continues to experience its present growth rate (Population Reference Bureau, 1994).

It will be Iran's birthrate, of course, that will determine whether the 3% growth rate will persist. At the present time the crude birthrate in Iran is very high, at 44; the total fertility rate is 6.6, which means that at present fertility levels an Iranian woman would be having an average of nearly seven children during her reproductive lifetime (Population Reference Bureau, 1994).

Over the last three decades the birthrate of Iran has fluctuated, often in response to the population policies of changing political regimes. The steady decline in fertility that usually occurs as a society experiences economic development seems to have been at least temporarily stalled by Iran's political events (Aghajanian, 1991).

During the reign of the Shah of Iran, in 1967, Iran adopted an explicit family planning policy aimed at reducing fertility. The main elements of Iran's program conformed to the standard family planning program: education, a promotional campaign, and readily available contraceptive services. The program was strengthened in 1970, and again in 1976 when abortion was made legal. The legalization of abortion was partially disguised through a legal loophole: A new law allowed women to have *"any kind of operation . . . provided the consent of all persons was obtained"* (Aghajanian, 1992, p. 4, emphasis added). Even though abortion was not mentioned specifically, Iran had adopted a *disguised* abortion policy.

The official family planning policy was also significantly augmented by a policy "enhancing the legal and social status of women and increasing their participation in the social and economic domains outside the household" (Aghajanian, 1991, p. 709). This relatively liberal policy with regard to the rights and status of women was part of a general policy of modernizing and westernizing Iran that was pursued during the Shah's regime.

The combined effects of the family planning program and the more liberal treatment of Iranian women had the expected effect on fertility. The crude birthrate declined from 49 in 1966 to 43 in 1976.[5] Among urban women during this decade, the crude birthrate declined from 45 to 33.

When the Shah of Iran was deposed by the Islamic Revolution in 1979, the family planning policy and the more liberal policy toward women were both reversed. The government's family planning council was dissolved immediately after the revolution and many family planning clinics were closed.

After the revolution, the role of women in Iranian society became, once again, very narrowly circumscribed. This pattern of reinstating a stringent patriarchal system, which subordinates women, has occurred wherever Islamic religious fundamentalists have gained control of countries (Obermeyer, 1992). In Iran, political and religious leaders (often the same men) stressed that women should be largely restricted to the domestic area and they "were praised above all for being good mothers and wives" (Aghajanian, 1991, p. 713). Women were segregated in public places, at universities, and at work. The permissible age for marriage was lowered to 9 for females and to 14 for males—early marriage and procreation were encouraged.

Again, as would be expected, these changes in family planning policy

---

[5]The 1966 and 1976 dates do not correspond precisely to the times of political change, but to the dates of decennial censuses. The same is true for the census of 1986, which was only 7 years into the Islamic fundamentalist regime.

and in the social status of women had the anticipated effect of increasing childbearing. Between 1976 and 1986, the population growth rate of Iran started to accelerate again, after having declined in the earlier decade. The crude birthrate went up to 48, which was just 1 point short of the 1966 level.

By the early 1990s, the political leaders of Iran—for reasons best known to them—were shifting their views on population growth once again. The new view was that the rate of growth should be slowed and fertility should decline. The government was importing the contraceptive implant Norplant and making it available, along with the pill and the IUD, in public health clinics (Aghajanian, 1992). Male sterilization was being encouraged for the first time, and the government was using radio and television to promote two- or three-child families. In 1991, the government announced that the fourth child in a family "would not be eligible for food rationing or nutritional supplements and other public child benefits" (Aghajanian, 1992, p. 4).

This review of Iran's population policies (both direct and indirect) in the last few decades illustrates how political changes can bring about increases in fertility as well as decreases. A similar pattern may be seen in the case of Singapore, but for different reasons.

Singapore's Too Successful Population Policy

Singapore is an island society off the tip of the Malayan Peninsula, with a current population of slightly under 3 million people. In 1966, the population was growing rapidly for a tiny country that had limited room for expansion. Singapore at that time adopted a population policy that was aiming to "reach replacement reproduction by 1990 and to achieve zero population growth by 2030" (Singh, Fong, & Ratnam, 1991, p. 73).

Singapore established an effective family planning program that combined free contraceptive services, the availability of abortion, and the promotion of voluntary sterilization for mothers with more than one child. By 1975, some 15 years *before* the 1990 target date, the goal of replacement-level fertility had already been reached. Declines in fertility continued into the 1980s and population projections indicated that Singapore would soon have a high proportion of old people in its population.

By 1984 the Singapore government, led by Prime Minister Lee Kwan Yew, became concerned about the low level of fertility, and especially about the small number of children being born to the more educated and affluent members of the society (Singh et al., 1991). Prime Minister Lee became very concerned that the low level of childbearing by the most educated women and the higher levels of childbearing by the least educated women would lower the intellectual level of the population (Gould, 1987). He therefore initiated a policy that would reward the most educated women for having children and would give payments to the least educated women for having sterilization operations. The women with the highest educations could receive a maximum payment of up to $10,000 for any child they might have. (The payment for having a child was based on a percentage of a woman's earned income, e.g., the payment for a third child was 15% of a woman's income.) Women at the lowest educational levels, by contrast, could receive $10,000 if they had a sterilization operation. Low educated women were only eligible for this money if they had no more than two children, and they had to meet a number of other requirements, including very low incomes. Furthermore, if one of these poor, low educated women did, for some reason, have another child after her sterilization operation, she was required to return the $10,000 with 10% compound interest (Singh et al., 1991, p. 74).

This policy, with its clear elitist bias, was only marginally successful in raising Singapore's birthrate, and was changed after 2 years. The new 1986 policy encouraged women of all social and educational levels to have children. This new pronatalist policy also led to a revision of the family planning slogan of Singapore. In the population control era the slogan had been "Two is enough," but now it has been changed to "Three children or more, if you can afford it" (Singh et al., 1991, p. 75).

The 1986 policy also included additional financial allowances for second, third, and fourth children; preferential school choices for mothers with three children or more; medical insurance for pregnancy care and the delivery of third or later children; and greater work-time flexibility and childcare services for employed mothers. Abortion continued to be available in Singapore, but counseling became mandatory before a pregnancy could be terminated.

Within 2 years after the 1986 policy went into effect the birthrate in Singapore had increased from 16.7 to 20.0. By 1994, however, the birthrate had dropped somewhat to 17.0 (Population Reference Bureau, 1994).

The population policies of India, Iran, and Singapore illustrate the reality that policies designed to reduce fertility can be successful, if governmental efforts are focused and vigorous. These cases also show that fertility can be increased, at least somewhat, when government policies support childbearing. But, most importantly, these three cases show clearly that political motivations are often very important determinants of changes and shifts in a nation's population policy.

THE U.S. INTERNAL POPULATION POLICY

The United States does not have an overall, explicit population policy. In general, throughout most of U.S. history, the country's implicit policy has favored population growth. Certainly, U.S. public health policies have been aimed at decreasing the death rate, which would, of course, increase the population. With regard to fertility, there has been no direct U.S. policy to stimulate fertility, at least not on the order of the child support payments Singapore, many European countries, and Canada hand out to encourage childbearing (Hyatt & Milne, 1991). On the other hand, fertility in the United States has never been discouraged, at least not as an official government policy. There are, in fact, a number of policies that indirectly encourage fertility, such as income tax deductions for dependent children (Whittington, 1992).

In recent decades, some social scientists, demographers, and environmentalists have called attention to the fact that a child born in an industrial country, and especially in the United States, will use far more resources and cause more pollution than a child born in a less technologically advanced society (Ehrlich & Ehrlich, 1990). The implication of this view is that Americans should control their fertility just as much as the people in the less developed parts of the world. But, up to this time, no major political leaders have proposed any policies or programs that would lower U.S. fertility. Certainly there is no official policy to do so.

Only in the case of immigration has the United States had an *explicit* population policy. As we noted earlier, until the last quarter of the 19th century, people from all countries of the world were allowed to come to the United States without restriction. When President Lincoln addressed Congress in 1863, he explicitly called for more immigrants, saying, "There is a still great deficiency of laborers in every field of industry, especially in agriculture and our mines" (Hutchinson, 1981, p. 48). He therefore recommended to the Congress that it adopt an official policy encouraging immigration.

While the need for foreign labor brought tens of millions of immigrants to the United States between the 1860s and the 1920s, both state legislatures and the U.S. Congress placed more and more restrictions on who might come into this country during this period. These restrictions, and U.S. immigration policy in general, were strongly influenced by the changing demand for workers, but the increasingly restrictive policies were also influenced by ethnic and racial prejudices and chauvinistic nationalism (LeMay, 1987).

In more recent decades foreign policy considerations have also played a part in shaping immigration policy, especially in regard to our admission of political refugees (Bouvier, 1992). In the section that follows, we will take a closer look at some of the details of U.S. immigration policy, and also discuss the issues and problems associated with illegal immigration into this country.

U.S. Immigration Policy

Three basic questions have dominated U.S. immigration policy since the 19th century (Murphy & Espenshade, 1990, p. 140):

1. How many immigrants should the United States accept?
2. From which countries should the immigrants be drawn?
3. What criteria should be employed in selecting immigrants?

At any given time, one or another of these questions may prevail in the debate about specific immigration legislation, but all three emerge again and again as the United States creates and re-creates its immigration policy.

Early Immigration Legislation

The earliest U.S. immigration legislation was generally prompted by the second question, From which countries should immigrants be drawn? Americans were having concerns and fears about the newest immigrants, concerns that were often rooted in racism and xenophobia (fear and hatred of foreigners).

Beginning in the 1850s, Chinese laborers were brought to this country by various commercial interests, especially those that were building the railroads in the West. From the time they arrived, Chinese immigrants were subjected to discrimination and prejudice. The California legislature passed special tax laws that effectively excluded the Chinese from gold mining. Their children were barred from the public schools, they were denied the right to testify in court against whites, and they were forbidden to marry whites (LeMay, 1987, p. 53).

In the 1870s congressmen and senators from western states (particularly California and Oregon) started introducing legislation in Congress designed to keep Chinese out of the country or deprive them of almost all rights if they did enter (Hutchinson, 1981). Finally, in 1882, Congress passed the Chinese Exclusion Act, which marked the first time the United States placed restrictions on the immigration of a specific nationality group. In the year before the law went into effect, nearly 40,000 Chinese entered the country. By 1885 the number was reduced to 23 (LeMay, 1987).

Japanese immigration to the United States started somewhat later than Chinese immigration, reaching a peak in the first decade of the 20th century. Again, state and local laws and judicial rulings deprived the Japanese of rights that other Americans had, including the right to own property or to lease it for more than 3 years. The Japanese, like the Chinese, could not marry whites or become naturalized citizens. In 1907–1908, the U.S. government pressured the Japanese government to voluntarily restrict emigration to the United States, and thereafter Japanese immigration declined dramatically (LeMay, 1987, p. 67).

As the American industrial economy grew in the 19th century, there was a need for many unskilled and manual workers: Immigrants helped to fill this need. Through most of the 19th century the immigrants came primarily from northwestern European countries, especially from England, Ireland, Germany, and Scandinavia. Near the end of the 19th century, the places of European origins for immigrants started to change. By the last decade of the century the immigrants were coming in ever increasing numbers from the countries of southern and eastern Europe.

Not only were the nationalities of the new immigrants different from those of the older immigrants, but the total number of immigrants reached a level that has never since been equalled. In the decade between 1900 and 1910, nearly 9 million people entered the country as immigrants. Since the total population of the United States was only 92 million in 1910, *new* immigrants made up about 10% of the population. (The numbers of legal immigrants coming into the United States each decade, from 1820 to 1990, are shown in Table 10.1.)

The large numbers of immigrants coming into the country around the turn of the century led to more and more demands for restricting immigration. Support for restricting immigrants came from many quarters, including an organization called the Immigration Restriction League, a variety of charitable and law enforcement agencies, and many sociologists and biologists who argued that some nationality groups were naturally inferior to others. Various patriotic groups and labor unions, each reflecting their own particular concerns, also called for restrictions on immigration (LeMay, 1987, p. 65).

In 1904 President Theodore Roosevelt urged Congress to deny admission to "masses of men whose standards of living and whose personal customs and habits are such that they tend to lower the level of the American wage-worker" (Hutchinson, 1981, p. 494). Roosevelt's statement combines the economic argument for restricting immigration with a suggestion that recent immigrants were people of less desirable "personal customs and habits." In the first decade of the 20th century more than 70% of the immigrants were coming from southern and eastern Europe. Many Americans, including many scientists and politicians, considered the people from these countries "racially" inferior. (In that era, Italians, Slavs, Greeks, and others with slightly darker skins than northern Europeans were considered separate races.) One U.S. Commissioner of Immigration, Francis A. Walker, characterized these "new immigrants" in the following way:

> They are beaten men from beaten races, representing the world's failures in the struggle for existence. They have none of the ideas and aptitudes which fit men to take up readily and easily the problems of self-care and self-government. (quoted in Saveth, 1948, p. 40)

An influential author, Madison Grant, advanced his racist theory by claiming that the United States was being inundated by

> a large and increasing number of the weak and broken and mentally crippled of all races being drawn from the lowest stratum of the Mediterranean basin and the Balkans, together with hordes of the wretched, submerged populations of the Polish ghettos. (quoted in LeMay, 1987, p. 79)

TABLE 10.1. Number of Immigrants to the United States and the
Immigration Rates, 1820–1990

Decade	Number	Rate
1820–1830	152,000	1.2
1831–1840	599,000	3.9
1841–1850	1.7 million	8.4
1851–1860	2.6 million	9.3
1861–1870	2.3 million	6.4
1871–1880	2.8 million	6.2
1881–1890	5.2 million	9.2
1891–1900	3.7 million	5.3
1901–1910	8.8 million	10.4
1911–1920	5.7 million	5.7
1921–1930	4.1 million	3.5
1931–1940	.5 million	0.4
1941–1950	1.0 million	0.7
1951–1960	2.5 million	1.5
1961–1970	3.3 million	1.7
1971–1980	4.5 million	2.1
1981–1990	7.3 million	3.1

Note. U.S. Bureau of the Census (1993a, Table 5, p. 10).

These bigoted statements were supplemented by a "Nordic race" supe-
riority theory, which claimed that the United States should be made up primarily of the descendants of immigrants from northern Europe. One con-
gressman from Maine said:

> God intended, I believe, [the U.S.] to be the home of a great people. En-
> glish speaking—a white race with great ideals, the Christian religion, one
> race, one country, and one destiny. (quoted in LeMay, 1987, p. 80)

A combination of views such as these, and the opposition to immigration by
various interest groups—especially labor organizations—led to the passage
in the 1920s of the National Origins Quota System.

The National Origins Quota System

During the early 1920s, Congress had made several attempts to limit and
control immigration. These efforts sought both to reduce the total number
of yearly immigrants and to cut off the large number of immigrants coming from the countries of southern and eastern Europe. In 1924, Congress
passed a law that nearly stopped the flow of immigrants from these countries and, incidentally, from most of the rest of the world. The new immi-

gration policy was called the quota system[6] because it gave a quota of immigrants to each country in proportion to their representation in the U.S.
population as of 1890. This system, of course, meant that many southern
and eastern European immigrants who had arrived after 1890 were not
counted as part of the total. As an example of how effectively this system
worked, Great Britain and Ireland were allotted 84,000 of the 150,000 immigrants admitted in a year, while Italy was allotted less than 6,000 (Hutchinson, 1981).

Since the quota system was adopted in the 1920s, U.S. immigration policy
has significantly changed several times: in 1952, with the passage of the
McCarren–Walter Act; in 1965, with the Immigration and Nationality Act;
and in 1986, with the Immigration Reform and Control Act. The most recent formulation of U.S. immigration policy is reflected in the Immigration
Act of 1990.

The McCarren–Walter Act

This act retained the national origins quota system, but it did open up immigration from some countries that had previously been excluded. Asians were
no longer excluded, since a minimal number of immigrants (typically 105
people) were allowed from each of the countries in the Asia-Pacific Triangle
(a triangular area that encompassed most Asian countries). Although the
McCarren–Walter Act was supposed to eliminate racial barriers to immigration, it introduced one curiously racist feature. Included in the quota were
any people whose ancestry could be traced to an Asian Pacific country. Thus,
any Asians who were living in Latin America or Canada (about 800,000 at
the time) had to be counted against the quota of their Asian country of origin. This made immigration more difficult for them, since immigration from
Western Hemisphere countries was essentially open. This provision was
clearly racist, because Asians were treated in terms of their racial heritage
rather than in terms of the country in which they lived (Hutchinson, 1981;
LeMay, 1987).

The McCarren–Walter Act also allowed the Immigration and Naturalization Service to exclude and deport various categories of people, including
criminals, subversives, and traffickers in illicit drugs. This provision has remained in effect, and has often been invoked to exclude or deport individuals who have politically unpopular views. The number of immigrants to the
United States for selected years since 1955, according to region of birth, are
shown in Figure 10.1.

[6]The law passed in 1924 firmly established the national origins principle for U.S. immigration policy. In 1929, the National Origins Act was passed, which made the quota system
fully operational (LeMay, 1987).

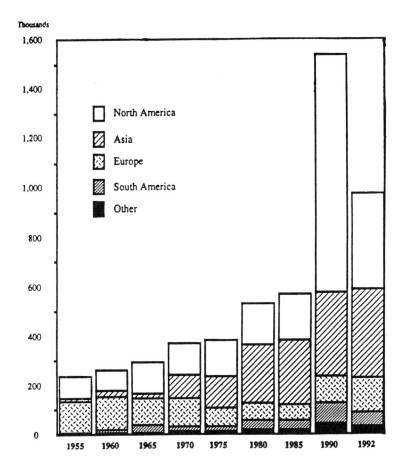

FIGURE 10.1. Immigrants to the United States for selected years, 1955–1992. Note: (1) North America included Canada and Mexico, and (2) the large number of 1990 immigrants from North America result from the admission of spouses and minor children of aliens legalized under the Immigration Reform and Control Act of 1986. From *Statistical Yearbook of the Immigration and Naturalization Service, 1993* (1993, p. 12).

The Immigration and Nationality Act of 1965

The national origins quota system adopted in the 1920s was finally abandoned in 1965. This act also did away with special rules for immigrants from the Asia-Pacific Triangle and thus eliminated race or ancestry as a basis for admitting immigrants. It also placed a numerical limit on the number of immigrants coming from Western Hemisphere countries.

The 1965 law had what has been called "a humanitarian dimension" in that it gave preference to immigrants who would be reuniting with family members already living in the United States. Family reunification continues to the present day to be the principal basis for admitting immigrants to this country. A second dimension, which was introduced by the 1965 act and also continues to be an important criterion for allowing immigrants into this country, is the preference given to people who have training or experience in occupations and professions needed by the United States.

The 1965 bill also provided openings for people who had been displaced by political upheavals, communist aggression, and natural disasters. This part of U.S. immigration policy is usually called the *refugee* provision.

The Immigration Reform and Control Act of 1986

The 1986 immigration act focused primarily on the issue of illegal aliens, or, to use a more neutral description, undocumented immigrants to this country. Most Americans, including many U.S. political leaders, have become very concerned about what they perceive as a flood of illegal aliens coming into the United States. The rhetoric on this issue includes assertions that the United States has "lost control of its borders" (Espenshade, 1990a), or, in more vivid language, that there is a "hemorrhaging at our borders." The objective of the 1986 legislation was to stop, or at least slow, the amount of illegal immigration into the United States. Much of the content of the new law was intended to address the special problem of immigration from Mexico, since the majority of undocumented immigrants cross the Mexican–U.S. border.[7]

One of the key provisions of this law was to punish employers who hired undocumented immigrants; the punishments included fines and even, in extreme cases, jail sentences. The assumption was that if employers faced real penalties, they would no longer hire illegal aliens, and thus there would be no motivation for people to enter the country illegally. A second important provision of the 1986 bill allowed illegal aliens, if they had lived in this country since 1982, to legitimize their status as permanent residents (Espenshade et al., 1990). A third feature of the 1986 law increased border patrols and immigration law enforcement.

Has the Flow of Illegal Aliens Been Slowed?

It is generally agreed that no legislation, or other governmental actions, can completely stop all illegal immigration into the United States. A more realistic objective, and one that the 1986 Immigration Control and Reform Act attempted, is to reduce the numbers of undocumented immigrants. The 1986 law has now been in effect for a number of years, giving researchers time to

[7]Other points of considerable illegal entry into the United States include Florida and major international airports.

assess whether or not the law has been achieving its goal (Bean & Fix, 1992; Bean et al., 1990; Cornelius, 1989; Donato, Durand, & Massey, 1992; White, 1990; Woodrow, 1992).

Some researchers have tried to assess the impact of the 1986 law by examining the number of apprehensions of illegal immigrants made by the Immigration and Naturalization Service in the years since the law went into effect (Bean et al., 1990; Espenshade, 1990b; White, Bean, & Espenshade, 1990). These studies have been consistent in showing that the law has had the desired effect of reducing the number of undocumented aliens attempting to enter the country, especially in the first 2 years after its passage (Bean & Fix, 1992). One study found a 38% decline in the number of undocumented migrants crossing the Mexican–U.S. border between November 1986 and September 1988 (Espenshade, 1990b, p. 175). But this same study also concluded that the number of undocumented aliens coming across the border into the United States still may have been as many as 4 million during the 2-year period after the passage of the law. This does not mean that 4 million illegal aliens were added to the U.S. population, since there was also a steady counterflow of migrants back to Mexico.

Other researchers have gone into Mexican communities to see if the number of residents leaving for the United States has been changing (Cornelius, 1989). One such study, in which respondents from three Mexican communities were interviewed, showed that many Mexicans were aware of the new U.S. law, and recognized that the U.S. labor market might therefore be more difficult to enter, but they were not significantly deterred from considering migration. The author states: "We found no evidence that the 1986 immigration law has reduced the traditionally heavy flow of workers to the United States from rural communities and small towns in west-central Mexico" (Cornelius, 1989, p. 701).

Another team of researchers interviewed respondents in seven Mexican communities *and* in several U.S. communities that commonly receive migrants from Mexico, and they too reached the conclusion that the 1986 immigration law had not significantly deterred undocumented migration from Mexico (Donato et al., 1992). These researchers asked respondents if they had ever gone into the United States and, if so, when they did so. This allowed them to make decade-to-decade and year-to-year comparisons with regard to such things as the likelihood of making a first trip, a repeat trip, or of getting apprehended. They also measured trends over time in the cost of making trips across the border. There were no apparent changes in these factors after the passage of the 1986 law. The researchers concluded that the millions of dollars and thousands of hours that have been spent to enforce the 1986 immigration law have done almost nothing "to stem the tide of Mexican migrants to the United States" (Donato et al., 1992, p. 156).

Another study has even suggested that the increased border control and apprehension efforts may actually be increasing the number of illegal immigrants residing in the United States (Koussoudji, 1992). Because of the threat of apprehension, many illegal immigrants are staying in the United States for longer periods of time once they do get into the country.

There is disagreement among researchers who have tried to assess the impact of the 1986 immigration law. Some see the decline in apprehensions as an indication of the law's effectiveness, while others say that apprehensions have decreased because the legalization part of the 1986 law offered legal status to undocumented immigrants (Donato et al., 1992). This part of the law reduced cross-border comings and goings, thereby reducing the number of apprehensions.

All researchers agree that the new law has not come close to stopping illegal immigration. Undocumented immigrants come into the United States because they believe they will find better economic opportunities here. Immigrants can only find such opportunities when employers provide jobs for those who come across the borders illegally and apparently there are still plenty of U.S. employers willing to break the law to do so.

The extensive debate about illegal immigration to the United States, which led to the passage of the 1986 immigration law, is based on the view that such immigration is a significant problem and has negative effects on the country. This is also an issue about which there is considerable debate and disagreement.

How Serious Is the Illegal Immigration Problem?

The considerable concern of Americans about illegal immigration revolves around two questions: How many illegal aliens are living in the United States? What effect do they have on the society and economy? Counting the number of illegal immigrants in the United States is, of course, difficult, since those who have entered this country illegally are inherently difficult to find and count. Assessing their impact, both positive and negative, on U.S. society and the U.S. economy is even more difficult.

Officials of the Immigration and Naturalization Service have, on various occasions, made estimates of the number of the illegal immigrants living in the United States; their estimates have generally been both high and unsubstantiated (Bean et al., 1987). In the 1970s the Immigration and Naturalization Service conjectured that the illegal immigrant population in the United States was between 8 and 12 million people. More careful analyses, based on 1980 census figures, placed the number between 3 and 6 million (Passel & Woodrow, 1984). Further estimates by Bean et al. (1987) suggest that the number in 1980 may have been at the low end of that range, or about 3.8 million. Since the 1986 law allowed 1.7 million aliens to legalize their status as U.S. residents, it is quite possible that the number of illegal immigrants residing in the United States in the early 1990s is no higher, and it may even be lower, than in 1980. Different estimations based on the 1990 census and other Census Bureau surveys lead to a 1990 estimate of 3.3 million (Himes & Clogg, 1992). If that estimate is correct, only 1.3% of the U.S. population is in this country illegally.

One of the most vocal advocates for immigrants, both legal and illegal, is Julian Simon (1990). The thrust of Simon's argument is that immigrants, again both legal and illegal, are often the least likely to use government services; illegal immigrants, in particular, receive few tax-supported services, and yet a high proportion contribute to the Social Security system and pay federal taxes. Further, Simon says, immigrants take jobs that many American citizens do not want, and by working, they are producing new demands for goods and services (and thus new jobs).

There have been many empirical studies of the economic impact of immigrants, conducted in different places, with different approaches and methodologies, and, as one might expect, they have produced some contradictory results. These many studies have been reviewed and assessed and the conclusion reached by one set of analysts is that "the effects of immigrants (both legal and undocumented) on the wages and earnings of other labor force groups are either *nonexistent or small (and sometimes positive)* (Bean et al., 1987, p. 685, emphasis added). So while U.S. immigration policy, especially with respect to undocumented aliens, is often based on the assumption that the immigrants are detrimental to other U.S. workers, that view is not generally supported by the evidence.

As an adjunct to the research on illegal immigrants, demographers and economists have frequently studied the economic effects of *legal* immigration. Researchers generally seek the answers to two questions: Do immigrants take jobs away from Americans who are already living in the country? Do immigrants receive exceptionally large amounts of money from governmental social programs, such as public welfare, food stamps, unemployment compensation, and so on? While these are difficult questions to answer conclusively, there has been enough research to provide us with tentative answers (Borjas & Tienda, 1987). A review of the available research has concluded that legal immigrants have not generally taken jobs away from native Americans, and may only require marginally more government assistance programs. The U.S. economy has generally been able to absorb immigrant workers without displacing native workers, and the negative impact on native workers' wages has been quite small. Some recent immigrants, specifically those who are Hispanic and Asian, have received slightly more public assistance than statistically similar native families, although that is not true of non-Hispanic white and black immigrants who received less (Borjas & Tienda, 1987). It is true that the economies of certain areas of the country have been more negatively impacted than other regions, because they are the places where new immigrants are more likely to settle. For the economy as a whole, the impact of immigrants has not been nearly as detrimental as some critics of immigration claim. On the other hand, as long as the immigration policy of the United States is influenced more by political and humanitarian factors than by purely economic considerations, immigration may have some limited negative impacts on the economy (Borjas & Tienda, 1987; Briggs, 1992).

U.S. Immigration Policy Today

The most recent immigration law, the Immigration Act of 1990, went into effect on October 1, 1991. The new immigration law did not substantially alter the fundamental principles of existing immigration policy. The primary reasons for admitting immigrants to the United States are to reunify family members and to add people with occupational characteristics needed in the United States. The 1990 legislation did, however, substantially increase the number of immigrants admitted each year. For the fiscal years 1992 through 1994, the legal limit for immigrants was set at 700,000 persons. This is an increase of 35% over previous levels (Briggs, 1992). Beginning in 1995, the number will be reduced slightly, to 675,000. The major features of current U.S. immigration policy are summarized in Table 10.2.

The numerical limits set by the 1990 act were exceeded by the total number of actual immigrants in both 1991 and 1992 because of the legalizations produced by the 1986 Immigration Control and Reform Act. In 1992, a total of 973,977 immigrants were admitted to the United States. However, by the end of 1992 almost all of the aliens eligible for permanant residence under the 1986 act had attained that status (U.S. Immigration and Naturalization Service, 1993, p. 13). The total number of immigrants for 1992 (973,977) is shown in Table 10.3, according to their country of origin (only the top 15 countries are shown).

An Overall U.S. Population Policy

It is sometimes argued that the United States should have an explicit population policy. An ideal population policy would establish goals for growth (or nongrowth), and specify how the birthrate and immigrant quotas might contribute to any agreed-upon rate of growth. It might also set goals for lowering the death rate, not just in old age and infancy but at all ages. An ideal population policy might also set targets for redistributing the population from one region of the country to another.

The advantages of having a national population policy are obvious. Public and private institutions and organizations could carry out their missions more effectively and efficiently if they knew in advance what the demographic trends of the future were likely to be. Educational systems could plan better for the numbers of classrooms and teachers needed. The Social Security system and pension systems generally could assess the needs of the future long before they were upon us. Businesses and industries could redi-

TABLE 10.2. The Major Features of U.S. Immigration Policy

U.S. law gives preferential immigration status to:

1. Aliens with a close family relationship with a U.S. citizen or legal permanent resident
2. Aliens with needed job skills
3. Aliens who qualify as refugees

Two categories of immigrants are *subject to direct numerical limitations* (366,000):

1. Family relationship preference: 226,000
2. Employment-based preference: 140,000

Other categories of immigrants are *not subject to direct numerical limitations*:

1. Immediate relatives of U.S. citizens: spouses, children (including orphans),
   and parents (of citizens 21 years or older)
2. Refugees and asylees
3. Aliens from countries "adversely affected" by the Immigration and Nationality Act Amendments of 1965[a]
4. Spouses and children of aliens legalized under the Immigration Reform and
   Control Act of 1986
5. Amerasians born in Vietnam
6. Certain parolees from the [former] Soviet Union and Indochina

Total number of immigrants per year:
 1992–1994 = 700,000
 1995+ = 675,000

Note. From U.S. Immigration and Naturalization Service (1993).

[a]Thirty-four countries were judged to have been "adversely affected" by the Immigration and
Nationality Act Amendments of 1965 (their numbers of immigrants decreased after 1965).
Among the 120,000 visas issued under this provision over a period from 1992 to 1994, 48,000
(40%) are reserved for natives of Ireland.

rect and modify their production systems to meet the needs of the population. A population policy for the nation would not resolve every issue and
avert every problem, but it could aid in doing so.

So why do we not have a national population policy if there are so many
obvious advantages? The answer can be found in many of the cases we have
considered in this chapter. The population policies of nations are shaped by
political considerations; by economic interests; by racial, ethnic, and social
class biases; and by ideological and moral beliefs. With all of the potential
competing interests and views that are found in the United States, it is doubtful
that this nation could ever arrive at a single population policy.

If we were just to consider the issue of whether the United States should

TABLE 10.3. Immigrants Admitted to the United States in 1992:
Numbers and Percentage from the Top 15 Countries

	Number	Percent
Mexico	213,802	22.0
Vietnam	77,735	8.0
Philippines	61,022	6.3
Soviet Union	43,590	4.5
Dominican Republic	41,969	4.3
China, Mainland	38,907	4.0
India	36,755	3.8
El Salvador	26,191	2.7
Poland	25,504	2.6
United Kingdom	19,973	2.1
Korea	19,359	2.0
Jamaica	18,915	1.9
Taiwan	16,344	1.7
Canada	15,205	1.6
Iran	13,233	1.4
Other	305,449	31.4
Total	973,977	100.0

Note. From U.S. Immigration and Naturalization Service (1993,
Table D, p. 21).

have a growth or a no-growth policy, it is easy to see that there would be
many different and opposing views. Environmentalists, for example, would
generally, though not always, favor a no-growth (or slow-growth) policy.
Other interest groups, perhaps manufacturers or service industries, might feel
that a nongrowing population would lead to a stagnating economy. Still other
groups and individuals in our society would favor either growth or no growth
for various political, economic, ideological, religious, or other reasons.

The United States is too heterogeneous in its composition and too varied in its ideas for a national population policy to gain acceptance. While an
overall population policy would probably be advantageous, and while demographers could probably provide the information needed to make it work
moderately well, it is unlikely that the United States will have such a policy
in the near future.

SUMMARY

Population politics and policies are closely related. Population policies are
governmental actions aimed at influencing the growth, decline, size, composition, or distribution of a population. Population policies may be *direct*, as
when a government has a specific policy to lower the birthrate of the population. Some policies are *indirect*, as when a speed-limit policy designed to

conserve gasoline also reduces automobile accident deaths. Some population policies are *disguised*, as when an abortion policy, which reduces fertility, is justified for maternal health reasons.

Population policies may be internal or external. Internal population policies are national policies that primarily affect a country's own population. External policies are national policies that affect the population of another nation or group of people.

The U.N. debate about world population growth during the 1950s and 1960s revealed the external population policies of many nations. The communist and Catholic nations effectively blocked U.N. actions on population for many years. The U.N. World Conferences on Population in 1974, 1984, and 1994 were all affected by national and international politics.

Internal population policies are often affected by political considerations. The population policies of India, Iran, and Singapore all illustrate, in different ways, how politics can play an important part in population policy and how populations are affected.

The internal population policy of the United States has not been explicit, except with regard to immigration policy. Beginning in the 19th century, the United States started to place limitations on immigration, especially from Asian countries. After immigration reached a peak in the first decade of the 20th century, the national origins quota system was adopted in the 1920s. This legislation, along with the Great Depression of the 1930s and World War II, effectively reduced immigration for several decades. The major immigration legislation since the mid-20th century included the McCarren–Walter Act of 1952, the Immigration and Nationality Act of 1965, the Immigration Reform and Control Act of 1986, and the Immigration Act of 1990.

These pieces of immigration legislation have shaped current immigration policy, which gives priority to immigrants who are family members of U.S. residents and to those who have occupational and professional skills needed in the United States. Refugees from political repression are also given preference.

Illegal immigration, a major national concern, was addressed specifically by the 1986 law. Illegal immigration persists despite this legislation.

The United States is unlikely to ever adopt a comprehensive population policy because it has too many competing economic and political interest groups with different objectives.

Appendix A
U.S. CENSUS FORM, 1990

This booklet shows the content of the two main questionnaires being used in the 1990 U.S. Census. See the explanatory notes on page 2.

CENSUS '90

OFFICIAL 1990 U.S. CENSUS FORM

Thank you for taking time to complete and return this census questionnaire. It's important to you, your community, and the Nation.

The law requires answers but guarantees privacy.

By law (Title 13, U.S. Code), you're required to answer the census questions to the best of your knowledge. However, the same law guarantees that your census form remains confidential. For 72 years–or until the year 2062–only Census Bureau employees can see your form. No one else–no other government body, no police department, no court system or welfare agency–is permitted to see this confidential information under any circumstances.

How to get started–and get help.

Start by listing on the next page the names of all the people who live in your home. Please answer all questions with a black lead pencil. You'll find detailed instructions for answering the census in the enclosed guide. If you need additional help, call the toll-free telephone number to the left, near your address.

Please answer and return your form promptly.

Complete your form and return it by April 1, 1990 in the postage-paid envelope provided. Avoid the inconvenience of having a census taker visit your home.

Again, thank you for answering the 1990 Census. Remember: Return the completed form by April 1, 1990.

Para personas de habla hispana –
(For Spanish-speaking persons)

Si usted desea un cuestionario del censo en español, llame sin cargo alguno al siguiente número: **1-800-XXXXXXX**
(o sea 1-800-XXX-XXXX)

U.S. Department of Commerce
BUREAU OF THE CENSUS

FORM **D-61**

OMB No. 0607-0628
Approval Expires 07/31/91

If wrong apartment identification, please write the correct number of apartment above.

INFORMATION COPY

Page 1

The 1990 census must count every person at his or her "usual residence." This means the place where the person lives and sleeps most of the time.

1a. List on the numbered lines below the name of each person living here on Sunday, April 1, including all persons staying here who have no other home. If EVERYONE at this address is staying here temporarily and usually lives somewhere else, follow the instructions given in question 1b below.

Include

- Everyone who usually lives here such as family members, housemates and roommates, foster children, roomers, boarders, and live-in employees
- Persons who are temporarily away on a business trip, on vacation, or in a general hospital
- College students who stay here while attending college
- Persons in the Armed Forces who live here
- Newborn babies still in the hospital
- Children in boarding schools below the college level
- Persons who stay here most of the week while working even if they have a home somewhere else
- Persons with no other home who are staying here on April 1

Do NOT include

- Persons who usually live somewhere else
- Persons who are away in an institution such as a prison, mental hospital, or a nursing home
- College students who live somewhere else while attending college
- Persons in the Armed Forces who live somewhere else
- Persons who stay somewhere else most of the week while working

Print last name, first name, and middle initial for each person. Begin on line 1 with the household member (or one of the household members) in whose name this house or apartment is owned, being bought, or rented. If there is no such person, start on line 1 with any adult household member.

	LAST	FIRST	INITIAL		LAST	FIRST	INITIAL
1				7			
2				8			
3				9			
4				10			
5				11			
6				12			

1b. If EVERYONE is staying here only temporarily and usually lives somewhere else, list the name of each person on the numbered lines above, fill this circle ⟶ ○ and print their usual address below. DO NOT PRINT THE ADDRESS LISTED ON THE FRONT COVER.

House number	Street or road/Rural route and box number	Apartment number
City	State	ZIP Code
County or foreign country	Names of nearest intersecting streets or roads	

NOW PLEASE OPEN THE FLAP TO PAGE 2 AND ANSWER ALL QUESTIONS FOR THE FIRST 7 PEOPLE LISTED. USE A BLACK LEAD PENCIL ONLY.

QUESTIONS ASKED OF ALL PERSONS

Page 2

	PERSON 1	PERSON 2
Please fill one column ➡ for each person listed in Question 1a on page 1.	Last name First name Middle initial	Last name First name Middle initial

2. How is this person related to PERSON 1?
 	
Fill ONE circle for each person.
 	
If **Other relative** of person in column 1, fill circle and print exact relationship, such as mother-in-law, grandparent, son-in-law, niece, cousin, and so on.
	START in this column with the household member (or one of the members) in whose name the home is owned, being bought, or rented.
	
	If there is no such person, start in this column with any adult household member.
	
	■
		If a RELATIVE of Person 1:
		○ Husband/wife ○ Brother/sister
		○ Natural-born ○ Father/mother
		  or adopted ○ Grandchild
		  son/daughter ○ Other relative ⌐
		○ Stepson/
		  stepdaughter
		If NOT RELATED to Person 1:
		○ Roomer, boarder, ○ Unmarried
		  or foster child partner
		○ Housemate, ■ ○ Other
		  roommate nonrelative

3. Sex Fill ONE circle for each person.	○ Male ○ Female	○ Male ○ Female
4. Race Fill ONE circle for the race that the person considers himself/herself to be. If **Indian (Amer.)**, print the name of the enrolled or principal tribe. ➡ If **Other Asian or Pacific Islander (API)**, print one group, for example: Hmong, Fijian, Laotian, Thai, Tongan, Pakistani, Cambodian, and so on. ➡ If **Other race**, print race. ➡	○ White ○ Black or Negro ○ Indian (Amer.) (Print the name of the enrolled or principal tribe.) ⌐ ○ Eskimo ○ Aleut Asian or Pacific Islander (API) ○ Chinese ○ Japanese ○ Filipino ■ ○ Asian Indian ○ Hawaiian ○ Samoan ○ Korean ○ Guamanian ○ Vietnamese ○ Other API ⌐ ○ Other race (Print race) ⌐	○ White ○ Black or Negro ○ Indian (Amer.) (Print the name of the enrolled or principal tribe.) ⌐ ○ Eskimo ○ Aleut Asian or Pacific Islander (API) ○ Chinese ○ Japanese ○ Filipino ■ ○ Asian Indian ○ Hawaiian ○ Samoan ○ Korean ○ Guamanian ○ Vietnamese ○ Other API ⌐ ○ Other race (Print race) ⌐

5. Age and year of birth
 	
a. Print each person's age at last birthday. Fill in the matching circle below each box.
 	
b. Print each person's year of birth and fill the matching circle below each box.
	a. Age b. Year of birth
	     / /
	0 0 0 0 1 ● 8 0 0
	1 1 1 1 9 1 1
	2 2 2 2
	3 3 3 3
	4 4 ■ 4 4
	5 5 5 5
	6 6 6 6
	7 7 7 7
	8 8 8 8
	9 9 9 9
		a. Age b. Year of birth
		     / /
		0 0 0 0 1 ● 8 0 0
		1 1 1 1 9 1 1
		2 2 2 2
		3 3 3 3
		4 4 ■ 4 4
		5 5 5 5
		6 6 6 6
		7 7 7 7
		8 8 8 8
		9 9 9 9

6. Marital status Fill ONE circle for each person.	○ Now married ○ Separated ○ Widowed ○ Never married ○ Divorced	○ Now married ○ Separated ○ Widowed ○ Never married ○ Divorced

7. Is this person of Spanish/Hispanic origin?
Fill ONE circle for each person.
 	
If **Yes, other Spanish/Hispanic**, print one group. ➡
	○ No (not Spanish/Hispanic)
	○ Yes, Mexican, Mexican-Am., Chicano
	○ Yes, Puerto Rican ■
	○ Yes, Cuban
	○ Yes, other Spanish/Hispanic
	  (Print one group, for example: Argentinean, Colombian, Dominican, Nicaraguan, Salvadoran, Spaniard, and so on.) ⌐
		○ No (not Spanish/Hispanic)
		○ Yes, Mexican, Mexican-Am., Chicano
		○ Yes, Puerto Rican
		○ Yes, Cuban
		○ Yes, other Spanish/Hispanic
		  (Print one group, for example: Argentinean, Colombian, Dominican, Nicaraguan, Salvadoran, Spaniard, and so on.) ⌐

FOR CENSUS USE ➡	○ ○	○ ○

PLEASE ALSO ANSWER HOUSING QUESTIONS ON PAGE 3 →

PERSON 3	PERSON 4	PERSON 5	PERSON 6
Last name	Last name	Last name	Last name
First name — Middle initial	First name — Middle initial	First name — Middle initial	First name — Middle initial

If a RELATIVE of Person 1:

○ Husband/wife ○ Natural-born or adopted son/daughter ○ Stepson/stepdaughter	○ Brother/sister ○ Father/mother ○ Grandchild ○ Other relative ⌐	○ Husband/wife ○ Natural-born or adopted son/daughter ○ Stepson/stepdaughter	○ Brother/sister ○ Father/mother ○ Grandchild ○ Other relative ⌐

(repeated for Person 5 and Person 6)

If NOT RELATED to Person 1:

○ Roomer, boarder, or foster child ■ Housemate, roommate	○ Unmarried partner ○ Other nonrelative

(repeated for Person 4, 5, 6)

Male ■ Female ○ (repeated across persons)

Race / origin section (Person 3 column, partially visible):

- White
- Black or Negro
- Indian (Amer.) (Print the name of enrolled or principal tribe.) ⌐
- Eskimo
- Aleut — Asian or Pacific Islander
- Chinese — Jap...
- Filipino ■ — Asi...
- Hawaiian — Sa...
- Korean — Gu...
- Vietnamese — Oth...
- Other race (Print race) ⌐

a. Age

0 0 0	1 ● 8
1 1 1	9
2 2 2	
3 3 3	
4 4 4	
5 5 5 ■	
6 6 6	
7 7 7	
8 8 8	
9 9 9	

b. Year of birth (Person 3) / Year of birth (Person 6 side)

Marital status:
- Now married — Sep...
- Widowed — Ne...
- Divorced

Spanish/Hispanic origin:
- No (not Spanish/Hispanic)
- Yes, Mexican, Mexican-Am...
- Yes, Puerto Rican
- Yes, Cuban
- Yes, other Spanish/Hispanic (Print one group, for example: Argentinean, Colombian, Dominican, Nicaraguan, Salvadoran, Spaniard, and so on.) ⌐

(Race section for Person 6, partially visible, right side:)
- Print the name of the ...cipal tribe.) ⌐
- ...cific Islander (API)
- Japanese
- Asian Indian
- Samoan
- Guamanian
- Other API ⌐
- ...race) ⌐
- Year of birth
- 8 0 0
- 9 1 1
- 2 2
- 3 3
- 4 4 ■
- 5 5
- 6 6
- 7 7
- 8 8
- 9 9
- ○ Separated
- ○ Never married
- /Hispanic)
- ...exican-Am., Chicano
- Yes, Puerto Rican
- Yes, Cuban
- Yes, other Spanish/Hispanic (Print one group, for example: Argentinean, Colombian, Dominican, Nicaraguan, Salvadoran, Spaniard, and so on.) ⌐

EXPLANATORY NOTES

This booklet shows the content of the two 1990 census questionnaires being delivered by mail. The content of these forms was determined after review of the 1980 census experience, extensive consultation with many government and private users of census data, and a series of experimental censuses and surveys in which various alternatives were tested.

Two principal types of data-collection forms — a 100-percent questionnaire (or "short form") and a sample questionnaire (or "long form") — are being used in the census. Each household receives one of the two questionnaires.

Short form — This questionnaire contains 7 population questions and 7 housing questions, shown on pages 1–3 of this booklet. On average, about 5 in every 6 households will receive the short form. For the average household, this form will take an estimated 14 minutes to complete.

Long form — This questionnaire has all of the short-form questions plus housing questions H8 through H26, shown on pages 4 and 5, and population questions 8 through 33, shown on pages 6 and 7. The population questions are repeated for each member of the household but these pages were not reproduced in this booklet. A statistical sample of approximately 1 in every 6 households will receive the long form. For the average household, this form will take an estimated 43 minutes to complete.

An instruction guide accompanies each questionnaire to help the respondents complete the form, and a preaddressed envelope is provided for returning the questionnaire.

For additional information about the 1990 U.S. Census, please write the Director, Bureau of the Census, Washington, DC 20233.

QUESTIONS ASKED OF ALL HOUSEHOLDS

Page 3

NOW PLEASE ANSWER QUESTIONS H1a—H26 FOR YOUR HOUSEHOLD

PERSON 7

Last name

First name   Middle initial

If a RELATIVE of Person 1:

Husband/wife	Brother/sister
Natural-born	Father/mother
or adopted	Grandchild
son/daughter	Other relative ⌐
Stepson/	
stepdaughter	

If NOT RELATED to Person 1:

Roomer, boarder,	Unmarried
or foster child	partner
Housemate, ■	Other
roommate	nonrelative

Male   Female

White
Black or Negro
Indian (Amer.) (Print the name of the enrolled or principal tribe.) ⌐

Eskimo
Aleut   Asian or Pacific Islander (API)

Chinese	Japanese
Filipino ■	Asian Indian
Hawaiian	Samoan
Korean	Guamanian
Vietnamese	Other API ⌐

Other race (Print race) ⌐

a. Age   **b. Year of birth**

				/		
0	0	0	1 ● 8	0	0
1	1	1	9	1	1
2	2			2	2
3	3			3	3
4	4	■		4	4
5	5			5	5
6	6			6	6
7	7			7	7
8	8			8	8
9	9			9	9

Now married	Separated
Widowed	Never married
Divorced	

No (not Spanish/Hispanic)
Yes, Mexican, Mexican-Am., Chicano
Yes, Puerto Rican ■
Yes, Cuban
Yes, other Spanish/Hispanic
(Print one group, for example: Argentinean, Colombian, Dominican, Nicaraguan, Salvadoran, Spaniard, and so on.) ⌐

H1a. Did you leave anyone out of your list of persons for Question 1a on page 1 because you were not sure if the person should be listed — for example, someone temporarily away on a business trip or vacation, a newborn baby still in the hospital, or a person who stays here once in a while and has no other home?

Yes, please print the name(s)   No ■
and reason(s) ⌐

---

b. Did you include anyone in your list of persons for Question 1a on page 1 even though you were not sure that the person should be listed — for example, a visitor who is staying here temporarily or a person who usually lives somewhere else?

Yes, please print the name(s)   No
and reason(s) ⌐

H2. Which best describes this building? Include all apartments, flats, etc., even if vacant.

A mobile home or trailer
A one-family house detached from any other house
A one-family house attached to one or more houses
A building with 2 apartments
A building with 3 or 4 apartments
A building with 5 to 9 apartments
A building with 10 to 19 apartments
A building with 20 to 49 apartments
A building with 50 or more apartments
Other

H3. How many rooms do you have in this house or apartment? Do NOT count bathrooms, porches, balconies, foyers, halls, or half-rooms.

1 room ■	4 rooms	7 rooms
2 rooms	5 rooms	8 rooms
3 rooms	6 rooms	9 or more rooms

H4. Is this house or apartment —

Owned by you or someone in this household with a mortgage or loan?
Owned by you or someone in this household free and clear (without a mortgage)?
Rented for cash rent?
Occupied without payment of cash rent?

If this is a ONE-FAMILY HOUSE —

H5a. Is this house on ten or more acres?

Yes   No

b. Is there a business (such as a store or barber shop) or a medical office on this property?

Yes   No

Answer only if you or someone in this household OWNS OR IS BUYING this house or apartment —

H6. What is the value of this property; that is, how much do you think this house and lot or condominium unit would sell for if it were for sale?

Less than $10,000	$70,000 to $74,999
$10,000 to $14,999	$75,000 to $79,999
$15,000 to $19,999	$80,000 to $89,999
$20,000 to $24,999	$90,000 to $99,999
$25,000 to $29,999	$100,000 to $124,999
$30,000 to $34,999	$125,000 to $149,999
$35,000 to $39,999	$150,000 to $174,999
$40,000 to $44,999	$175,000 to $199,999
$45,000 to $49,999	$200,000 to $249,999
$50,000 to $54,999	$250,000 to $299,999
$55,000 to $59,999	$300,000 to $399,999
$60,000 to $64,999	$400,000 to $499,999
$65,000 to $69,999	$500,000 or more

Answer only if you PAY RENT for this house or apartment —

H7a. What is the monthly rent?

Less than $80	$375 to $399
$80 to $99	$400 to $424
$100 to $124	$425 to $449
$125 to $149	$450 to $474
$150 to $174	$475 to $499
$175 to $199	$500 to $524
$200 to $224 ■	$525 to $549
$225 to $249	$550 to $599
$250 to $274	$600 to $649
$275 to $299	$650 to $699
$300 to $324	$700 to $749
$325 to $349	$750 to $999
$350 to $374	$1,000 or more

b. Does the monthly rent include any meals?

Yes   No

FOR CENSUS USE

A. Total persons	B. Type of unit		D. Months vacant	G. DO	ID ■
	Occupied **Vacant**		Less than 1 6 up to 12		
	First form Regular		1 up to 2 12 up to 24		
	Cont'n Usual home		2 up to 6 24 or more		
		elsewhere	**E. Complete after**		
	C1. Vacancy status		LR TC QA JIC 1		
	For rent For seas/		P/F RE I/T		
	For sale only rec/occ		MV ED EN ■		
	Rented or For migrant		P0 P3 P6		
	sold, not workers		P1 P4 IA JIC 2		
	occupied Other vacant		P2 P5 SM		
	C2. Is this unit boarded up?	**F. Cov.**			
	Yes No	1b 1a 7 H1		

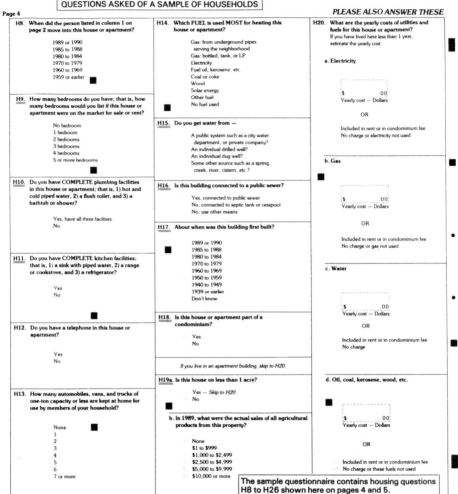

QUESTIONS ASKED OF A SAMPLE OF HOUSEHOLDS

Page 4

H8. When did the person listed in column 1 on page 2 move into this house or apartment?

- 1989 or 1990
- 1985 to 1988
- 1980 to 1984
- 1970 to 1979
- 1960 to 1969
- 1959 or earlier

H9. How many bedrooms do you have; that is, how many bedrooms would you list if this house or apartment were on the market for sale or rent?

- No bedroom
- 1 bedroom
- 2 bedrooms
- 3 bedrooms
- 4 bedrooms
- 5 or more bedrooms

H10. Do you have COMPLETE plumbing facilities in this house or apartment; that is, 1) hot and cold piped water, 2) a flush toilet, and 3) a bathtub or shower?

- Yes, have all three facilities
- No

H11. Do you have COMPLETE kitchen facilities; that is, 1) a sink with piped water, 2) a range or cookstove, and 3) a refrigerator?

- Yes
- No

H12. Do you have a telephone in this house or apartment?

- Yes
- No

H13. How many automobiles, vans, and trucks of one-ton capacity or less are kept at home for use by members of your household?

- None
- 1
- 2
- 3
- 4
- 5
- 6
- 7 or more

H14. Which FUEL is used MOST for heating this house or apartment?

- Gas: from underground pipes serving the neighborhood
- Gas: bottled, tank, or LP
- Electricity
- Fuel oil, kerosene, etc.
- Coal or coke
- Wood
- Solar energy
- Other fuel
- No fuel used

H15. Do you get water from —

- A public system such as a city water department, or private company?
- An individual drilled well?
- An individual dug well?
- Some other source such as a spring, creek, river, cistern, etc.?

H16. Is this building connected to a public sewer?

- Yes, connected to public sewer
- No, connected to septic tank or cesspool
- No, use other means

H17. About when was this building first built?

- 1989 or 1990
- 1985 to 1988
- 1980 to 1984
- 1970 to 1979
- 1960 to 1969
- 1950 to 1959
- 1940 to 1949
- 1939 or earlier
- Don't know

H18. Is this house or apartment part of a condominium?

- Yes
- No

If you live in an apartment building, skip to H20.

H19a. Is this house on less than 1 acre?

- Yes — *Skip to H20*
- No

b. In 1989, what were the actual sales of all agricultural products from this property?

- None
- $1 to $999
- $1,000 to $2,499
- $2,500 to $4,999
- $5,000 to $9,999
- $10,000 or more

PLEASE ALSO ANSWER THESE

H20. What are the yearly costs of utilities and fuels for this house or apartment?
If you have lived here less than 1 year, estimate the yearly cost.

a. Electricity

$ _____ .00
Yearly cost — Dollars

OR

Included in rent or in condominium fee
No charge or electricity not used

b. Gas

$ _____ .00
Yearly cost — Dollars

OR

Included in rent or in condominium fee
No charge or gas not used

c. Water

$ _____ .00
Yearly cost — Dollars

OR

Included in rent or in condominium fee
No charge

d. Oil, coal, kerosene, wood, etc.

$ _____ .00
Yearly cost — Dollars

OR

Included in rent or in condominium fee
No charge or these fuels not used

The sample questionnaire contains housing questions H8 to H26 shown here on pages 4 and 5.

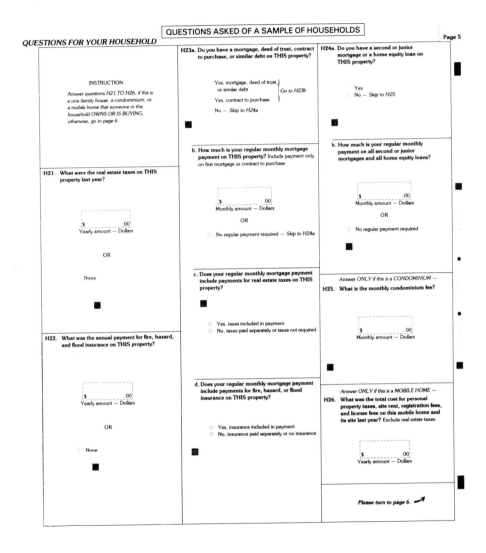

QUESTIONS FOR YOUR HOUSEHOLD

QUESTIONS ASKED OF A SAMPLE OF HOUSEHOLDS

INSTRUCTION:

Answer questions H21 TO H26, if this is a one-family house, a condominium, or a mobile home that someone in this household OWNS OR IS BUYING; otherwise, go to page 6.

H21. What were the real estate taxes on THIS property last year?

$ _____ .00
Yearly amount — Dollars

OR

None

H22. What was the annual payment for fire, hazard, and flood insurance on THIS property?

$ _____ .00
Yearly amount — Dollars

OR

○ None

H23a. Do you have a mortgage, deed of trust, contract to purchase, or similar debt on THIS property?

Yes, mortgage, deed of trust, or similar debt ⎫
Yes, contract to purchase ⎬ Go to H23b

No — Skip to H24a

b. How much is your regular monthly mortgage payment on THIS property? Include payment only on first mortgage or contract to purchase.

$ _____ .00
Monthly amount — Dollars

OR

○ No regular payment required — Skip to H24a

c. Does your regular monthly mortgage payment include payments for real estate taxes on THIS property?

○ Yes, taxes included in payment
○ No, taxes paid separately or taxes not required

d. Does your regular monthly mortgage payment include payments for fire, hazard, or flood insurance on THIS property?

○ Yes, insurance included in payment
○ No, insurance paid separately or no insurance

H24a. Do you have a second or junior mortgage or a home equity loan on THIS property?

○ Yes
○ No — Skip to H25

b. How much is your regular monthly payment on all second or junior mortgages and all home equity loans?

$ _____ .00
Monthly amount — Dollars

OR

○ No regular payment required

Answer ONLY if this is a CONDOMINIUM —

H25. What is the monthly condominium fee?

$ _____ .00
Monthly amount — Dollars

Answer ONLY if this is a MOBILE HOME —

H26. What was the total cost for personal property taxes, site rent, registration fees, and license fees on this mobile home and its site last year? Exclude real estate taxes.

$ _____ .00
Yearly amount — Dollars

Please turn to page 6.

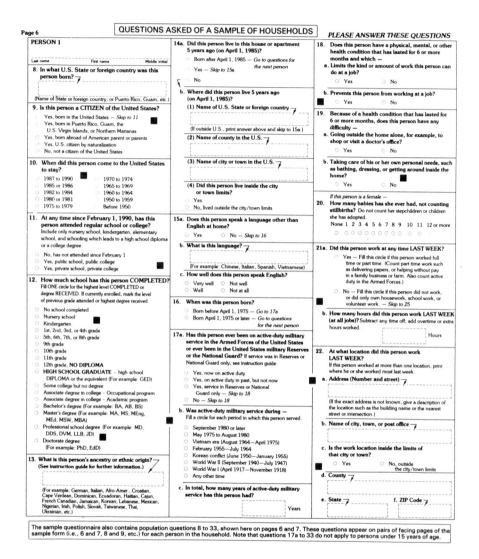

The sample questionnaire also contains population questions 8 to 33, shown here on pages 6 and 7. These questions appear on pairs of facing pages of the sample form (i.e., 6 and 7, 8 and 9, etc.) for each person in the household. Note that questions 17a to 33 do not apply to persons under 15 years of age.

FOR PERSON 1 ON PAGE 2   **QUESTIONS ASKED OF A SAMPLE OF HOUSEHOLDS**   Page 7

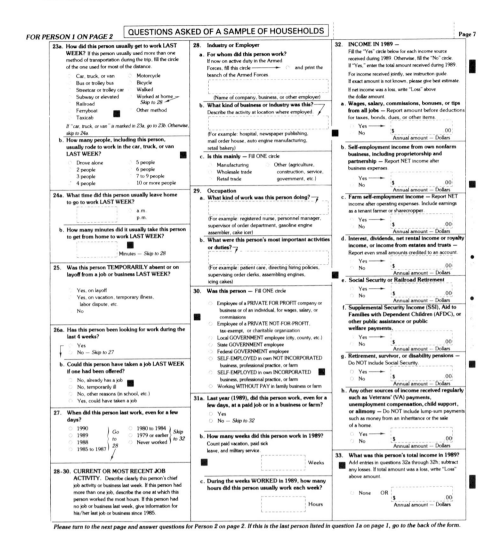

23a. How did this person usually get to work LAST WEEK? If this person usually used more than one method of transportation during the trip, fill the circle of the one used for most of the distance.

- Car, truck, or van
- Bus or trolley bus
- Streetcar or trolley car
- Subway or elevated
- Railroad
- Ferryboat
- Taxicab
- Motorcycle
- Bicycle
- Walked
- Worked at home — Skip to 28
- Other method

If "car, truck, or van" is marked in 23a, go to 23b. Otherwise, skip to 24a.

b. How many people, including this person, usually rode to work in the car, truck, or van LAST WEEK?

- Drove alone
- 2 people
- 3 people
- 4 people
- 5 people
- 6 people
- 7 to 9 people
- 10 or more people

24a. What time did this person usually leave home to go to work LAST WEEK?

- a.m.
- p.m.

b. How many minutes did it usually take this person to get from home to work LAST WEEK?

Minutes — Skip to 28

25. Was this person TEMPORARILY absent or on layoff from a job or business LAST WEEK?

- Yes, on layoff
- Yes, on vacation, temporary illness, labor dispute, etc.
- No

26a. Has this person been looking for work during the last 4 weeks?

- Yes
- No — Skip to 27

b. Could this person have taken a job LAST WEEK if one had been offered?

- No, already has a job
- No, temporarily ill
- No, other reasons (in school, etc.)
- Yes, could have taken a job

27. When did this person last work, even for a few days?

- 1990
- 1989
- 1988
- 1985 to 1987
- 1980 to 1984
- 1979 or earlier
- Never worked

Go to 28 / Skip to 32

28-30. CURRENT OR MOST RECENT JOB ACTIVITY. Describe clearly this person's chief job activity or business last week. If this person had more than one job, describe the one at which this person worked the most hours. If this person had no job or business last week, give information for his/her last job or business since 1985.

28. Industry or Employer

a. For whom did this person work? If now on active duty in the Armed Forces, fill this circle ———— and print the branch of the Armed Forces.

(Name of company, business, or other employer)

b. What kind of business or industry was this? Describe the activity at location where employed.

(For example: hospital, newspaper publishing, mail order house, auto engine manufacturing, retail bakery)

c. Is this mainly — Fill ONE circle

- Manufacturing
- Wholesale trade
- Retail trade
- Other (agriculture, construction, service, government, etc.)

29. Occupation

a. What kind of work was this person doing?

(For example: registered nurse, personnel manager, supervisor of order department, gasoline engine assembler, cake icer)

b. What were this person's most important activities or duties?

(For example: patient care, directing hiring policies, supervising order clerks, assembling engines, icing cakes)

30. Was this person — Fill ONE circle

- Employee of a PRIVATE FOR PROFIT company or business or of an individual, for wages, salary, or commissions
- Employee of a PRIVATE NOT-FOR-PROFIT, tax-exempt, or charitable organization
- Local GOVERNMENT employee (city, county, etc.)
- State GOVERNMENT employee
- Federal GOVERNMENT employee
- SELF-EMPLOYED in own NOT INCORPORATED business, professional practice, or farm
- SELF-EMPLOYED in own INCORPORATED business, professional practice, or farm
- Working WITHOUT PAY in family business or farm

31a. Last year (1989), did this person work, even for a few days, at a paid job or in a business or farm?

- Yes
- No — Skip to 32

b. How many weeks did this person work in 1989? Count paid vacation, paid sick leave, and military service.

Weeks

c. During the weeks WORKED in 1989, how many hours did this person usually work each week?

Hours

32. INCOME IN 1989 —
Fill the "Yes" circle below for each income source received during 1989. Otherwise, fill the "No" circle. If "Yes," enter the total amount received during 1989.
For income received jointly, see instruction guide. If exact amount is not known, please give best estimate. If net income was a loss, write "Loss" above the dollar amount.

a. Wages, salary, commissions, bonuses, or tips from all jobs — Report amount before deductions for taxes, bonds, dues, or other items.

- Yes
- No

$ .00
Annual amount — Dollars

b. Self-employment income from own nonfarm business, including proprietorship and partnership — Report NET income after business expenses.

- Yes
- No

$ .00
Annual amount — Dollars

c. Farm self-employment income — Report NET income after operating expenses. Include earnings as a tenant farmer or sharecropper.

- Yes
- No

$ .00
Annual amount — Dollars

d. Interest, dividends, net rental income or royalty income, or income from estates and trusts — Report even small amounts credited to an account.

- Yes
- No

$ .00
Annual amount — Dollars

e. Social Security or Railroad Retirement

- Yes
- No

$ .00
Annual amount — Dollars

f. Supplemental Security Income (SSI), Aid to Families with Dependent Children (AFDC), or other public assistance or public welfare payments.

- Yes
- No

$ .00
Annual amount — Dollars

g. Retirement, survivor, or disability pensions — Do NOT include Social Security.

- Yes
- No

$ .00
Annual amount — Dollars

h. Any other sources of income received regularly such as Veterans' (VA) payments, unemployment compensation, child support, or alimony — Do NOT include lump-sum payments such as money from an inheritance or the sale of a home.

- Yes
- No

$ .00
Annual amount — Dollars

33. What was this person's total income in 1989? Add entries in questions 32a through 32h; subtract any losses. If total amount was a loss, write "Loss" above amount.

- None   OR $ .00
Annual amount — Dollars

Please turn to the next page and answer questions for Person 2 on page 2. If this is the last person listed in question 1a on page 1, go to the back of the form.

Please make sure you have . . .

1. FILLED this form completely.

2. ANSWERED Question 1a on page 1.

3. ANSWERED Questions 2 through 7 for each person you listed in Question 1a.

4. ANSWERED Questions H1a through H26 on pages 3, 4, and 5.

5. ANSWERED the questions on pages 6 through 19 for each person you listed in Question 1a.

Also . . .

6. PRINT here the name of a household member who filled the form, the date the form was completed, and the telephone number at which a person in this household can be called.

Name		Date
Telephone number ⟶	Area code | Number	○ Day ○ Night

Then . . .

7. FOLD the form the way it was sent to you.

8. MAIL it back by April 1, or as close to that date as possible, in the envelope provided; no stamp is needed. When you insert your completed questionnaire, please make sure that the address of the U.S. Census Office can be seen through the window on the front of the envelope.

NOTE — If you have listed more than 7 persons in Question 1a, please make sure that you have filled the form for the first 7 people. Then mail back this form. A census taker will call to obtain the information for the other people.

Thank you very much.

☆U.S. GOVERNMENT PRINTING OFFICE: 1988—239-809

Appendix B
ABRIDGED LIFE TABLE

T he life table shown on pages 284–291 is an abridged life table. It is an abridged rather than a complete table because, except for the first year of life (age interval 0–1), the age intervals after age 5 are combined into 5-year categories. (The last age interval is open-ended, since it is 85 and over.) An abridged life table yields nearly the same information as a complete table, and it has the advantages of being much shorter and involving fewer calculations.

The fundamental idea of a life table is that prevailing death rates for each age (or age interval) are applied to a hypothetical cohort of 100,000 newborns. As these age-specific death rates are applied to each successive age interval, the cohort is gradually diminished until, at last, the entire cohort is gone. But, of course, during that process the cohort accumulates many years of living. In the life table, the cohort of 100,000 persons accumulates 7,525,922 years of living under the prevailing age-specific death rates (see the top number of Column 6). If that total number of years is divided by 100,000, it averages out to 75.3 years per person (see the top number of Column 7). That number is the average number of years a person will live from birth or, more commonly, life expectancy at birth.

With this general description in mind, a step-by-step examination of the seven columns will reveal more details of how the calculations are made and the final numbers in Column 7 are obtained. The numbered columns are also labeled with the traditional column headings used by demographers.

Column 1: Age interval (x to x + n). The age interval shown in Column 1 is the interval between the two exact ages indicated. For instance, 20–25 means the 5-year interval between the 20th birthday and the 25th.

Column 2: Proportion dying $(_nq_x)$. This column shows the proportion of the cohort who are alive at the beginning of an indicated age interval who will die before reaching the end of that interval. For example, for people in the age interval 20–25, the proportion dying is .0056—out of every 1,000 people alive and exactly 20 years old at the beginning of the period, between 5 and 6 (5.6) will die before reaching their 25th birthdays. In other words, the $_nq_x$ values represent *probabilities* that persons who are alive at the beginning of a specific age interval will die before reaching the beginning of the next age interval. The "Proportion Dying" column forms the basis of the life table; the life table is so constructed that all other columns are derived from it.

Column 3: Number surviving (l_x). This column shows the number of persons, starting with the hypothetical cohort of 100,000 at birth, who survive to the exact age marking the beginning of each interval. The l_x values are computed from the $_nq_x$ values, which are successively applied to the remainder of the original 100,000 persons still alive at the beginning of each age interval. Thus out of the original 100,000 newborns, 99,014 will complete the first year of life and enter the second, 98,822 will begin the sixth year, 98,145 will reach age 20, and 31,231 will live to age 85.

Column 4: Number dying $(_nd_x)$. This column shows the number dying in each successive age interval out of 100,000 live births. Out of the original 100,000 newborns, 986 die in the first year of life, 192 in the succeeding 4 years, 548 in the 5-year period between exact ages 20 and 25, and 31,231 die after reaching 85. Each figure in Column 4 is the difference between two successive figures in Column 3.

Column 5: Number of person-years lived in age interval $(_nL_x)$. This column shows the number of person-years that would be lived between the two indicated birthdays by all those reaching the earlier birthday. Thus the figure 489,382 for persons in the age interval 20–25 is the total number of years lived between the 20th and 25th birthdays by the 98,145 (Column 3) who reached their 20th birthday. This calculation is as follows: Of the 98,145 who reach age 20, there will be 97,597 who reach age 25. For these survivors, we can multiply 5 years times 97,597 (5 × 97,597 = 487,985 person-years). However, the 548 persons who died during this age interval will also have lived some years. Assuming their deaths were distributed fairly evenly over the 5-year period, they would have lived on an average about 2½ years. (Because the age-specific death rates are not exactly the same for each year, the average is, in fact, 2.549 years.) Multiplying 2.549 times 548 adds another 1,397 years to the 487,985 years lived by the survivors of the entire interval between 20 and 25. The total number of person-years lived during this age interval is 489,382.

Column 6: Person-years lived in this and all subsequent age intervals (T_x). This column is the cumulative number of person-years that will be lived by those who start any age interval. The numbers in this column can only be obtained after all the calculations for Column 5 have been completed. Then, working from the bottom of the column, the person-years lived in each age interval can be calculated. Notice that the 193,200 years lived by those who reach age 85 are added to the 195,668 years that will be lived between the years 80 and 85. The total of 388,868 is the total number of person-years lived by those who reach age 80.

When Column 6 has been calculated, the total is the 7,525,922 years discussed earlier. Again, since the entire cohort of 100,000 lived a total of 7,525,922 years, the average number of years, or life expectancy at age zero, is said to be 75.3 years.

Column 7: Average number of years of life remaining at the beginning of age interval $(\mathring{e}_x)$. This column tells us more than the average life expectancy at birth (75.3 years). The numbers down Column 7 are the expectations for years yet to live at any age. For example, for those persons reaching age 20, it can be seen in Column 6 that these survivors will still live a total of 5,552,208 years. Since 98,145 persons reached age 20, 5,552,208 can be divided by 98,145 to see how many more years each person can be expected to live. The answer, which can be read from

Column 7, is 56.6 years. In other words, persons who survive to age 20 can expect to live to 76.6 years on the average. It should not be surprising that life expectancy is greater at 20 than it is at birth (76.6 years vs. 75.3 years), since those who reach age 20 have already lived through infancy and early childhood when death rates are higher.

There is an alternate interpretation of a life table, which is associated with the concept of a stationary population (see Chapter 4). The application of this interpretation is for making comparative measurements of mortality and for use in studies of population structure (see Shryock & Siegel, 1976).

Since death rates vary by gender and race, it is obvious that life expectancies will also differ. The remainder of the abridged life table shows life expectancies for white males, white females, black males, and black females for 1989. This table can now be examined for the information they yield about differential life expectancy by race and sex.

LIFE TABLES

Abridged Life Tables by Race and Sex: United States, 1989

Age interval	Proportion dying	Of 100,000 born alive		Stationary population		Average remaining lifetime
Period of life between two exact ages stated in years, race, and sex	Proportion of persons alive at beginning of age interval dying during interval	Number living at beginning of age interval	Number dying during age interval	In the age interval	In this and all subsequent age intervals	Average number of years of life remaining at beginning of age interval
(1)	(2)	(3)	(4)	(5)	(6)	(7)
x to $x+n$	nq_x	l_x	nd_x	nL_x	T_x	$\overset{\circ}{e}_x$
ALL RACES						
0-1	0.0099	100,000	986	99,154	7,525,922	75.3
1-5	.0019	99,014	192	395,606	7,426,768	75.0
5-10	.0012	98,822	117	493,791	7,031,162	71.1
10-15	.0013	98,705	132	493,269	6,537,371	66.2
15-20	.0043	98,573	428	491,894	6,044,102	61.3
20-25	.0056	98,145	548	489,382	5,552,208	56.6
25-30	.0062	97,597	604	486,474	5,062,826	51.9
30-35	.0076	96,993	734	483,173	4,576,352	47.2
35-40	.0097	96,259	932	479,097	4,093,179	42.5
40-45	.0127	95,327	1,213	473,819	3,614,082	37.9
45-50	.0186	94,114	1,753	466,508	3,140,263	33.4
50-55	.0292	92,361	2,694	455,493	2,673,755	28.9
55-60	.0463	89,667	4,154	438,547	2,218,262	24.7
60-65	.0707	85,513	6,044	413,274	1,779,715	20.8
65-70	.1026	79,469	8,156	377,836	1,366,441	17.2
70-75	.1522	71,313	10,851	330,381	988,605	13.9
75-80	.2235	60,462	13,511	269,356	658,224	10.9
80-85	.3348	46,951	15,720	195,668	388,868	8.3
85 and over	1.0000	31,231	31,231	193,200	193,200	6.2

MALE

Age					
0-1	.0109	100,000	1,086	99,065	7,182,240	71.8
1-5	.0021	98,914	211	395,167	7,083,175	71.6
5-10	.0014	98,703	134	493,149	6,688,008	67.8
10-15	.0016	98,569	163	492,548	6,194,859	62.8
15-20	.0062	98,406	606	490,674	5,702,311	57.9
20-25	.0085	97,800	827	486,977	5,211,637	53.3
25-30	.0091	96,973	885	482,628	4,724,660	48.7
30-35	.0109	96,088	1,047	477,853	4,242,032	44.1
35-40	.0138	95,041	1,308	472,100	3,764,179	39.6
40-45	.0172	93,733	1,608	464,923	3,292,079	35.1
45-50	.0245	92,125	2,255	455,407	2,827,156	30.7
50-55	.0375	89,870	3,368	441,466	2,371,749	26.4
55-60	.0598	86,502	5,174	420,303	1,930,283	22.3
60-65	.0912	81,328	7,419	389,050	1,509,980	18.6
65-70	.1316	73,909	9,726	346,142	1,120,930	15.2
70-75	.1947	64,183	12,499	290,381	774,788	12.1
75-80	.2839	51,684	14,675	221,862	484,407	9.4
80-85	.4115	37,009	15,229	146,096	262,545	7.1
85 and over	1.0000	21,780	21,780	116,449	116,449	5.3

FEMALE

Age					
0-1	.0088	100,000	882	99,247	7,860,815	78.6
1-5	.0017	99,118	171	396,063	7,761,568	78.3
5-10	.0010	98,947	101	494,460	7,365,505	74.4
10-15	.0010	98,846	100	494,014	6,871,045	69.5
15-20	.0025	98,746	243	493,161	6,377,031	64.6
20-25	.0027	98,503	264	491,867	5,883,870	59.7
25-30	.0032	98,239	317	490,419	5,392,003	54.9
30-35	.0043	97,922	416	488,618	4,901,584	50.1
35-40	.0056	97,506	551	486,250	4,412,966	45.3
40-45	.0084	96,955	817	482,884	3,926,716	40.5
45-50	.0130	96,138	1,251	477,790	3,443,832	35.8
50-55	.0213	94,887	2,022	469,694	2,966,042	31.3
55-60	.0338	92,865	3,141	456,935	2,496,348	26.9
60-65	.0523	89,724	4,694	437,568	2,039,413	22.7
65-70	.0778	85,030	6,613	409,483	1,601,845	18.8
70-75	.1183	78,417	9,280	370,076	1,192,362	15.2
75-80	.1815	69,137	12,547	315,798	822,286	11.9
80-85	.2898	56,590	16,402	243,090	506,488	9.0
85 and over	1.0000	40,188	40,188	263,398	263,398	6.6

LIFE TABLES

Abridged Life Tables by Race and Sex: United States, 1989—Con.

Age interval	Proportion dying	Of 100,000 born alive		Stationary population		Average remaining lifetime
Period of life between two exact ages stated in years, race, and sex	Proportion of persons alive at beginning of age interval dying during interval	Number living at beginning of age interval	Number dying during age interval	In the age interval	In this and all subsequent age intervals	Average number of years of life remaining at beginning of age interval
(1)	(2)	(3)	(4)	(5)	(6)	(7)
x to $x + n$	nq_x	l_x	nd_x	nL_x	T_x	$\overset{\circ}{e}_x$
WHITE						
0-1	0.0081	100,000	814	99,299	7,599,881	76.0
1-5	.0017	99,186	170	396,346	7,500,582	75.6
5-10	.0011	99,016	108	494,786	7,104,236	71.7
10-15	.0013	98,908	124	494,304	6,609,450	66.8
15-20	.0041	98,784	409	492,989	6,115,146	61.9
20-25	.0049	98,375	487	490,673	5,622,157	57.2
25-30	.0054	97,888	526	488,111	5,131,484	52.4
30-35	.0064	97,362	625	485,282	4,643,373	47.7
35-40	.0081	96,737	784	481,838	4,158,091	43.0
40-45	.0110	95,953	1,055	477,322	3,676,253	38.3
45-50	.0165	94,898	1,569	470,875	3,198,931	33.7
50-55	.0265	93,329	2,474	460,874	2,728,056	29.2
55-60	.0434	90,855	3,942	445,022	2,267,182	25.0
60-65	.0675	86,913	5,863	420,746	1,822,160	21.0
65-70	.0993	81,050	8,048	386,053	1,401,414	17.3
70-75	.1494	73,002	10,906	338,750	1,015,361	13.9
75-80	.2214	62,096	13,747	277,027	676,611	10.9
80-85	.3331	48,349	16,105	201,747	399,584	8.3
85 and over	1.0000	32,244	32,244	197,837	197,837	6.1

WHITE, MALE

Age						
0-1	.0091	100,000	908	99,216	7,267,465	72.7
1-5	.0019	99,092	187	395,936	7,168,249	72.3
5-10	.0012	98,905	123	494,189	6,772,313	68.5
10-15	.0016	98,782	154	493,635	6,278,124	63.6
15-20	.0057	98,628	565	491,869	5,784,489	58.6
20-25	.0075	98,063	731	488,512	5,292,620	54.0
25-30	.0080	97,332	775	484,685	4,804,108	49.4
30-35	.0093	96,557	898	480,562	4,319,423	44.7
35-40	.0115	95,659	1,104	475,677	3,838,861	40.1
40-45	.0147	94,555	1,394	469,545	3,363,184	35.6
45-50	.0216	93,161	2,013	461,176	2,893,639	31.1
50-55	.0338	91,148	3,080	448,564	2,432,463	26.7
55-60	.0560	88,068	4,933	428,754	1,983,899	22.5
60-65	.0875	83,135	7,275	398,491	1,555,145	18.7
65-70	.1278	75,860	9,694	356,045	1,156,654	15.2
70-75	.1919	66,166	12,696	299,881	800,609	12.1
75-80	.2827	53,470	15,115	229,747	500,728	9.4
80-85	.4108	38,355	15,756	151,476	270,981	7.1
85 and over	1.0000	22,599	22,599	119,505	119,505	5.3

WHITE, FEMALE

Age						
0-1	.0071	100,000	715	99,387	7,923,460	79.2
1-5	.0015	99,285	151	396,777	7,824,073	78.8
5-10	.0009	99,134	93	495,418	7,427,296	74.9
10-15	.0009	99,041	93	495,008	6,931,878	70.0
15-20	.0025	98,948	245	494,166	6,436,870	65.1
20-25	.0024	98,703	237	492,927	5,942,704	60.2
25-30	.0027	98,466	267	491,673	5,449,777	55.3
30-35	.0035	98,199	341	490,181	4,958,104	50.5
35-40	.0046	97,858	454	488,234	4,467,923	45.7
40-45	.0073	97,404	710	485,380	3,979,689	40.9
45-50	.0116	96,694	1,119	480,886	3,494,309	36.1
50-55	.0195	95,575	1,862	473,525	3,013,423	31.5
55-60	.0316	93,713	2,957	461,629	2,539,898	27.1
60-65	.0493	90,756	4,478	443,269	2,078,269	22.9
65-70	.0746	86,278	6,439	416,182	1,635,000	19.0
70-75	.1155	79,839	9,220	377,392	1,218,818	15.3
75-80	.1788	70,619	12,628	323,122	841,426	11.9
80-85	.2877	57,991	16,685	249,482	518,304	8.9
85 and over	1.0000	41,306	41,306	268,822	268,822	6.5

LIFE TABLES

Abridged Life Tables by Race and Sex: United States, 1989—Con.

Age interval	Proportion dying	Of 100,000 born alive		Stationary population		Average remaining lifetime
Period of life between two exact ages stated in years, race, and sex	Proportion of persons alive at beginning of age interval dying during interval	Number living at beginning of age interval	Number dying during age interval	In the age interval	In this and all subsequent age intervals	Average number of years of life remaining at beginning of age interval
(1)	(2)	(3)	(4)	(5)	(6)	(7)
x to $x+n$	nq_x	l_x	nd_x	nL_x	T_x	$\overset{\circ}{e}_x$
BLACK						
0-1	0.0186	100,000	1,863	98,399	6,923,552	69.2
1-5	.0031	98,137	303	391,837	6,825,153	69.5
5-10	.0017	97,834	167	488,707	6,433,316	65.8
10-15	.0018	97,667	176	487,961	5,944,609	60.9
15-20	.0056	97,491	550	486,237	5,456,648	56.0
20-25	.0095	96,941	921	482,549	4,970,411	51.3
25-30	.0118	96,020	1,134	477,367	4,487,862	46.7
30-35	.0163	94,886	1,544	470,715	4,010,495	42.3
35-40	.0221	93,342	2,067	462,017	3,539,780	37.9
40-45	.0288	91,275	2,625	450,200	3,077,763	33.7
45-50	.0380	88,650	3,366	435,325	2,627,563	29.6
50-55	.0531	85,284	4,529	415,635	2,192,238	25.7
55-60	.0747	80,755	6,031	389,270	1,776,603	22.0
60-65	.1067	74,724	7,971	354,275	1,387,333	18.6
65-70	.1445	66,753	9,648	310,162	1,033,058	15.5
70-75	.1956	57,105	11,171	258,031	722,896	12.7
75-80	.2672	45,934	12,273	199,018	464,865	10.1
80-85	.3847	33,661	12,949	135,555	265,847	7.9
85 and over	1.0000	20,712	20,712	130,292	130,292	6.3

BLACK, MALE

Age						
0-1	.0200	100,000	2,004	98,275	6,484,736	64.8
1-5	.0034	97,996	332	391,215	6,386,461	65.2
5-10	.0020	97,664	192	487,788	5,995,246	61.4
10-15	.0022	97,472	214	486,923	5,507,458	56.5
15-20	.0088	97,258	855	484,424	5,020,535	51.6
20-25	.0150	96,403	1,449	478,636	4,536,111	47.1
25-30	.0177	94,954	1,681	470,701	4,057,475	42.7
30-35	.0239	93,273	2,229	460,963	3,586,774	38.5
35-40	.0332	91,044	3,023	447,995	3,125,811	34.3
40-45	.0414	88,021	3,644	431,479	2,677,816	30.4
45-50	.0534	84,377	4,507	411,229	2,246,337	26.6
50-55	.0737	79,870	5,887	385,264	1,835,108	23.0
55-60	.0980	73,983	7,253	352,333	1,449,844	19.6
60-65	.1336	66,730	8,917	311,848	1,097,511	16.4
65-70	.1816	57,813	10,501	263,083	785,663	13.6
70-75	.2444	47,312	11,563	207,666	522,580	11.0
75-80	.3261	35,749	11,658	149,139	314,914	8.8
80-85	.4524	24,091	10,900	92,358	165,775	6.9
85 and over	1.0000	13,191	13,191	73,417	73,417	5.6

BLACK, FEMALE

Age						
0-1	.0172	100,000	1,717	98,527	7,351,437	73.5
1-5	.0028	98,283	272	392,479	7,252,910	73.8
5-10	.0014	98,011	141	489,661	6,860,431	70.0
10-15	.0014	97,870	135	489,043	6,370,770	65.1
15-20	.0024	97,735	237	488,135	5,881,727	60.2
20-25	.0043	97,498	424	486,497	5,393,592	55.3
25-30	.0065	97,074	626	483,882	4,907,095	50.6
30-35	.0095	96,448	912	480,082	4,423,213	45.9
35-40	.0127	95,536	1,215	474,846	3,943,131	41.3
40-45	.0181	94,321	1,709	467,622	3,468,285	36.8
45-50	.0252	92,612	2,337	457,586	3,000,663	32.4
50-55	.0363	90,275	3,275	443,612	2,543,077	28.2
55-60	.0547	87,000	4,757	423,662	2,099,465	24.1
60-65	.0841	82,243	6,913	394,592	1,675,803	20.4
65-70	.1142	75,330	8,600	355,876	1,281,211	17.0
70-75	.1597	66,730	10,656	307,828	925,335	13.9
75-80	.2280	56,074	12,783	248,925	617,507	11.0
80-85	.3445	43,291	14,913	179,295	368,582	8.5
85 and over	1.0000	28,378	28,378	189,287	189,287	6.7

LIFE TABLES

Abridged Life Tables by Race and Sex: United States, 1989—Con.

Age interval	Proportion dying	Of 100,000 born alive		Stationary population		Average remaining lifetime
Period of life between two exact ages stated in years, race, and sex	Proportion of persons alive at beginning of age interval dying during interval	Number living at beginning of age interval	Number dying during age interval	In the age interval	In this and all subsequent age intervals	Average number of years of life remaining at beginning of age interval
x to $x+n$	nq_x	l_x	nd_x	nL_x	T_x	$\overset{\circ}{e}_x$
(1)	(2)	(3)	(4)	(5)	(6)	(7)
ALL OTHER						
0-1	0.0163	100,000	1,630	98,610	7,124,869	71.2
1-5	.0028	98,370	279	392,825	7,026,259	71.4
5-10	.0016	98,091	155	490,026	6,633,434	67.6
10-15	.0017	97,936	163	489,339	6,143,408	62.7
15-20	.0052	97,773	509	487,734	5,654,069	57.8
20-25	.0085	97,264	827	484,369	5,166,335	53.1
25-30	.0103	96,437	992	479,777	4,681,966	48.5
30-35	.0136	95,445	1,298	474,092	4,202,189	44.0
35-40	.0183	94,147	1,722	466,664	3,728,097	39.6
40-45	.0235	92,425	2,167	457,035	3,261,433	35.3
45-50	.0317	90,258	2,863	444,567	2,804,398	31.1
50-55	.0455	87,395	3,975	427,537	2,359,831	27.0
55-60	.0653	83,420	5,450	404,033	1,932,294	23.2
60-65	.0941	77,970	7,340	372,110	1,528,261	19.6
65-70	.1286	70,630	9,085	330,995	1,156,151	16.4
70-75	.1760	61,545	10,832	281,159	825,156	13.4
75-80	.2420	50,713	12,272	223,032	543,997	10.7
80-85	.3530	38,441	13,568	158,038	320,965	8.3
85 and over	1.0000	24,873	24,873	162,927	162,927	6.6

ALL OTHER, MALE

Age interval					
0-1	.0176	100,000	1,757	98,495	6,712,872	67.1
1-5	.0031	98,243	305	392,266	6,614,377	67.3
5-10	.0018	97,938	175	489,208	6,222,111	63.5
10-15	.0020	97,763	198	488,419	5,732,903	58.6
15-20	.0080	97,565	776	486,126	5,244,484	53.8
20-25	.0132	96,789	1,282	480,934	4,758,358	49.2
25-30	.0153	95,507	1,460	473,976	4,277,424	44.8
30-35	.0197	94,047	1,857	465,716	3,803,448	40.4
35-40	.0269	92,190	2,477	455,040	3,337,732	36.2
40-45	.0335	89,713	3,002	441,477	2,882,692	32.1
45-50	.0439	86,711	3,810	424,590	2,441,215	28.2
50-55	.0617	82,901	5,114	402,316	2,016,625	24.3
55-60	.0860	77,787	6,688	372,780	1,614,309	20.8
60-65	.1194	71,099	8,490	334,833	1,241,529	17.5
65-70	.1620	62,609	10,141	288,041	906,696	14.5
70-75	.2202	52,468	11,555	233,598	618,655	11.8
75-80	.2951	40,913	12,075	174,097	385,057	9.4
80-85	.4182	28,838	12,061	113,306	210,960	7.3
85 and over	1.0000	16,777	16,777	97,654	97,654	5.8

ALL OTHER, FEMALE

Age interval					
0-1	.0150	100,000	1,499	98,729	7,522,063	75.2
1-5	.0026	98,501	253	393,400	7,423,334	75.4
5-10	.0014	98,248	134	490,870	7,029,934	71.6
10-15	.0013	98,114	126	490,287	6,539,064	66.6
15-20	.0024	97,988	232	489,408	6,048,777	61.7
20-25	.0039	97,756	386	487,869	5,559,369	56.9
25-30	.0057	97,370	552	485,528	5,071,500	52.1
30-35	.0080	96,818	775	482,252	4,585,972	47.4
35-40	.0108	96,043	1,040	477,795	4,103,720	42.7
40-45	.0150	95,003	1,421	471,704	3,625,925	38.2
45-50	.0214	93,582	2,002	463,227	3,154,221	33.7
50-55	.0318	91,580	2,912	451,010	2,690,994	29.4
55-60	.0479	88,668	4,247	433,243	2,239,984	25.3
60-65	.0731	84,421	6,169	407,330	1,806,741	21.4
65-70	.1015	78,252	7,945	372,121	1,399,411	17.9
70-75	.1425	70,307	10,017	327,335	1,027,290	14.6
75-80	.2054	60,290	12,385	271,065	699,955	11.6
80-85	.3138	47,905	15,031	202,207	428,890	9.0
85 and over	1.0000	32,874	32,874	226,683	226,683	6.9

Note. From National Center for Health Statistics (1992).

GLOSSARY

Age effect. Social changes that may be attributable to a specific age group.

Age heaping. The tendency of some people to round their ages to numbers ending in 5 or 0.

Age–sex pyramid. *See* Population Pyramid.

Age-specific fertility rate. The number of live births in 1 year per 1,000 women in a specific age category.

Availability. A component in the decision-making model of migration, emphasizing that migration must be physically and cognitively possible.

Canvasser method. A method of census taking in which a person representing the census-taking agency visits each residence, asks questions, and records answers; also known as *direct interviewer*.

Census. A complete count or enumeration of a population conducted under the auspices of some governmental authority.

Chain migration. When one family member follows another to the same destination.

Cohort. The entire set of people who experience the same demographic event (e.g., birth) at the same period of time (e.g., 1940 birth cohort).

Cohort effect. Social changes that may be attributable to a specific cohort of people.

Constrictive pyramid. A population pyramid showing an age structure with declining growth rates.

Crude birthrate. The number of births in a single year per 1,000 population.

Crude death rate. The number of deaths in a single year per 1,000 population.

Deductive historical demography. The estimation of what demographic facts might have been on the basis of known historical, economic, social, medical, or other facts.

De facto census. A census of the population in which people are counted as being residents of the place where they happen to be at the time the census is taken.

De jure census. A census of the population in which people are counted as being residents of the place where they usually live.

293

Demographic transition theory. The process of change from high birthrates and death rates to low birthrates and death rates; related to changes in economic development.

Demography. The scientific study of human populations.

Dependency ratio. The ratio of the dependent population (under age 15 and ages 65 and over) to the productive population (aged 15–64).

Dependent population. The population under age 15 and ages 65 and over who typically are not active in the labor force.

Dependent variable. Effect variables; variables that are affected or changed by other, independent, variables.

Differential fertility. Comparisons of childbearing behavior by different social groups or categories: nationalities, races, and so on.

Direct interviewer. *See* Canvasser Method.

Double count. An error in data collection in which some people in the population are counted more than once; same as overcount.

Doubling time. The amount of time it takes for a population to double its size.

Emigrant. A person who moves away from a country of origin.

European marriage pattern. A term used to describe the tendency to delay marriage or not to marry at all, based on data from historical records throughout Europe.

Excessive survivors effect. When the number of children in one generation who survive to adulthood is greater than expected, causing the next generation to lower its fertility.

Expansive pyramid. A population pyramid showing an age structure with a growing population.

Expectancy. A component in the decision-making model of migration, emphasizing that individuals must believe migration will help them to achieve their goals.

Experimental science. A science in which the data are generally produced by experiments conducted in laboratories under conditions that are closely controlled by the scientist.

Family demography. A subfield of demography, concerned with the composition and sizes of households and families, and with family processes.

Family reconstitution. A data-producing method used by historical demographers; a method of obtaining demographic data by assembling information from family life-cycle events.

Fecundity. The biological capacity to bear children.

Fertility. One of the demographic processes referring to actual childbearing behavior.

Field experiments. A type of experimental study in which the researcher controls or manipulates some factors, but subjects are in natural settings such as communities.

Formal demography. A type of population study in which population variables are used to explain variations in other population variables.

General fertility rate. The number of children ever born per 1,000 women aged 15–49 in a given year. The ages 15–49 are used to represent women in their childbearing years.

Genocide. The process of exterminating an entire social group or category of people.

Growth rate. Increase in a population during a year as determined by births, deaths, and net migration.

Historical demography. The study of the demographic events and conditions of the past.

Householder method. A method of census taking in which members of the household must read and answer the questions; also known as *self-enumeration.*

Immigrant. A person who moves into a new country.

Incentive. A component of the decision-making model of migration, emphasizing that individuals must evaluate the specific characteristics of their current places of residence and of potential destinations.

Incrementalists. A population policy position that views population growth as the key inhibitor of economic development.

Independent variables. Causal variables; variables that have the potential for producing changes in other, dependent, variables.

Infant mortality. Deaths occurring during the first year after birth.

Insurance strategy. *See* Replacement strategy.

Internal migration. The movement of people within the boundaries of a nation-state.

International migration. The movement of people from one nation-state to another.

Life expectancy. The average number of years of life remaining to individuals of a specified age, under prevailing age-specific death rates.

Life span. The biological maximum number of years a person can live.

Life table. A statistical method of estimating the average life expectancy at any given age.

Longevity. The ability to resist death and thus the length of time one can live.

Macroscopic level of analysis. The study of the larger structures of societies such as institutions, cities, the labor force, and so on.

Marriage squeeze. An imbalance between the numbers of males and females in the prime marriage ages.

Maternal mortality. Deaths to women as a result of childbirth.

Median age. A measure of a central tendency whereby 50% of a population is older and 50% is younger.

Metropolitan Statistical Area. A term developed by the Bureau of the Census to classify central cities and their interrelated surrounding areas.

Microscopic level of analysis. The study of individuals, often at the social–psychological level.

Migrant. A person who makes a permanent change in his or her regular place of residence.

Migration. The movement of individuals or groups from one place of residence to another when they have the intention of remaining in the new place for some substantial period of time.

Migration interval. The amount of time specified to determine the occurrence of migration.

Migration stream. The flow of people from one area to another designated area.

Misclassification. An error in census data in which an individual's characteristics, such as educational level, are in error.

Morbidity. The incidence and prevalence of disease and illness in a population.

Mortality. The demographic process related to death.

Motive. A component in the decision-making model of migration, emphasizing that decisions are based on values.

Natural fertility. Unlimited fertility where no effort is made to control childbearing.

Natural increase. Population growth produced by the excess of births over deaths in any given year.

Neolocal residence. A nuclear family form where the newly married couple sets up a residence separate from that of both their parents.

Neo-Malthusians. Contemporary demographers who follow Malthus in believing that population growth is often the cause of societal and human problems and that population must be curbed.

Neonatal mortality rate. The number of infants who die within the first 28 days of life per 1,000 live births.

Net migration. In a given population, the difference between the number of persons entering and leaving a given area through migration.

Net reproduction rate. The average number of daughters who would be born to a female birth cohort, given current age-specific fertility and mortality rates.

Observational science. A science in which the data are obtained by observing and recording events that occur naturally in the world.

Period effect. Social changes that may be attributable to a specific historical time period (e.g., the Vietnam War era).

Personal status characteristics. The positions or statuses which individuals may have in the economic system and family.

Political arithmeticians. A group of 17th-century researchers and scientists noted for their application of the scientific method to demographic issues.

Population policy. A legitimate act by a governmental authority that aims to influence the growth, decline, size, composition, or distribution of a population.

Population projection. A procedure to estimate, or project, the size and characteristics of a population at some future time.

Population pyramid. A graphic method to show the age and sex composition of a population.

Population register. A system of continuous registration of data on the major vital events for each individual in a population.

Population study. A study in which nondemographic factors are introduced as either independent or dependent variables in a relationship with demographic phenomena.

Population studies, type I. A relationship where nondemographic variables are used as independent variables to explain or predict demographic phenomena.

Population studies, type II. A relationship where demographic variables are used as independent variables to explain or predict nondemographic phenomena.

Positive checks. Term used by Malthus to identify factors that increase mortality and thus check the rate of population growth (e.g., war, famine, and disease).

Postenumeration survey. Reinterviewing a sample of the population for the purpose of determining the accuracy of the initial census-taking method.

Postneonatal mortality. Infant deaths occurring after the first 28 days of life, but before the end of the first year of life.

Posttransition stage. Final stage in the demographic transition where the birthrates and death rates are relatively low.

Pretransition stage. First stage in the demographic transition where the birthrates and death rates are relatively high.

Preventive checks. A term used by Malthus to identify factors that decrease fertility, but especially delayed marriage.

Productive population. People in the population aged 15 to 64 who typically are active in the labor force.

Rate. A measure that reflects the frequency of an event relative to some population that may experience that event.

Rationalism. Individual actions based on self-interest, efficiency, and effectiveness.

Redistributionists. A population policy position that views population problems as the result of economic underdevelopment and the unequal distribution of economic resources and power.

Replacement strategy. The idea that couples have additional children to replace infants who die so that a certain number of children are alive when the parents reach old age.

Return migrant. A person who moves back to his or her previous place of residence.

Self-designation method. Individuals designate their own race or ethnicity, as opposed to census takers making that determination.

Self-enumeration. *See* Householder Method.

Sex ratio. The number of males per 100 females in a population.

Social psychology. The study of individual attitudes and orientations and their influence on behavior.

Stable population. A population whose rate of growth is constant and in which the birthrate, death rate, and age–sex structure are also constant.

Standardization. A statistical technique used to adjust rates so that populations may be compared as if they were the same.

Stationary population. A population that is not increasing or decreasing where the age structure is unchanging.

Stationary pyramid. A population pyramid showing a stationary population, where the population is not growing.

Stem family. An intermediate family form that shares characteristics of the extended family and the nuclear family. A family home remains as a base for all of the other family members.

Surrogate measures of fertility. An indirect measure of fertility, determined by the childbearing intentions of individuals and couples.

Total fertility rate. The average number of children that would be born alive to a cohort of women given current age-specific fertility rates.

Traditionalism. Individual action based primarily on the customary ways of doing things in a group or society.

Transition stage. The second stage in the demographic transition where the birth rate remains high and the death rate is falling rapidly.

Triangulation. The use of three different methods in scientific research to make an estimate or to reach a conclusion about some fact or issue.

Undercount. An error in data collection in which some people in the population are not located and therefore not counted.

Urban. According to the U.S. Census, any community of 2,500 or more residents.

Variable. Any event or occurrence that can change or take on different values.

Vital registration system. A system in which vital events, such as births, deaths, marriage, and divorce are registered with governmental authorities.

Vital statistics. Data on births, deaths, marriages, and divorces.

Zero population growth. The number of births plus immigration equals the number of deaths plus emigration.

REFERENCES

Abbott, Margaret H., Helen Abbey, David R. Bolling, and Edmond A. Murphy
  1978 "The Familial Component in Longevity—A Study of Offspring of Non-
      agenarians: III. Intrafamilial Studies." *American Journal of Medical
      Genetics* 2: 105–120.
Abbott, Margaret H., Edmond Murphy, David R. Bolling, and Helen Abbey
  1974 "The Familial Component of Longevity—A Study of Offspring of Non-
      agenarians: II. Preliminary Analysis of the Completed Study." *Johns
      Hopkins Medical Journal* 134(1): 1–16.
Adams, John, and Alice Bel Kasakoff
  1991 "Estimates of Census Undernumeration Based on Genealogies." *Social
      Science History* 15(4): 527–543.
Aghajanian, Akbar
  1991 "Population Change in Iran, 1966–86: A Stalled Demographic Transi-
      tion." *Population and Development Review* 17: 703–715.
  1992 "Slow Growth Efforts Renewed in Iran." *Population Today* 20: 4–5.
Ahlburg, Dennis A., and Carol J. DeVita
  1992 "New Realities of the American Family." *Population Bulletin* 47(2):
      1–44.
Ahlburg, Dennis A., and James W. Vaupel
  1990 "Alternative Projections of the U.S. Population." *Demography* 27:
      639–652.
Alter, George
  1990 "New Perspectives on European Marriage in the Nineteenth Century."
      Journal of Family History 16: 1–5.
Altman, Lawrence K.
  1992 "Researchers Report a Much Grimmer AIDS Outlook." *New York
      Times*, June 4, pp. A1, B10.
Antonovsky, Aaron
  1967 "Social Class, Life Expectancy and Overall Mortality." *Milbank Me-
      morial Fund Quarterly* 45(1): 31–73.
Arensberg, Conrad M., and Solon T. Kimball
  1968 *Family and Community in Ireland.* 2d ed. Cambridge, Mass.: Harvard
      University Press.
Arioka, Jiro
  1991 "Fewer Babies: A Private Matter?" *Japan Quarterly* 38: 50–56.

Aristotle
    1932   *The Politics*. Translated by H. Rackham. London: William Heinemann.
Ashford, Lori S.
    1995   "New Perspectives on Population: Lessons from Cairo." *Population Bulletin* 50(1): 1–44.
Axinn, William G.
    1992   "Family Organization and Fertility Limitation in Nepal." *Demography* 29(4): 503–522.
Bachu, Amara
    1991   "Profile of the Foreign-Born Population in the United States." In U.S. Bureau of the Census, *Studies in American Fertility*, Current Population Reports, Series P-23, no. 176.
Back, Kurt
    1967   "New Frontiers in Demography and Social Psychology." *Demography* 4: 90–97.
Back, Robert L., and Joel Smith
    1977   "Community Satisfaction, Expectations of Moving and Migration." *Demography* 14(2): 147–167.
Banerji, Debabar
    1992   "Family Planning in the Nineties: More of the Same?" *Economic and Political Weekly* 27: 883–887.
Banister, Judith
    1987   *China's Changing Population*. Stanford, Calif.: Stanford University Press.
Banister, Judith, and Samuel H. Preston
    1981   "Mortality in China." *Population and Development Review* 7(1): 98–110.
Banks, J. A.
    1954   *Prosperity and Parenthood*. London: Routledge and Kegan Paul.
Baringa, Marcia
    1991   "How Long Is the Human Life-Span?" *Science* 254: 936–938.
Barlow, Robin
    1994   "Population Growth and Economic Growth: Some More Correlations." *Population and Development Review* 20(1): 153–165.
Basu, Alaka Malwade
    1992   *Culture, the Status of Women, and Demographic Behaviour*. Oxford, England: Clarendon Press.
Bean, Frank D., Thomas J. Espenshade, Michael J. White, and Robert F. Dymmowksi
    1990   "Post-IRCA Changes in the Volume and Composition of Undocumented Migration to the United States: An Assessment Based on Apprehensions Data." In Frank D. Bean, Barry Edmonston, and Jeffrey S. Passel (eds.), *Undocumented Migration to the United States: ICRA and the Experience of the 1980's*, 111–158. Washington, D.C.: Urban Institute.
Bean, Frank D., and Michael Fix
    1992   "The Signifigance of Recent Immigration Policy Reforms in the United States." In Gary P. Freeman and James Jupp (eds.), *Nations of Immigrants: Australia, the United States, and International Migration*, 41–55. Melbourne, Australia: Oxford University Press.

Bean, Frank D., Edward E. Telles, and B. Lindsay Lowell
1987 "Undocumented Migration to the United States: Perceptions and Evidence." *Population and Development Review* 13: 671–690.

Beaver, Steven E.
1975 *Demographic Transition Theory Reinterpreted.* Lexington, Mass.: Lexington Books.

Becker, G., E. Landes, and R. Michael
1977 "An Economic Analysis of Marital Instability." *Journal of Political Economy* 85: 1141–1187.

Berger, Mark
1989 "Demographic Cycles, Cohort Size, and Earnings." *Demography* 26(2): 311–321.

Berkov, Beth
1971 "Illegitimate Fertility in California's Population." In Kingsley Davis and Frederick G. Styles (eds.), *Symposium on California's Population Problems and Policies.* Berkeley and Los Angeles: University of California Press.

Birdsell, Nancy
1980 "Population Growth and Poverty in the Developing World." *Population Bulletin* 35(5): 1–46.

Blake, Judith
1972 "Coercive Pronatalism and American Population Policy." In Charles Westoff and Robert Parke (eds.), *Social and Demographic Aspects of Population Growth and the American Future.* Washington, D.C.: U.S. Government Printing Office.

Blanc, Ann K.
1991 "Demographic and Health Surveys World Conference." In *Executive Summary IRD/Macro International, Inc.,* 1–28. Columbia, Md.: IRD/ Macro International.

Bollen, Kenneth A., and David P. Phillips
1981 "Suicide Motor Vehicle Fatalities in Detroit: A Replication." *American Journal of Sociology* 87(2): 404–412.

Bongaarts, John
1983 "The Formal Demography of Families and Households: An Overview." *IUSSP Newsletter* 17: 27–42.

Booth, Allan, and John N. Edwards
1985 "Age at Marriage and Marital Stability." *Journal of Marriage and the Family* 47(1): 67–75.

Borjas, George J., and Marta Tienda
1987 "The Economic Consequences of Immigration." *Science* 235: 645–651.

Bouvier, Leon F.
1992 *Peaceful Invasions: Immigration and Changing America.* Lanham, Md: University Press of America.

Boyd, Robert
1992 "Population Decline from Two Epidemics on the Northwest Coast." In John W. Verano and Douglas H. Ubelaker (eds.), *Disease and Demography in the Americas,* 249–255. Washington, D.C.: Smithsonian Institution Press.

Boyer, Richard, and David Savageau
    1989   *Places Rated Almanac.* Skokie, Ill.: Rand-McNally.
Briggs, Vernon N. Jr.
    1992   *Mass Immigration and the National Interest.* New York: M. E. Sharpe.
Brown, David
    1992a   "Smoking Deaths Seen Climbing." *Washington Post*, May 22, p. A8.
    1992b   "Infectious Disease Reemerging as a Major Cause of U.S. Illness."
             Washington Post, October 16, p. A2.
Brown, James S., Harry K. Schwarzweller, and Joseph J. Mangalam
    1963   "Kentucky Mountain Migration and the Stem Family: An American
            Variation of a Theme by Leplay." *Rural Sociology* 28: 48–69.
Brown, Lester R.
    1991   *Building a Sustainable Society.* New York: W. W. Norton.
Bumpass, Larry L., and James A. Sweet
    1972   "Differentials in Marital Instability." *American Sociological Review* 37:
            754–766.
Burch, Thomas K.
    1979   "Household and Family Demography: A Bibliographic Essay." *Population Index* 45: 173–195.
Caldwell, John C.
    1982   *Theory of Fertility Decline.* New York: Academic Press.
    1990   "Cultural and Social Factors Influencing Mortality Levels in Developing Countries." *Annals of the American Academy of Political and Social Science* 510: 44–59.
Caldwell, John C., I. O. Orubuloye, and Pat Caldwell
    1992   "Fertility Decline in Africa: A New Type of Transition?" *Population and Development Review* 18(2): 211–242.
Carnes, Bruce A., and S. Jay Olshansky
    1993   "Evolutionary Perspectives on Human Senescence." *Population and Development Review* 19: 793–806.
Cassen, R. H.
    1978   *India: Population, Economy and Society.* New York: Holmes and Meir.
Castaño, Gabriel Murillo.
    1988   "Effects of Emigration and Return on Sending countries: The Case of Cubumbia." In Charles Stahl (ed.), *International Migration Today*, 2: 191–203. Leige, Belgium: UNESCO.
Centers for Disease Control
    1992   "Addendum to the Proposed Expansion of the AIDS Surveillance Case Definition," October 22. Atlanta, Ga.: Author.
    1993   *HIV/AIDS Surveillance Report.* Atlanta, Ga.: Author.
    1994   *HIV/AIDS Surveillance Report.* 5: p. 8. Atlanta, Ga.: Author.
    1995   *HIV/AIDS Surveillance Report.* Atlanta, Ga.: Author.
Chaudhry, Mahinder D.
    1992   "Population Growth Trends in India: 1991 Census." *Population and Environment* 14: 31–48.
Chen, P. C., and A. Kols
    1982   *Population and Birth Planning in the People's Republic of China.* Population Reports, Series J, no. 25. Baltimore: Johns Hopkins University Press.

Cherlin, Andrew
    1977   "The Effects of Children on Marital Dissolution." *Demography* 14(3):
           266–272.
Choldin, Harvey
    1988   "Government Statistics: The Conflict between Research and Privacy."
           Demography 25(1): 145–154.
1 Chronicles 21, 1–6
    1953   In *The Holy Bible* (Rev. Standard Version, p. 357). New York: Thomas
           Nelson and Sons.
Clarkson, L. A.
    1981   "Irish Population Revisited, 1687–1821." In J. M. Goldstrom and L. A.
           Clarkson (eds.), *Irish Population, Economy and Society.* Oxford, En-
           gland: Clarendon Press.
Cleland, John, and Christopher Wilson
    1987   "Demand Theories of the Fertility Transition: An Iconoclastic View."
           Population Studies 41: 5–30.
"Clinton Overturns Mexico City Policy with Stroke of Pen."
    1993   *Popline* 15:1.
Coale, Ansley J.
    1973   "The Demographic Transition Reconsidered." In *International Popu-
           lation Conference*, 1: 53–72. Leige, Belgium: International Union for
           the Scientific Study of Population.
    1989   "Marriage and Childbearing in China since 1940." *Social Forces* 67(4):
           833–850.
    1991a "Excess Female Mortality and the Balance of the Sexes in the Popula-
           tion: An Estimate of the Number of 'Missing Females.'" *Population
           and Development Review* 17: 517–523.
    1991b "People Over Age 100: Fewer Than We Think." *Population Today*
           19: 6–8.
Coale, Ansley J., and Susan Watkins (eds.)
    1986   *The Decline of Fertility in Europe.* Princeton, N.J.: Princeton Univer-
           sity Press.
Coale, Ansley J., and M. Zelnik
    1963   *New Estimates of Fertility and Population in the United States.* Prince-
           ton, N.J.: Princeton University Press.
Cochrane, Susan Hill
    1979   *Fertility and Education: What Do We Really Know?* Baltimore: Johns
           Hopkins University Press.
Coleman, David, and John Salt
    1992   *The British Population: Patterns, Trends and Processes.* New York:
           Oxford University Press.
Condorcet, Marie Jean Antoine Nicolas, Marquis de
    1795/1970   *Esquisse d'un tableau historique des progrès de l'esprit humain.*
                Paris: J. Urin.
Connell, Kenneth H.
    1950   *The Population of Ireland, 1740–1845.* Oxford, England: Clarendon Press.
    1962   "Peasant Marriage in Ireland: Its Structure and Development since the
           Famine." *Economic History Review*, 2d. ser., 14: 502–523.

Cook, Robert C.
  1967   "Soviet Population Theory from Marx to Kosygin." *Population Bulletin* 23: 85–115.
Cooney, Rosemary Santana, and Jiali Li
  1994   "Household Registration Type and Compliance with the 'One Child' Policy in China, 1979–1988," *Demography* 31(1): 21–32.
Cornelius, Wayne A.
  1989   "Impacts of the 1986 U.S. Immigration Law on Emigration from Rural Mexican Sending Communities." *Population and Development Review* 15: 689–705.
"Correspondence."
  1991   *New England Journal of Medicine* 325: 1041–1043.
Cowgill, Ursala
  1970   "The People of York: 1538–1812." *Scientific American* 222(1): 104–112.
Cramer, James
  1980   "Fertility and Female Employment: Problems of Causal Direction." *American Sociological Review* 45: 167–190.
Crimmins, Elaine M.
  1981   "The Changing Pattern of American Mortality Decline 1970–77 and Its Implications for the Future." *Population and Development Review* 7(2): 229–254.
Dailey, George H. Jr., and Rex R. Campbell
  1980   "The Ozark-Ouachita Uplands: Growth and Consequences." In David L. Brown and John M. Wardwell (eds.), *New Directions in Urban–Rural Migration.* New York: Academic Press.
Das Gupta, Monica
  1987   "Selective Determination against Female Children in Rural Punjab, India." *Population and Development Review* 13: 77–100.
DaVanzo, Julie S., and Peter A. Morrison
  1981   "Return and Other Sequences of Migration in the United States." *Demography* 18(1): 85–101.
  1982   *Migration Sequences: Who Moves Back and Who Moves On?* Santa Monica, Calif.: Rand Corporation.
David, Henry P.
  1982   "Eastern Europe: Pronatalist Policies and Private Behavior." *Population Bulletin* 36(6): 1–48.
Davis, F. James
  1991   *Who Is Black?* University Park: Pennsylvania State University Press.
Davis, George A., and O. Fred Donaldson
  1975   *Blacks in the United States: A Geographic Perspective.* Boston: Houghton Mifflin.
Davis, Kingsley
  1955   "Institutional Patterns Favoring High Fertility in Underdeveloped Countries." *Eugenics Quarterly* 2: 33–39.
Day, Jennifer Cheeseman
  1993   *Population Projections of the United States, by Age, Sex, Race, and Hispanic Origin: 1993 to 2050.* U.S. Bureau of the Census, Current Population Reports, P25-1104. Washington, D.C.: U.S. Government Printing Office.

de Castro, Josue
    1952 *Geography of Hunger*. Boston: Little, Brown.

DeJong, Gordon
    1980 "Nonmetropolitan Area Migrants: Preference and Satisfaction." *Intercom*, November/December, pp. 8–10.

DeJong, Gordon, and Robert W. Gardner (eds.)
    1981 *Migration Decision Making*. New York: Pergamon Press.

Della Pergola, Sergio
    1980 "Patterns of American Jewish Fertility." *Demography* 17(3): 261–73.

del Pinal, Jorge
    1992 *Exploring Alternative Race–Ethnic Comparison Groups in Current Population Surveys*. U.S. Bureau of the Census, Current Population Reports Series p23–182 Special Studies. Washington, D.C.: U.S. Government Printing Office.

Demeny, Paul
    1968 "Early Fertility Decline in Austria-Hungary: A Lesson in Demographic Transition." *Daedalus* 97: 502–522.

Demerath, Nicholas
    1976 *Birth Control and Foreign Policy: The Alternatives to Family Planning*. New York: Harper & Row.

Dixon, R. B.
    1975 *Women's Rights and Fertility*. Reports on Population/Family Planning, no. 17. New York: Population Council.

Dixon-Mueller, Ruth
    1993 *Population Policy and Women's Rights*. Westport, Conn.: Praeger.

Donaldson, Loraine
    1991 *Fertility Transition; The Social Dynamics of Population Change*. Cambridge, Mass.: Basil Blackwell.

Donaldson, Peter J.
    1990 "On the Origins of the United States Government's International Population Policy." *Population Studies* 44: 385–399.

Donato, Katharine M., Jorge Durand, and Douglas S. Massey
    1992 "Stemming the Tide? Assessing the Deterrent Effects of the Immigration Reform and Control Act." *Demography* 29(2): 139–157.

Doubleday, Thomas
    1853 *The True Law of Population, Shown to Be Connected with the Food of the People*. 3d ed. London: Smith, Elder.

Douglas, Mary
    1966 "Population Control in Primitive Groups." *British Journal of Sociology* 17: 263–273.

Drake, Michael
    1963 "Marriage and Population Growth in Ireland, 1740–1845." *Economic History Review* 16: 311.
    1969 *Population and Society in Norway, 1735–1865*. Cambridge: Cambridge University Press.

D'Souza, Stan, and Lincoln C. Chen
    1980 "Sex Differences in Mortality in Rural Bangladesh." *Population and Development Review* 6: 257–270.

Durand, John D.

    1960   "Mortality Estimates from Roman Tombstone Inscriptions." *American Journal of Sociology* 65: 365–373.

    1977   "Historical Estimates of World Popuation." *Population and Development Review* 3(3): 253–296.

Durkheim, Emile

    1897/1951   *Suicide.* Translated by John A. Spaulding and George Simpson. New York: Free Press.

Easterlin, Richard A.

    1980   *Birth and Fortune.* New York: Basic Books.

Ehrlich, Paul

    1971   *The Population Bomb.* 2d ed. New York: Ballantine Books.

Ehrlich, Paul, and Anne Ehrlich

    1972   *Population, Resources, Environment.* 2d ed. New York: W. H. Freeman.

    1990   *The Population Explosion.* New York: Simon & Schuster.

El-Badry, Mohammed A.

    1991   "The Growth of World Population: Past, Present and Future." In United Nations, *Consequences of Rapid Population Growth in Developing Countries,* Proceedings of the United Nations/Institut National d'Etudes Demographiques Expert Group Meeting, New York, 23–26, August 1988. New York: United Nations.

Enos, Darryl D., and Paul Sultan

    1977   *The Sociology of Health Care: Social, Economic, and Political Perspectives.* New York: Praeger.

Enstrom, James E.

    1978   "Cancer and Total Mortality among Active Mormons." *Cancer* 42: 1943–1951.

    1989   "Health Practices and Cancer Mortality among Active California Mormons." *Journal of the National Cancer Institute* 81: 1379–1384.

Espenshade, Thomas J.

    1990a   "A Short History of U.S. Policy toward Illegal Immigration." *Population Today* 18(2): 6–9.

    1990b   "Undocumented Migration to the United States: Evidence from a Repeated Trials Model." In Frank D. Bean, Barry Edmonston, and Jeffrey S. Passel (eds.), *Undocumented Migration to the United States: ICRA and the Experience of the 1980's.* Washington, D.C.: Urban Institute.

Espenshade, Thomas J., Frank D. Bean, Tracy Ann Goodis, and Michael J. White

    1990   "Immigration Policy in the United States: Future Prospects for the Immigration Reform and Control Act of 1986." In Godfrey Roberts (ed.), *Population Policy: Contemporary Issues,* 59–84. New York: Praeger.

Espenshade, Thomas J., and Joseph J. Minarik

    1987   "Demographic Implications of the 1986 U.S. Tax Reform." *Population and Development Review* 13: 115–127.

Farley, Reynolds

    1991   "The New Census Question about Ancestry: What Did It Tell Us?" *Demography* 28(3): 411–429.

Fassmann, Heinz, and Rainer Munz
  1992   "Patterns and Trends of International Migration in Western Europe."
         Population and Development Review 18(3): 457–480.
Feeney, Griffith, and Wang Feng
  1993   "Parity Progression and Birth Intervals in China." *Population and
         Development Review* 19(1): 61–102.
Feiler, Gil
  1991   "Migration and Recession: Arab Labor Mobility in the Middle East,
         1982–89." *Population and Development Review* 17(1): 134–156.
Feinleib, Manning
  1992   "The Demographic Setting: Trends in Rankings and Levels of Perinatal
         and Infant Mortality, Low Birth Weight, and Other Outcome Mea-
         sures." In *Proceedings of the International Collaborative Effort on
         Perinatal and Infant Mortality, Volume III. Papers presented at the
         International Symposium of Perinatal and Infant Mortality, 1990,
         Bethesda, Maryland, Sponsored by the National Center for Health Sta-
         tistics.* Hyattsville, Md.: National Center for Health Statistics.
Fingerhut, Lois A., and Joel C. Kleinman
  1990   "International and Interstate Comparisons of Homicide Rates among
         Young Males." *Journal of the American Medical Association* 263:
         3292–3295.
Fingerhut, Lois A., Joel C. Kleinman, Elizabeth Godfrey, and Harry Rosenberg
  1991   "Firearm Mortality among Children, Youth, and Young Adults 1–34
         Years of Age, Trends and Current Status: United States, 1979–88."
         Monthly Vital Statistics Report 39(11, Suppl.). Hyattsville, Md.: Na-
         tional Center for Health Statistics.
Finkle, Jason L., and Barbara R. Crane
  1975   "The Politics of Bucharest: Population Development and the New In-
         ternational Economic Order." *Population and Development Review*
         1(1): 87–114.
Forbes, Douglas, and W. Parker Frisbie
  1991   "Spanish Surname and Anglo Infant Mortality: Differentials over Half
         a Century." *Demography* 28(4): 639–660.
Fox, A. J.
  1990   "Socio-Economic Differences in Mortality and Morbidity." *Scandina-
         vian Journal of Social Medicine* 18: 1–8.
Fox, A. J., and A. M. Adelstein
  1978   "Occupational Mortality: Work or Way of Life." *Journal of Epidemol-
         ogy and Community Health* 32(2): 73–78.
Fraser, Steward E.
  1982   "How China Conducted Its 1982 Census." *Intercom* 10(11–12): 3–5.
Freedman, Ronald
  1990   "Family Planning Programs in the Third World." *Annals of the Ameri-
         can Academy of Political and Social Science* 510: 33–43.
Frey, William, and Arden Speare Jr.
  1992   "The Revival of Metropolitan Population Growth in the United States:
         An Assessment of Findings from the 1990 Census." *Population and
         Development Review* 18(1): 129–146.

Fries, J. F.
  1980   "Aging, Natural Death and the Compression of Morbidity." *New England Journal of Medicine* 303: 130–135.
Frisbie, W. Parker, and D. L. Poston Jr.
  1975   "Components of Sustenance Organization and Nonmetropolitan Population Change: A Human Ecological Investigation." *American Sociological Review* 40: 773–784.
  1976   "The Structure of Sustenance Organization and Population Change in Nonmetropolitan America." *Rural Sociology* 41: 354–370.
Frisch, Rose E.
  1980   "Fatness, Puberty, and Fertility." *Natural History* 89(10): 16–27.
Fuguitt, Glenn V., David L. Brown, and Calvin L. Beale
  1989   *Rural and Small Town America.* New York: Russell Sage.
Garreau, Joel
  1992   "These Cows Now Moo in Washington." *Washington Post*, October 10, pp. A1, A16.
Garson, Lea Keil
  1991   "The Centenarian Question: Old-Age Mortality in the Soviet Union, 1897–1970." *Population Studies* 45: 265–278.
Gee, E. M., and J. E. Veevers
  1983   "Accelerating Sex Differences in Mortality: An Analysis of Contributing Factors." *Social Biology* 30: 75–85.
Gertler, Paul J., and John W. Molyneaux
  1994   "How Economic Development and Family Planning Programs Combined to Reduce Indonesian Fertility." *Demography* 31(1): 33–63.
Gladwell, Malcolm
  1991   "Left-Handers Die Younger, Study Says." *Washington Post*, April 4, pp. A1, A18.
Glenn, Norval, and Beth Ann Shelton
  1985   "Regional Differences in Divorce in the United States." *Journal of Marriage and the Family* 47(3): 641–652.
Glenn, Norval, and Michael Supancic
  1984   "The Social and Demographic Correlates of Divorce and Separation in the United States: An Update and Reconsideration." *Journal of Marriage and the Family* 46(3): 563–575.
Glick, Paul C.
  1988   "Fifty Years of Family Demography: A Record of Social Change." *Journal of Marriage and the Family* 50: 861–873.
Glick, Paul C., and Arthur J. Norton
  1977   "Marrying, Divorcing and Living Together in the U.S. Today." *Population Bulletin* 32: 3–39.
Gober-Myers, P.
  1978   "Employment-Motivated Migration and Economic Growth in Post-Industrial Market Economics." *Progress in Human Geography* 2: 207–229.
Godwin, William
  1793/1946   *Enquiry Concerning Political Justice and Its Influence on Morals and Happiness.* 3 vols. Edited by F.E.L. Priestly. Toronto: University of Toronto Press.

Goldman, Noreen
    1993   "Marriage Selection and Mortality Patterns: Inferences and Fallacies." *Demography* 30: 189–208.

Goldman, Noreen, Charles F. Westoff, and Charles Hammerslough
    1984   "Demography of the Marriage Market in the United States." *Population Index* 50(1): 5–25.

Goldscheider, Calvin, and William D. Mosher
    1991   "Patterns of Contraceptive Use in the United States: The Importance of Religious Factors." *Studies in Family Planning* 22: 102–115.

Goodman, Louis J.
    1975   "The Longevity and Mortality of American Physicians, 1969–1973." *Milbank Memorial Fund Quarterly/Health and Society*, summer, pp. 353–375.

Goodstadt, Leo
    1982   "China's One-Child Family: Policy and Public Response." *Population and Development Review* 8: 37–58.

Goubert, Pierre
    1968   "Legitimate Fecundity and Infant Mortality in France during the Eighteenth Century: A Comparison." *Daedalus* 97: 593–603.

Gould, Stephen J.
    1987   *The Flamingo's Smile: Reflections in Natural History.* New York: W. W. Norton.

Graunt, John
    1662/1964   Observations Upon the Bills of Mortality. In Charles Henry Hull (ed.), *The Economic Writings of Sir William Petty, Together with the Observations upon the Bills of Mortality.* New York: Augustus M. Kelley.
    1662/1939   *Natural and Political Observations Mentioned in a Following Index and Made upon the Bills of Mortality.* Baltimore: Johns Hopkins University Press.

Guinnane, Timothy
    1990   "Re-thinking the Western European Marriage Pattern: The Decision to Marry in Ireland at the Turn of the Twentieth Century." *Journal of Family History* 16: 47–64.

Guttentag, Marcia, and Paul Secord
    1983   *Too Many Women? The Sex Ratio Question.* Beverly Hills, Calif.: Sage.

Gwatkin, Davidson R.
    1979   "Political Will and Family Planning: The Implications of India's Emergency Experience." *Population and Development Review* 5(1): 29–59.

Hackey, Melissa K.
    1991   "Injuries and Illnesses in the Workplace, 1989." *Monthly Labor Review*, May, pp. 34–36.

Haines, Michael
    1979   *Fertility and Occupation: Population Patterns in Industrialization.* New York: Academic Press.

Hajnal, J.
    1965   "European Marriage Patterns in Perspective." In D. V. Glass and D. E. C. Eversley (eds.), *Population in History*, 101–143. Chicago: Aldine.

Halpern, Diane F., and Stanley Coren
1988 "Do Right-Handers Live Longer?" *Nature* 332: 213.
1991 "Handedness and Life Span." *New England Journal of Medicine* 324: 998.
Harbison, Sarah F.
1981 "Family Structure and Family Strategy in Migration Decision Making." In Gordon F. DeJong and Robert W. Gardner (eds.), *Migration Decision Making*. New York: Pergamon Press.
Hardy, Ann M.
1992 "AIDS Knowledge and Attitudes for January–March 1991." *Advance Data from Vital and Health Statistics*, no. 216, 1–15. Hyattsville, Md.: National Center for Health Statistics.
Hassold, T., S. D. Quillen, and J. A. Yamane
1983 "Sex Ratios in Spontaneous Abortions." *Annales of Human Genetics* 47: 39–47.
Haub, Carl
1991a "World's Largest Head Count Ever." *Population Today* 19(1): 3.
1991b "Homeless Count Hits Street." *Population Today* 19(6): 5.
Haub, Carl, and Machiko Yanagishita
1991 "Infant Mortality: Who's Number One?" *Population Today* 19: 6–8.
Haupt, Arthur
1983 "The Shadow of Female Infanticide." *Intercom* 11: 13–14.
1984 "They Came to Mexico City . . ." *Population Today* 12(10): 8–12.
Haupt, Arthur, and Thomas T. Kane
1991 *The Population Reference Bureau's Population Handbook—U.S. Edition*. 3d ed. Washington, D.C.: Population Reference Bureau.
Hauser, Philip M., and Otis Dudley Duncan
1959 *The Study of Demography: An Inventory and Appraisal*. Chicago: University of Chicago Press.
Hawkins, Margaret, Edmond A. Murphy, and Helen Abbey
1965 "The Familial Component in Longevity: A Study of the Offspring of Nonagenarians: I. Methods and Preliminary. Report." *Bulletin of the Johns Hopkins Hospital* 117(1): 24–36.
Hawley, Amos
1950 *Human Ecology*. New York: Ronald Press.
Haynes, Suzanne C., and Manning Feinleib
1980 "Women, Work and Coronary Disease: Prospective Findings from the Framingham Heart Study." *American Journal of Public Health* 70: 133–141.
Heaton, Tim B., William B. Clifford, and Glenn W. Fuguitt
1981 "Temporal Shifts in the Determinants of Young and Elderly Migration in Nonmetropolitan Areas." *Social Forces* 60(1): 41–60.
Heaton, Tim B., Carl Fredrickson, Glenn Fuguitt, and James J. Zuiches
1979 "Residential Preference, Community Satisfaction and the Intention to Move." *Demography* 16(4): 565–573.
Hecht, Jacqueline
1987 "Johann Peter Sussmilch: A German Prophet in Foreign Countries." *Population Studies* 41(1): 31–58.

Heilig, Gerhard
  1991 "The Possible Impact of AIDS on Future Mortality." In Wolfgang Lutz (ed.), *Future Demographic Trends in Europe and North America*, 71–95. New York: Academic Press.
Heleniak, Timothy E.
  1991 "Glasnost and the Publication of Soviet Census Data." *Population Today* 19(12): 9.
Henry, Louis
  1968 "Historical Demography." *Daedalus* 97(2): 385–396.
Hershey, Sharon
  1993 "New U.N. Report Tracks Migration," *Population Today* 21(4): 5.
Hess, Peter N.
  1988 *Population Growth and Socioeconomic Progress in Less Developed Countries: Determinants of Fertility Transition*. New York: Praeger.
Himes, Christine L., and Clifford C. Clogg
  1992 "An Overview of Demographic Analysis as a Method for Evaluating Census Coverage in the United States." *Population Index* 58: 587–607.
Hodgson, Dennis
  1991 "Benjamin Franklin on Population: From Policy to Theory." *Population and Development Review* 17(4): 639–661.
Hoem, Jan M.
  1990 "Social Policy and Recent Fertility Change in Sweden." *Population and Development Review* 16: 735–748.
Horner, Edith R. (ed.)
  1992 *Almanac of the Fifty States 1992*. Palo Alto, Calif.: Information Publications.
Hoskin, Alan F.
  1991 "1990 Statistics Show No Change in Work Deaths." *Saftey and Health*, May, p. 74.
Hu, Yuanreng, and Noreen Goldman
  1990 "Mortality Differentials by Marital Status: An International Comparison." *Demography* 27(2): 233–250.
Hull, Charles Henry
  1899/1963 *The Economic Writings of Sir William Petty*, vol. 1. New York: Augustus M. Kelley.
Hull, Terence H.
  1990 "Recent Trends in Sex Ratios at Birth in China." *Population and Development Review* 16: 63–83.
Hull, Terence H., Valerie J. Hull, and Masri Singarimbun
  1977 "Indonesia"s Family Planning Story: Success and Challenge." *Population Bulletin* 32(6): 1–52.
Hummer, Robert A., Isaac W. Eberstein, and Charles B. Nam
  1992 "Infant Mortality Differentials among Hispanic Groups in Florida." *Social Forces* 70: 1055–1075.
Hutchinson, Edward P.
  1967 *The Population Debate*. New York: Houghton Mifflin.
  1981 *Legislative History of American Immigration Policy, 1798–1865*. Philadelphia: University of Pennsylvania Press.

Hyatt, Douglas E., and William J. Milne
  1991   "Can Public Policy Affect Fertility?" *Canadian Public Policy* 17: 77–85.
Isaac, Stephen L., and Renee J. Holt
  1987   "Redefining Procreation: Facing the Issues." *Population Bulletin* 42(3):
         1–37.
Janssen, Susan G., and Robert Hauser
  1981   "Religion, Socialization, and Fertility." *Demography* 18(4): 511–528.
Jerome, Harry
  1926   *Migration and Business Cycles.* New York: National Bureau of Econom-
         ics Research.
Johansson, S. Ryan
  1991   "'Implicit' Policy and Fertility during Development." *Population and
         Development Review* 17: 377–414.
Johansson, Sten, and Olga Nygren
  1991   "The Missing Girls of China: A New Demographic Account." *Popula-
         tion and Development Review* 17(1): 35–52.
Jones, James H.
  1981   *Bad Blood: The Tuskegee Syphilis Experiment.* New York: Free Press.
Jowett, B. (trans.)
  1937   *The Dialogue of Plato.* 3d ed. New York: Random House.
Kahn, H. A., R. L. Phillips, D. A. Snowden, and W. Choe
  1984   "Association between Reported Diet and All-Cause Mortality: Twenty-
         One Year Follow-up on 27,530 Adult Seventh-Day Adventists." *Ameri-
         can Journal of Epidemiology* 119: 775–787.
Kahn, Joan R.
  1994   "Immigrant and Native Fertility during the 1980s: Adaptation and
         Expectations for the Future." *International Migration Review* 28(3);
         501–519.
Kaku, Kanae
  1975   "Were Girl Babies Sacrificed to a Folk Superstitution in 1966 in Japan?"
         Annals of Human Biology 2(4): 391–393.
Kalish, Susan
  1992   "Multiculturalism Grows, but Segregation Lingers." *Population Today*
         20(7–8): 3–4.
  1994   "Cairo Built Momentum for Change, Say Advocates," *Global Steward-
         ship*, October/November, pp. 1–2.
Kammeyer, Kenneth C. W.
  1976   "The Dynamics of Population." In Harold Orel (ed.), *Irish History and
         Culture*, 189–223. Lawrence: University of Kansas Press.
Kammeyer, Kenneth, and Helen Ginn
  1986   *An Introduction to Population.* Chicago: Dorsey Press.
Kaplan, Charles, and Thomas Van Valey (eds.)
  1980   *Census '80: Continuing the Factfinder Tradition.* Washington, D.C.: U.S.
         Government Printing Office.
Karasek, Robert A., D. Baker, F. Marxer, A. Ahlbom, and Tores Thorsell
  1981   "Job Decision Latitude, Job Demands and Cardiovascular Disease: A
         Prospective Study of Swedish Men." *American Journal of Public Health*
         71: 694–705.

Karasek, Robert A., Tores Thorsell, Joseph E. Schwartz, Peter L. Schnall, Carl F. Pieper, and John L. Michela
  1988   "Job Characteristics in Relation to the Prevalence of Myocardial Infarction in the U.S. Health Examination Survey (HES) and the Health and Nutrition Examination Survey (HANES)." *American Journal of Public Health* 78: 910–918.

Kennedy, Robert E. Jr.
  1973   *The Irish: Emigration, Marriage, and Fertility.* Berkeley and Los Angeles: University of California Press.

Kephart, William M., and William W. Zellner
  1991   *Extraordinary Groups.* New York: St. Martin's Press.

Kertzler, David I., and Dennis P. Hogan
  1990   "Reflections on the European Marriage Pattern: Sharecropping and Proletarianization in Casalecchio, Italy, 1861–1921, *Journal of Family History* 16: 31–45.

Keyfitz, Nathan
  1981   "The Limits of Population Forecasting." *Population and Development Review* 7: 579–593.

Kiser, Clyde, and Myrna E. Frank
  1967   "Factors Associated with the Low Fertility of Nonwhite Women of College Attainment." *Milbank Memorial Fund Quarterly* 45(4): 427–449.

Kiser, Clyde V., and Pascal K. Whelpton
  1958   "Summary of Chief Findings and Implications for Future Studies." *Milbank Memorial Fund Quarterly* 36(3): 282–329.

Kishor, Sunita
  1993   "'May God Give Sons to All': Gender and Child Mortality in India." *American Sociological Review* 58: 247–265.

Kitagawa, Evelyn M., and Philip M. Hauser
  1964   "Trends in Differential Mortality in a Metropolis: Chicago." In Ernest W. Burgess and Donald J. Bogue (eds.), *Contributions to Urban Sociology*, 59–85. Chicago: University of Chicago Press.
  1968   "Educational Differentials in Mortality by Cause of Death: United States 1960." *Demography* 5(2): 318–353.
  1973   *Differential Mortality in the United States: A Study of Socioeconomic Epidemiology.* Cambridge, Mass.: Harvard University Press.

Knodel, John
  1987   "Starting, Stopping, and Spacing during the Early Stages of Fertility Transition: The Experience of German Village Populations in the 18th and 19th Centuries." *Demography* 24(2): 143–162.

Knodel, John, Napaporn Chayovan, and Siriwan Siriboon
  1992   "The Impact of Fertility Decline on Familial Support for the Elderly: An Illustration from Thailand." *Population and Development Review* 18(1): 79–103.

Knodel, John, and Etienne van de Walle
  1978   "European Populations in the Past: Family Level Relations." In Samuel H. Preston (ed.), *The Effects of Infant and Child Mortality on Fertility*, 21–45. New York: Academic Press.

1979   "Lessons from the Past: Policy Implications of Historical Fertility Stud-
        ies." *Population and Development Review* 5(2): 217–245.
Knodel, John, and Malinee Wongsith
1991   "Family Size and Children's Education in Thailand: Evidence from a
        National Sample." *Demography* 28(1): 119–131.
Kocher, James E.
1980   "Population Policy in India: Recent Developments and Current Pros-
        pects." *Population and Development Review* 6(2): 299–310.
Koussoudji, Sherrie A.
1992   "Playing Cat and Mouse at the U.S.–Mexican Border." *Demography*
        29(2): 159–180.
Kreager, Philip
1988   "New Light on Graunt," *Population Studies* 42: 129–140
Kristof, Nicholas D.
1993   "China's Crackdown on Births: A Stunning, and Harsh, Success." *New
        York Times*, April 25, pp. 1, 12.
Larsen, Ulla, and James W. Vaupel
1993   "Hutterite Fecundability by Age and Parity: Strategies for Frailty Mod-
        eling of Event Histories." *Demography* 30(1): 81–102.
Lazerwitz, Bernard
1980   "Religiosity and Fertility: How Strong a Connection?" *Contemporary
        Jewry* 6: 3–8.
Lavely, William, and Ronald Freedman
1990   "The Origins of the Chinese Fertility Decline." *Demography* 27(3):
        357–367.
Lee, Barrett A.
1989   "Stability and Change in an Urban Homeless Population." *Demography*
        26(2): 323–334.
Lee, Everett
1961   "The Turner Thesis Re-examined." *American Quarterly* 13: 77–83.
1966   "A Theory of Migration." *Demography* 3(1): 47–59.
1970   "Migration in Relation to Education, Intellect and Social Structure."
        Population Index 36(4): 437–444.
LeMay, Michael C.
1987   *From Open Door to Dutch Door: An Analysis of U.S. Immigration
        Policy since 1820.* New York: Praeger.
LePlay, Frederic
1872   *The Organization of Labor.* Translated by G. Emerson. Philadelphia:
        Claxton, Remsen and Haffelfinger.
Lesthaeghe, Ron
1992   "Beyond Economic Reductionism: The Transformation of the Repro-
        ductive Regimes in France and Belgium in the 18th and 19th Centu-
        ries." In Calvin Goldscheider (ed.), *Fertility Transitions, Family Struc-
        ture, and Population Policy*, 1–44. Boulder, Colo.: Westview Press.
Levine, Richard J., Ravi M. Mathew, C. Brandon Chenault, Michelle H. Brown, Mark
        E. Hurtt, Karin S. Bentley, Kathleen L. Mohr, and Peter K. Working
1990   "Differences in the Quality of Semen in Outdoor Workers during Sum-
        mer and Winter." *New England Journal of Medicine* 323: 12–16.

Lewis, G. J.
1982   *Human Migration*. New York: St. Martin's Press.
Linstead, K. D., S. Tonstad, and J. W. Kuzma
1991   "Self-Report of Physical Activity and Patterns of Mortality in Seventh-Day Adventist Men." *Journal of Clinical Epidemiology* 44: 355–364.
London, Bruce
1988   "Dependence, Distorted Development, and Fertility Trends in Noncore Nations: A Structural Analysis of Cross-National Data." *American Sociological Review* 53: 606–618.
1992   "School-Enrollment Rates and Trends, Gender, and Fertility: A Cross-National Analysis." *Sociology of Education* 65: 306–316.
London, Bruce, and Kenneth Hadden
1988   "The Spread of Education and Fertility Decline: A Thai Province Level Test of Caldwell's 'Wealth Flows Theory.'" *Rural Sociology* 54: 17–36.
Long, Larry H.
1972   "The Influence of Numbers and Ages of Children on Residential Mobility." *Demography* 9(3): 371–382.
1988   *Migration and Residential Mobility in the United States*. New York: Russell Sage.
Lunde, Anders S.
1980   *The Person–Number Systems of Sweden, Norway, Denmark, and Israel*. DHHS Publication no. (PHS) 80-1358. Washington, D.C.: U.S. Government Printing Office.
Malthus, Thomas Robert
1872/1973   *An Essay on the Principle of Population*. Introduction by T. H. Hollingsworth. London: J. M. Dent & Sons.
1798/1960   *An Essay on the Principle of Population as It Affects the Future Improvement of Society, with Remarks on the Speculation of Mr. Godwin, M. Condercet and Other Writers*. New York: Random House.
Mamdani, Mahmood
1972   *The Myth of Population Control*. New York: Monthly Review Press.
Manton, Kenneth G., Eric Stallard, and H. Dennis Tolley
1991   "Limits to Human Life Expectancy: Evidence, Prospects, and Implications." *Population and Development Review* 17: 603–638.
Mare, Robert D.
1990   "Socio-Economic Careers and Differential Mortality among Older Men in the United States." In Jacques Vallin, Stan D'Souza, and Alberto Palloni (ed.), *Measurement and Analysis of Mortality: New Approaches*, 362–387. New York: Oxford University Press.
Mason, Karen Oppenheim, and Karen Kuhlthau
1992   "The Perceived Impact of Child Care Costs on Women's Labor Supply and Fertility." *Demography* 29(4): 523–543.
Massey, Douglas S., Joaquin Arango, Graeme Hugo, Ali Kouaouci, Adela Pellegrino, and J. Edward Taylor
1993   "Theories of International Migration: A Review and Appraisal." *Population and Development Review* 19(3): 431–466.

Mauldin, W. Parker, Nazli Choucri, Frank W. Notestein, and Michael Teitelbaum
  1974   "A Report on Bucharest." *Studies in Family Planning* 5(12): pp. 356–395.
Mauldin, W. Parker, and John A. Ross
  1991   "Family Planning Programs: Efforts and Results, 1982–89." *Studies in Family Planning* 22(6): 350–367.
May, John F.
  1992   "Census Taking in Niue, South Pacific." *Population Today* 20(2): 5.
Mayer, Kurt
  1962   "Developments in the Study of Population." *Social Research* 29(3): 293–320.
McAuley, William J., and Cheri L. Nutty
  1982   "Residential Preferences and Moving Behavior: A Family Life-Cycle Analysis." *Journal of Marriage and the Family* 44: 301–309.
McFalls, Joseph A. Jr.
  1979a *Psychopathology and Subfecundity.* New York: Academic Press.
  1979b "Frustrated Fertility: A Population Paradox." *Population Bulletin* 34(2): 1–43.
McFalls, Joseph A. Jr., and Marguerite Harvey McFalls
  1984   *Disease and Fertility.* New York: Academic Press.
McKeown, Thomas R.
  1976   *The Modern Rise of Population.* London: Edward Arnold.
  1979   *The Role of Medicine: Dream, Mirage or Nemesis?* Princeton, N.J.: Princeton University Press.
McKeown, Thomas R., and R. G. Record
  1962   "Reasons for the Decline in Mortality in England and Wales during the Nineteenth Century." *Population Studies* 16: 94–122.
McKinlay, John Bond, and Sonja M. McKinlay
  1977   "The Questionable Contribution of Medical Measures to the Decline of Mortality in the United States in the Twentieth Century." *Milbank Memorial Fund Quarterly/Health and Society* 55(3): 405–428.
McMillen, Marilyn M.
  1979   "Differential Mortality by Sex in Fetal and Neonatal Deaths." *Science* 204: 89–91.
McQuillan, Kevin
  1979   "Common Themes in Catholic and Marxist Thought on Population and Development." *Population and Development Review* 5(4): 689–698.
Meadows, D. H., D. L. Meadows, J. Randers, and W. Behrens III
  1972   *The Limits of Growth.* New York: New American Library.
Merrick, Thomas W., and Stephen J. Tordella
  1988   "Demographics: People and Markets." *Population Bulletin* 43: 3–46.
Miller, Barbara D.
  1981   *The Endangered Sex: Neglect of Female Children in Rural North India.* Ithaca, N.Y.: Cornell University Press.
Miller, Sheila
  1976   "Family Life Cycle, Extended Family Orientations and Economic Aspirations as Factors in the Propensity to Migrate." *Sociological Quarterly* 17: 323–335.

Miller, Warren B.
    1981   *The Psychology of Reproduction.* Washington, D.C.: National Techni-
            cal Information Service.
    1992   "Personality Traits and Developmental Experiences as Antecedents of
            Childbearing Motivation." *Demography* 29(2): 265–285.
Morgan, S. Philip, Diane N. Lye, and Gretchen A. Condran
    1988   "Sons and Daughters, and the Risk of Marital Disruption." *American
            Journal of Sociology* 94: 110–129.
Morrison, Peter
    1990   "Demographic Factors Reshaping Ties to Family and Peace." *Research
            on Aging* 12(4): 399–408.
Mosher, William D.
    1980   "Demographic Responses and Demographic Transitions." *Demography*
            17(4): 395–412.
Mosher, William D., and William F. Pratt
    1990a   *Contraceptive Use in the United States, 1973–88.* Advance Data for
            Vital and Health Statistics, no. 182. Hyattsville, Md.: National Cen-
            ter for Health Statistics.
    1990b   *Fecundity and Infertility in the United States, 1965–88.* Advance Data
            for Vital and Health Statistics, no. 192. Hyattsville, Md.: National
            Center for Health Statistics.
Mosher, William D., Linda B. Williams, and David P. Johnson
    1992   "Religion and Fertility in the United States: New Patterns." *Demography*
            29(2): 199–214.
Moskos, Charles C. Jr.
    1980   *Greek Americans.* Englewood Cliffs, N.J.: Prentice-Hall.
Mueller, Charles F.
    1982   *The Economics of Labor Migration: A Behavioral Analysis.* New York:
            Academic Press.
Murphy, Francis X.
    1981   "Catholic Perspectives on Population Issues, II." *Population Bulletin*
            35(6): 1–43.
Murphy, Joseph P., and Thomas J. Espenshade
    1990   "Immigrations Prism: Historical Continuities in the Kennedy–Simpson
            Legal Immigration Reform Bill." *Population and Environment* 12:
            139–158.
Narayana, G., and John F. Kantner
    1992   *Doing the Needful: The Dilemma of India's Population Policy.* Boul-
            der, Colo.: Westview Press.
National Center for Health Statistics
    1990   *Monthly Vital Statistics Report* 39(4, Suppl.). Hyattsville, Md.: U.S.
            Public Health Service.
    1991   *Health, United States, 1990.* Hyattsville, Md.: U.S. Public Health Ser-
            vice.
    1992   *Vital Statistics of the United States, 1989,* vol. II, section 6, Life Tables.
            Hyattsville, Md.: U.S. Public Health Service.
    1993   "Advance Report on Final Mortality Statistics, 1991." *Monthly Vital
            Statistics Report,* vol. 42. Hyattsville, Md.: U.S. Public Health Service.

1994   "Births, Marriages, Divorces, and Deaths for 1993." *Monthly Vital Statistics Report*, vol. 42. Hyattsville, Md.: U.S. Public Health Service.

Navarro, Vincent

1990   "Race or Class versus Race and Class: Mortality Differentials in the United States." *Lancet* 17: 1238–1240.

1991   "Class and Race: Life and Death Situations." *Monthly Review* 43: 1–13.

Newland, Kathleen

1981   *Infant Mortality and the Health of Societies*. Washington, D.C.: Worldwatch Institute.

Nicholson, Beryl

1990   "The Hidden Component in Census-Derived Migration Data: Assessing Its Size and Distribution." *Demography* 27(1): 111–120.

Nortman, Dorothy L.

1978   "India's New Birth Rate Target: An Analysis." *Population and Development Review* 4(2): 277–312.

Oakley, Deborah

1978   "American–Japanese Interaction in the Development of Population Policy in Japan, 1945–52." *Population and Development Review* 4(4): 617–643.

Obermeyer, Carla Makhlouf

1992   "Islam, Women, and Politics: The Demography of Arab Countries." *Population and Development Review* 18: 33–60.

"Occupational Health Neglected."

1991   *Occupational Hazards*, January, p. 27.

O'Connell, Martin

1991   "Studies in American Fertility." *Current Population Reports*, Special Studies, Series P-23, no. 176. Washington, D.C.: U.S. Bureau of the Census.

ÓGráda, Cormac

1991   "New Evidence on the Fertility Transition in Ireland, 1880–1911." *Demography* 28(4): 535–548.

O'Hare, William P., Kelvin M. Pollard, Taynia L. Mann, and Mary M. Kent

1991   "African Americans in the 1990s." *Population Bulletin* 46(1): 1–40.

Olshansky, S. Jay

1992   "Estimating the Upper Limits to Human Longevity." *Population Today* 20: 6–8.

Olshansky, S. Jay, and Bruce A. Carnes

1994   "Demographic Perspectives on Human Senescence." *Population and Development Review* 20: 57–80.

Olshansky, S. Jay, B. A. Carnes, and C. Cassel

1990   "In Search of Methuselah: Estimating the Upper Limits to Human Longevity." *Science* 250: 634–640.

Omran, A. R.

1977   "Epidemiologic Transition in the U.S.: The Health Factor in Population Change." *Population Bulletin* 32(3): 1–42.

Parkin, Tim G.

1992   *Demography and Roman Society*. Baltimore: Johns Hopkins University Press.

Passel, Jeffrey S., and Karen A. Woodrow
    1984   "Geographic Distribution of Undocumented Immigrants: Estimates of
            Undocumented Aliens in the 1980 Census by State." *International
            Migration Review* 18: 642–671.
Pearl, Raymond
    1925   *The Biology of Population Growth*. New York: Alfred A. Knopf.
    1939   *The Natural History of Population*. New York: Oxford University Press.
Pearl, Raymond, and Ruth DeWitt Pearl
    1934   *The Ancestry of the Long Lived*. Baltimore: Johns Hopkins University
            Press.
Personick, Martin E., and Ethel C. Jackson
    1992   "Injuries and Illnesses in the Workplace, 1990." *Monthly Labor Re-
            view*, April p. 37.
Petersen, William
    1979   *Malthus*. Cambridge, Mass.: Harvard University Press.
Phillips, David P.
    1972   "Deathday and Birthday: An Unexpected Connection." In J. M. Tanur (ed.),
            Statistics: A Guide to the Unknown, 52–65. San Francisco: Holden-Day.
    1974   "The Influence of Suggestion on Suicide: Substantive and Theoretical
            Implications of the Werther Effect." *American Sociological Review* 39:
            340–354.
    1980   "Airplane Accidents, Murder, and the Mass Media: Towards a Theory
            of Imitation and Suggestion." *Social Forces* 58(4): 1001–1024.
Phillips, Roland L., J. W. Kuzma, W. Lawrence Beesman, and Terry Lotz
    1980   "Influence of Selection versus Lifestyle on Risk of Fatal Cancer and
            Cardiovascular Disease among Seventh-Day Adventists." *American
            Journal of Epidemiology* 112(2): 296–314.
Pierson, George W.
    1962   "The Migration Factor in American History." *American Quarterly*
            14(Suppl.): 275–289.
Population Council
    1992   "Nigeria 1990: Results from the Demographic and Health Survey."
            Studies in Family Planning 23(3): 211.
    1993   "Cameroon 1991: Results from the Demographic and Health Survey."
            Studies in Family Planning 24(2): 132.
Population Reference Bureau
    1993   *World Population Data Sheet*. Washington, D.C.: Author.
    1994   *World Population Data Sheet*. Washington, D.C.: Author.
Portes, Alejandro, and Rubén G. Rumbaut
    1990   *Immigrant America: A Portrait*. Berkeley and Los Angeles: University
            of California Press.
Poston, Dudley L. Jr., and Ralph White
    1978   "Indigenous Labor Supply, Sustenance Organization, and Population
            Redistribution in Nonmetropolitan America: An Extension of the Eco-
            logical Theory of Migration." *Demography* 15(4): 637–641.
Potter, Lloyd B.
    1991   "Socioeconomic Determinants of White and Black Males' Life Expect-
            ancy Differentials, 1980." *Demography* 28: 303–321.

Preston, Samuel H. (ed.)
  1978   *The Effects of Infant and Child Mortality on Fertility.* New York: Academic Press.
  1982   "Review Symposium on Julian Simon, *The Ultimate Resource.*" *Population and Development Review* 8(1): 174–177.
Pritchett, Lant H.
  1994   "Desired Fertility and the Impact of Population Policies." *Population and Development Review* 20(1): 1–55.
Ravenholt, R. T.
  1990   "Tobacco's Global Death March." *Population and Development Review* 16: 213–240.
Ravenstein, E. G.
  1885   "The Laws of Migration." *Journal of the Royal Statistical Society* 48(Part 2): 61–235.
  1889   "The Laws of Migration." *Journal of the Royal Statistical Society* 52: 241–305.
Razzell, P. E.
  1974   "An Interpretation of the Modern Rise of Population in Europe—A Critique." *Population Studies* 28(1): 5–17.
Relethford, John H.
  1991   "Sex Differentials in Unintentional Injury Mortality in Relation to Age at Death." *American Journal of Human Biology* 3: 369–375.
Remini, Robert V.
  1984   *Andrew Jackson and the Course of American Democracy, 1833–1845.* New York: Harper & Row.
Rensberger, Boyce
  1994   "Cairo Conference Ends with Broad Consensus for Plan to Curb Growth," *Washington Post,* September 14, p. A13.
Rich, Spencer
  1992   "Surgeon General Sees Oral Cancer Epidemic." *Washington Post,* December 14, p. A16.
Riley, Ann P., Albert I. Hermalin, and Luis Rosero-Bixby
  1993   "A New Look at the Determinants of Nonnumeric Response to Desired Family size: The Case of Costa Rica." *Demography* 30(2): 159–174.
Rindfuss, Ronald R., S. Philip Morgan, and Gray Swicegood
  1988   *First Births in America: Changes in the Timing of Parenthood.* Berkeley and Los Angeles: University of California Press.
Robey, Bryant
  1989   "Two Hundred Years and Counting: The 1990 Census." *Population Bulletin* 44(1): 1–38.
Roenneberg, Till, and Jürgen Aschoff
  1990a "Annual Rhythm of Human Reproduction: I. Biology, Sociology, or Both?" *Journal of Biological Rhythms* 5: 195–216.
  1990b "Annual Rhythm of Human Reproduction: II. Environmental Correlation." *Journal of Biological Rhythms* 5: 217–239.
Rogers, Richard G.
  1992   "Living and Dying in the U.S.A.: Sociodemographic Determinants of Death among Blacks and Whites." *Demography* 29(2): 287–303.

Rosen, Sherwin, and Paul Taubman
   1979   "Changes in the Impact of Education and Income on Mortality in the
          U.S." In Linda DelBene and Foritz Schueren (eds.), *Statistical Uses
          of Administrative Records with Emphasis on Mortality and Disabil-
          ity Research*. Washington, D.C.: U.S. Department of Health, Educa-
          tion and Welfare, Social Security Administration, Office of Research
          Statistics.
Rossi, Peter
   1955/1980   *Why Families Move*. New York: Free Press.
Ryder, Norman B.
   1990   "What Is Going to Happen to American Fertility?" *Population and
          Development Review* 16(3): 433–454.
Sadik, Nafis (ed.)
   1991   *Population Policies and Programmes: Lessons Learned from Two De-
          cades of Experience*. New York: New York University Press.
Sadler, Michael Thomas
   1829   *Ireland: Its Evils and Their Remedies*. 2d ed. London: John Murray.
Salaman, Redcliffe N.
   1949   *The History and Social Influence of the Potato*. Cambridge: Cambridge
          University Press.
Salt, John
   1981   "International Labor Migration in Western Europe: A Geographical
          Review." In Mary M. Kritz, Charles Keely, and Lydio Tomasi (eds.),
          Global Trends in Migration, 133–157. New York: Center for Migra-
          tion Studies.
Samuelson, Robert J.
   1991, March 20   "As the Boomers Turn Fifty." *The Washington Post*, p. A19.
Sandefur, Gary, and Wilbur J. Scott
   1981   "A Dynamic Analysis of Migration: An Assessment of the Effects of Age,
          Family, and Career Variables." *Social Forces* 18(3): 355–369.
Saveth, Edward N.
   1948   *American Historians and European Immigrants*. New York: Columbia
          University Press.
Sell, Ralph R.
   1983   "Transferred Jobs: A Neglected Aspect of Migration and Occupational
          Change." *Work and Occupations* 10(2): 179–206.
Sell, Ralph R., and Gordon DeJong
   1978   "Toward a Motivational Theory of Migration Decision Making." *Jour-
          nal of Population* 1(4): 313–335.
Sen, Amartya
   1990   "More than 100 Million Women Are Missing." *New York Review of
          Books* 20: 61–66.
Sharpe, Pamela
   1991   "Literally Spinsters: A New Interpretation of Local Economy and
          Demography in Colyton in the Seventeenth and Eighteenth Centuries."
          Economic History Review 44(1): 46–65.
Shryock, Henry S. Jr.
   1991   "Letter to the Editor." *Population Today* 19: 10.

Shryock, Henry S., and Jacob S. Siegel
  1976   *The Methods and Materials of Demography.* New York: Academic Press.
Siegel, Jacob S.
  1974   "Estimates of Coverage of Population by Sex, Race, and Age: Demographic Analysis." In United States Bureau of the Census, *Census of Population and Housing: 1970, Evaluation and Research Program.* Washington, D.C.: U.S. Government Printing Office.
Simmons, George B., Deborah Balk, and Khodezatul K. Faiz
  1991   "Cost-Effectiveness Analysis of Family Planning Programs in Rural Bangladesh: Evidence from Matlub." *Studies in Family Planning* 22(2): 83–101.
Simmoms, George B., Celeste Smucker, Stan Bernstein, and Eric Jensen
  1982   "Post-Neonatal Mortality in Rural India: Implications of an Economic Model." *Demography* 19(3): 371–389.
Simon, Julian L.
  1981   *The Ultimate Resource.* Princeton, N.J.: Princeton University Press.
  1990   *Population Matters: People, Resources, Environment, and Immigration.* New Brunswick, N.J.: Transaction.
Singh, Kuldip, Yoke Fai Fong, and S. S. Ratnam
  1991   "A Reversal of Fertility Trends in Singapore." *Journal of Biosocial Science* 23: 73–78.
Sirageldin, Ismail, and John F. Kantner
  1982   "Review Symposium on Julian L. Simon, *The Ultimate Resource.*" *Population and Development Review* 8(1): 169–173.
Sly, David F.
  1972   "Migration and the Ecological Complex." *American Sociological Review* 37: 615–628.
Sly, David F., and Jeffrey Tayman
  1980   "Metropolitan Morphology and Population Mobility: The Theory of Ecological Expansion Reexamined." *American Journal of Sociology* 86(1): 119–135.
Smith, George D., Martin J. Shipley, and Geoffrey Rose
  1990   "Magnitude and Causes of Socioeconomic Differentials in Mortality: Further Evidence from the Whitehall Study." *Journal of Epidemiology and Community Health* 44: 265–270.
Smith, S. L.
  1990   "Nervous about Neurotoxins." *Occupational Hazards*, December, pp. 37–40.
Smith-Lovin, Lynn, and Ann R. Tickamyer
  1978   "Labor Force Participation, Fertility Behavior and Sex Role Attitudes." *American Sociological Review* 43: 541–556.
Snowdon, D. A.
  1988   "Animal Product Consumption and Mortality Because of All Causes Combined, Coronary Heart Disease, Stroke, Diabetes, and Cancer in Seventh-Day Adventists." *American Journal of Clinical Nutrition* 48: 739–748.
Sommers, David G., and Katherine R. Rowell
  1992   "Factors Differentiating Elderly Residential Movers and Nonmovers." *Population Research and Policy Review* 11: 249–262.

South, Scott J., and Kim M. Lloyd
1992 "Marriage Opportunities and Family Formation: Further Implications of Imbalanced Sex Ratios." *Journal of Marriage and the Family* 54(2): 440–451.

Speare, Alden Jr.
1970 "Home Ownership, Life Cycle Stage, and Residential Mobility." *Demography* 7(4): 449–457.
1974 "Residential Satisfaction as an Intervening Variable in Residential Mobility." *Demography* 11(2): 173–188.

Speare, Alden Jr., Frances Kobrin, and Ward Kingkade
1982 "The Influence of Socioeconomic Bonds and Satisfaction on Interstate Migration." *Social Forces* 61(2): 551–574.

Specter, Michael
1992 "Neglected for Years, TB Is Back with Strains that Are Deadlier." *New York Times*, October 11, pp. 1, 44.

Spencer, Herbert
1873 *The Principles of Biology*, vol. 2. New York: D. Appleton.

Squires, Sally
1990 "Study Traces More Deaths to Working than Driving." *Washington Post*, August 31, p. A7.

Stack, Steven
1987 "Celebrities and Suicide: A Taxonomy and Analysis, 1948–1983." *American Sociological Review* 52: 401–412.

Stahl, Charles (ed.)
1988 *International Migration Today: Vol. 2, Emerging Issues*. Liege, Belgium: UNESCO.

Statistical Yearbook of the Immigration and Naturalization Service, 1993
1993 Washington, D.C.: U.S. Government Printing Service.

Stephen, Elizabeth Hervey, and Frank D. Bean
1992 "Assimilation, Disruption and the Fertility of Mexican-Origin Women in the United States." *International Migration Review* 26(1): 67–88.

Stinner, W. R., and G. F. DeJong
1969 "Southern Negro Migration: Social and Economic Components of an Ecological Model." *Demography* 6(3): 455–473.

Stinner, W. R., and Micheal B. Toney
1980 "Migrant–Native Differences in Social Background and Community Satisfaction in Nonmetropolitan Utah Communities." In David L. Brown and John M. Wardwell (eds.), *New Directions in Urban–Rural Migration*. New York: Academic Press.

Stinner, William, Mollie Van Loon, and Yongchan Byun
1992 "Plans to Migrate In and Out of Utah." *Sociology and Social Research* 76(3): 131–137.

Stolzenberg, Ross M., and Linda J. Waite
1977 "Age Fertility Expectations and Plans for Employment." *American Sociological Review* 42: 769–782.

Stycos, J. Mayone
1963 "Obstacles to Programs of Population Control—Facts and Fancies." *Marriage and Family Living* 25(1): 5–13.
1974 "Demographic Chic at the UN." *Family Planning Perspectives* 6(3): 160–63.

Stycos, J. Mayone, and Robert H. Weller
    1967 "Female Working Roles and Fertility." *Demography* 4(1): 210–217.
Sutherland, Ian
    1963 "John Graunt: A Tercentenary Tribute." *Royal Statistical Journal*, series A, 126: 537–556.
Sweet, James A.
    1977 "Demography and the Family." In Alex Inkeles, James Coleman, and Neil Smelser (eds.), *Annual Review of Sociology*, 363–405. Palo Alto, Calif.: Annual Review.
Symonds, Richard, and Michael Carder
    1973 *The United Nations and the Population Question*. New York: McGraw-Hill.
Taeuber, Cynthia
    1992 "Sixty-Five Plus in America." *Current Population Reports*, Special Studies, P23-178. Washington, D.C.: U.S. Bureau of the Census.
Teachman, Jay
    1983 "Early Marriage, Premarital Fertility and Marital Dissolution: Results for Blacks and Whites." *Journal of Family Issues* 4: 105–126.
Teachman, Jay, Karen A. Polonko, and John Scanzoni
    1987 "Demography of the Family." In Marvin B. Sussman and Suzanne K. Steinmetz (eds.), *Handbook of Marriage and the Family*, 3–36. New York: Plenum Press.
Teachman, Jay D., and Paul T. Schollaert
    1989 "Gender of Children and Birth Timing." *Demography* 26(3): 411–423.
Teitelbaum, Michael S.
    1975 "Relevance of Demographic Transition Theory for Developing Countries." *Science* 188(2): 420–425.
Third African Population Conference
    1993 "Dakar Declaration on Population." In *Population and Development Review* 19(1): 209–215.
Thomas, D. S., and S. Kuznets (eds.)
    1957, 1960, 1964 *Population Redistribution and Economic Growth, United States: 1870–1950*. 3 vols. Philadelphia: American Philosophical Society.
Thompson, Warren S.
    1929 "Population." *American Journal of Sociology* 34: 959–975.
Tien, H. Yuan
    1963 "Birth Control in Mainland China: Ideology and Politics." *Milbank Memorial Fund Quarterly* 41(3): 269–290.
    1973 *China's Population Struggle: Demographic Decisions of the People's Republic, 1949–1969*. Columbus: Ohio State University Press.
    1983 "China: Demographic Billionaire." *Population Bulletin* 38(2): 1–42.
Tien, H. Yuan (ed.)
    1980 *Population Theory in China*. White Plains, N.Y.: M. E. Sharpe.
Tien, H. Yuan, Zhang Tianlu, Ping Yu, Li Jingneng, and Liang Zhongtang
    1992 "China's Demographic Dilemmas." *Population Bulletin* 47: 1–43.
Timmer, C. Peter
    1982 "Review Symposium on Julian L. Simon, *The Ultimate Resouce*." *Population and Development Review* 8(1): 163–168.

Tolnay, Stewart E.
  1981   "Trends in Total and Marital Fertility for Black Americans, 1886–1899."
         Demography 18(4): 443–464.
  1989   "A New Look at the Effect of Venereal Disease on Black Fertility: The
         Deep South in 1940." *Demography* 26(4): 679–690.
Toscano, Guy, and Janice Windau
  1991   "Further Test of a Census Approach to Compiling Data on Fatal Work
         Injuries." *Monthly Labor Review*, October, pp. 33–36.
United Nations
  1953   Department of Social Affairs, Population Division, *The Determinants
         and Consequences of Population Trends.* New York: United Nations.
  1992   *Demographic Yearbook, 1991.* New York: Author.
U.S. Bureau of the Census
  1907   *Heads of Families at the First Census of the United States Taken in the
         Year 1790.* Washington, D.C.: U.S. Government Printing Office.
  1979   *Statistical Abstract of the United States.* Washington, D.C.: U.S. Gov-
         ernment Printing Office.
  1980   "The Social and Economic Status of the Black Population in the United
         States: An Historical View, 1790–1978." *Current Population Reports,*
         Special Studies, P-23, no. 80. Washington, D.C.: U.S. Government Print-
         ing Office.
  1982   *Statistical Abstract of the United States 1982–3.* Washington, D.C.: U.S.
         Government Printing Office.
  1990a *Statistical Abstract of the United States, 1990.* Washington D.C.: U.S.
         Government Printing Office.
  1990b "Fertility of American Women: June 1990." *Current Population Re-
         ports,* Series P-20, no. 454. Washington D.C.: U.S. Government Print-
         ing Office.
  1991a "Population Profile of the United States 1991." *Current Population
         Reports,* Special Studies, Series P-23, no. 173. Washington, D.C.: U.S.
         Government Printing Office.
  1991b "Geographical Mobility: March 1987 to March 1990." *Current Popu-
         lation Reports, Population Characteristics,* Series P-20, no. 456. Wash-
         ington D.C.: U.S. Government Printing Office.
  1991c "School Enrollment—Social and Economic Characteristics of Students
         October 1989." *Current Population Reports, Population Character-
         istics,* Series P-20, no. 452. Washington, D.C.: U.S. Government Print-
         ing Office.
  1991d "Fertility of American Women: June 1990 Errata." *Current Popula-
         tion Reports, Population Characteristics,* Series P-20, no. 454, table 3.
         Washington D.C.: U.S. Government Printing Office.
  1991e "Money Income of Households, Families and Persons in the United States:
         1990." *Current Population Reports, Consumer Income,* Series P-60, no.
         174, tables A and D. Washington, D.C.: U.S. Government Printing Office.
  1992a "Marital Status and Living Arrangements: March 1991." *Current
         Population Reports, Population Characteristics,* Series P-20, no. 461.
         Washington D.C.: U.S. Government Printing Office.
  1992b "Household and Family Characteristics: March 1991." *Current Popu-*

lation Reports, Population Characteristics, Series P-20, no. 458. Washington, D.C.: U.S. Government Printing Office.

1992c   "Residents of Farms and Rural Areas 1990." *Current Population Reports*, Series P-20, no. 457. Washington D.C.: U.S. Government Printing Office.

1992d   "The Black Population in the United States: 1991." *Current Population Reports*, Series P-20, no. 464. Washington, D.C.: U.S. Government Printing Office.

1992e   *Statistical Abstract of the United States, 1992.* Washington, D.C.: U.S. Government Printing Office.

1992f   "Geographical Mobility: March 1990 to March 1991." *Current Population Reports, Population Characteristics*, Series P-20, no. 463. Washington D.C.: U.S. Government Printing Office.

1992g   *1990 Census of Population, General Population Characteristics, Alaska.* Washington, D.C.: U.S. Government Printing Office.

1992h   *General Population Characteristics for Virginia*, CP-1-48, Table 79. Washington, D.C.: U.S. Government Printing Office.

1992i   "Population Projections of the United States by Age, Sex, Race, and Hispanic Origin 1992–2050." *Current Population Reports*, P-25-1092, pp. xv–xviii. Washington, D.C.: U.S. Government Printing Office.

1993a   *Statistical Abstract of the United States, 1993.* Washington, D.C.: U.S. Government Printing Office.

1993b   "Population Profile of the United States 1993." *Current Population Reports, Special Studies*, P23-185. Washington, D.C.: U.S. Government Printing Office.

1993c   "Fertility of American Women: June 1992." *Current Population Reports*, P20-470. Population Characteristics. Washington D.C.: U.S. Government Printing Office.

1993d   "American Housing Survey for the United States in 1991." *Current Housing Reports*, Series H150, no. 91, p. 60. Washington, D.C.: U.S. Government Printing Office.

U.S. Department of Health and Human Services

1991a   "Advance Report of Final Marriage Statistics, 1988." *Monthly Vital Statistics Report* 40(4) August 26. Hyattsville, Md.: National Center for Health Statistics.

1991b   "Advance Report of Final Divorce Statistics, 1988." *Monthly Vital Statistics Report* 39(12) May 21. Hyattsville, Md.: National Center for Health Statistics.

1992   "Advance Report of New Data from the 1989 Birth Certificate." *Monthly Vital Statistics Report* 40(12) April 15. Hyattsville, Md.: National Center for Health Statistics.

1993a   "Advanced Report on Final Mortality Statistics, 1990," *Monthly Vital Statistics Report* 41(7, Suppl.). Hyattsville, Md.: National Center for Health Statistics.

1993b   "Advanced Report on Final Natality Statistics, 1990." Monthly Vital Statistics Report 41 (9, suppl.) February 25. Hyattsville, Md. National Center for Health Statistics.

1994   "Births, Marriages, Divorces, and Deaths for January 1994." *Monthly*

Vital Statistics Report 43(1) June 3. Hyattsville, Md.: National Center for Health Statistics.

U.S. Immigration and Naturalization Service
1992   *Immigrants Admitted By Country or Region of Birth*. Washington, D.C.: U.S. Government Printing Office.
1993   *Statistical Yearbook of Immigration and Naturalization Service, 1992*. Washington, D.C.: U.S. Government Printing Office.

van de Walle, Etienne
1992   "Fertility Transition, Conscious Choice, and Numeracy." *Demography* 29(4): 487–502.

van de Walle, Etienne, and John Knodel
1980   "Europe's Fertility Transition: New Evidence and Lessons for Today's Developing World." *Population Bulletin* 34(6): 1–43.

Veevers, Jean E.
1980   *Childless by Choice*. Toronto: Butterworths.
1988   "The 'Real' Marriage Squeeze: Mate Selection, Mortality, and the Marriage Gradient." *Sociological Perspective* 31: 169–189.

Vera, Hernan, Donna H. Berardo, and Felix Berardo
1985   "Age Heterogamy in Marriage." *Journal of Marriage and the Family* 47: 553–566.

Visaria, Pravin, and Leela Visaria
1981   "India's Population: Second and Growing." *Population Bulletin* 36(4): 1–55.

Von Drehle, David
1994   "Population Summit Has Pope Worried." *Washington Post*, June 16, pp. A1, A30.

Waite, Linda J., and Ross M. Stolzenberg
1976   "Intended Childbearing and Labor Force Participation of Young Women: Insights from Nonrecursive Models." *American Sociological Review* 41: 235–251.

Waldron, Ingrid
1983   "The Role of Genetic and Biological Factors in Sex Differences in Mortality." In A. D. Lopez and L. T. Ruzicka (eds.), *Sex Differential in Mortality: Trends, Determinants, and Consequences*, 141–164. Canberra, Australia: Department of Demography, Australian National University.

Waldrop, Judith
1992   "Live Long and Prosper." *American Demographics* 14: 40–45.

Watkins, John F.
1990   "Appalachian Elderly Migration: Patterns and Implications." *Research on Aging* 12(4): 409–429.

Watkins, Susan C.
1984   "Spinsters." *Journal of Family History* 9: 310–325.
1991   *From Provinces into Nations: Demographic Integration in Western Europe, 1870–1960*. Princeton, N.J.: Princeton University Press.

Weiner, Myron
1991   *The Child and the State in India: Child Labor and Education Policy in Comparative Perspective*. Princeton, N.J.: Princeton University Press.

Westoff, Charles, and Norman Ryder
    1977   *The Contraceptive Revolution*. Princeton, N.J.: Princeton University
           Press.
Westoff, Charles, Robert Potter, Philip C. Sagi, and Elliot Mishler
    1961   *Family Growth in Metropolitan America*. Princeton, N.J.: Princeton
           University Press.
Whelpton, P. K., and Clyde V. Kiser
    1943   "Differential Fertility among 41,498 Native-White Couples in India-
           napolis." *Milbank Memorial Fund Quarterly* 21(3): 221–280.
    1950   "Social and Psychological Factors Affecting Fertility." *Milbank Memo-
           rial Fund Quarterly* 2(iv–x): 139–466.
White, Michael J., Frank D. Bean, and Thomas J. Espenshade
    1990   "The U.S. 1986 Immigration Reform and Control Act and Undocu-
           mented Migration to the United States." *Population Research and Policy
           Review* 9: 93–116.
White, Stephen
    1983   "Return Migration to Appalachian Kentucky: A Typical Case of Non-
           metropolitan Migration Reversal." *Rural Sociology* 48(3): 471–491.
Whittington, Leslie A.
    1992   "Taxes and the Family: The Impact of the Tax Exemption for Depen-
           dents on Marital Fertility." *Demography* 29: 215–226.
Williams, Linda B., and Basil Zimmer
    1990   "The Changing Influence of Religion on U.S. Fertility: Evidence from
           Rhode Island." *Demography* 27(3): 475–481.
Willson, Peters D.
    1981   "'Pro Growth' Returns with 'The Ultimate Resource.'" *Intercom* 9(11):
           5–6.
Wilson, Thomas
    1992   "Urbanism, Migration, and Tolerance: A Reassessment." *American
           Sociological Review* 56(1): 117–123.
Wolpert, Julian
    1965   "Behavioral Aspects of the Decision to Migrate." *Papers of the Regional
           Science Association* 15: 159–169.
Woodrow, Karen A.
    1992   "A Consideration of the Effect of Immigration Reform on the Number
           of Undocumented Residents in the United States." *Population Research
           and Policy Reviews* 11: 249–262.
World Bank
    1984   *World Development Report*. New York: Oxford University Press.
Wrigley, Edward A.
    1966   *An Introduction to English Historical Demography*. New York: Basic
           Books.
    1968   "Mortality in Pre-Industrial England: The Example of Colyton, Devon,
           over Three Centuries." *Daedalus* 97(2): 546–580.
    1969   *Population and History*. New York: McGraw-Hill.
Wrong, Dennis
    1958   "Trends in Class Fertility in Western Nations." *Canadian Journal of
           Economics and Political Science* 24: 216–229.

WuDunn, Sheryl
    1993   "Births Punished by Fine, Beating or Ruined Home." *New York Times*, April 25, p. 12.

Ye, Wenzhen
    1992   "China's "Later" Marriage Policy and Its Demographic Consequences." *Population Research and Policy Review* 11: 51–71.

Yi, Zeng, Tu Ping, Guo Liu, and Xie Ying
    1991   "A Demographic Decomposition of the Recent Increase in Crude Birth Rates in China." *Population and Development Review* 17(3): 435–458.

Youssef, Nadia Haggag
    1974   *Women and Work in Developing Societies.* Westport, Conn.: Greenwood Press.

AUTHOR INDEX

SUBJECT INDEX